Innovations in GIS

selected papers from the First National Conference on GIS Research UK

Innovations in GIS

selected papers from the First National Conference on GIS Research UK

Editor

Michael F. Worboys
Department of Computer Science
Keele University
Keele ST5 5BG UK

CRC Press is an imprint of the
Taylor & Francis Group, an **informa** business
A TAYLOR & FRANCIS BOOK

First published 1994 by Taylor & Francis Ltd

Published 2020 by CRC Press
Taylor & Francis Group
6000 Broken Sound Parkway NW, Suite 300
Boca Raton, FL 33487-2742

First issued in paperback 2020

ISBN 13: 978-0-367-57983-8 (pbk)
ISBN 13: 978-0-7484-0140-6 (hbk)

Visit the Taylor & Francis Web site at
http://www.taylorandfrancis.com

and the CRC Press Web site at
http://www.crcpress.com

British Library Cataloguing in Publication Data

A catalogue record for this book is available from the British Library

ISBN 0-7484-0140-7 (cloth)
0-7484-0141-5 (paper)

Library of Congress Cataloging in Publication Data are available

Cover design by Hybert • Design & Type

Typeset by Euroset, Dover Close, Alresford, Hants SO24 9PG

Contents

Foreword vii
David Unwin

Preface ix

Contributors xi

GISRUK Committees xiii

Introduction 1

SECTION I COMPUTATIONAL SUPPORT FOR GIS 7

1 On the integration of database systems and computational support for high-level modelling of spatio-temporal phenomena 11
Terence Smith

2 Automatic GIS data capture and conversion 25
Colin Flanagan, Cullen Jennings and Neil Flanagan

3 Computer-assisted tools for cartographic data capture 39
Tim Mayo

4 An automatic generalization system for large-scale topographic maps 53
Gary Robinson and Fang Lee

5 A consistent user-model for a GIS incorporating remotely sensed data 65
Mark Gahegan

6 Computational support for spatial information systems: models and algorithms 75
Christopher Yearsley, Michael Worboys, Philip Story, Priyantha Jayawardena and Petros Bofakos

7 Incorporating the temporal dimension in a GIS 89
Zarine Kemp and Arek Kowalczyk

8 Towards temporality in GIS 105
Monica Wachowicz and Richard Healey

SECTION II SPATIAL ANALYSIS AND ERROR 119

9 A concepts-rich approach to spatial analysis, theory generation, and scientific discovery in GIS using massively parallel computing 123
Stan Openshaw

10 Exploratory spatial analysis in GIS: generating geographical hypotheses from spatial data 139
Richard Aspinall

11 Visualizing spatial association in area-value data 149
Jason Dykes

12 Probable and fuzzy models of the viewshed operation 161
Peter Fisher

13 Error simulation in vector GIS using neural computing methods 177
Chris Brunsdon and Stan Openshaw

14 An investigation into the spatial structure of error in digital elevation data 201
Colin Monckton

SECTION III APPLICATIONS OF GIS 215

Information sharing and the implementation of GIS: some key issues 217
Ian Masser and Heather Campbell

16 Modelling environmental systems with GIS: Theoretical barriers to progress 229
David Livingstone and Jonathan Raper

17 Coupling of process-based vegetation models to GIS and knowledge-based systems with reference to vegetation change 241
David Miller

18 Using GIS to target agri-environmental policy 251
Charlotte Norman, Clive Potter and Hadrian Cook

Index 263

Foreword

The GIS industry in UK and elsewhere is both large and diverse. Like all systems, it can be thought of in a number of ways. One possible model is as what ecologists call a 'pyramid of numbers' in which individual activities are considered to belong to distinct levels in a pyramid. At its base is a very large and rapidly growing number of people who use a GIS - or something like it - in their day-to-day work over an increasingly wide range of applications. At the next level up is a smaller number of specialists creating and maintaining the systems on which these applications depend. Above this lies a smaller layer of systems developers and, towards the apex of the pyramid, an even smaller, but happily growing, number of research scientists. As Michael Worboys makes clear in the Introduction to this volume, even this research community is rather diverse.

This model has a number of weaknesses, but it also has one very great advantage. It emphasizes the fact that individual layers of the pyramid depend on those above and below for their survival. Usually in ecology attention is directed to the process whereby higher layers of a pyramid feed off those below and so sustain a flow of energy upwards from base to apex. This view emphasizes the dependency of those in the upper levels on those below. An alternative view is to observe that the upper layers perform necessary functions without which lower layers could not exist. In this view, the lower layers are equally dependent on those above. In the case of the GIS pyramid, flows are not simply those of people or money. There are also flows of problems and solutions upwards and downwards through the pyramid. Just as in an ecological pyramid, the dependencies are two-way.

As anyone who has worked at grass roots with a GIS knows, real applications have a habit of generating problems that at first sight seem easy but on closer inspection turn out to be surprisingly difficult. They may find a solution at the next level in the pyramid in some clever fix using standard tools, or as some new function added to a system toolkit, but often they emerge as real research problems. In a sense, these problems are the food which sustains GIS research and this volume contains numerous examples of them. An obvious - but particularly searching - example might be someone's messy plot at 1:50 000 scale from a digital geography captured at 1:10 000. Some system fixes can correct this, but, ultimately, the issue of how we generalize geographic data asks important theoretical questions about how we conceptualize our geography.

The reverse flow is that of potential solutions, found by computer scientists, mathematicians and geographers, down the pyramid to be incorporated in everyday working practice. The cynical view is that this seldom happens and that research is almost totally decoupled from practice. Anyone who has watched the emergence of GIS from their early beginnings in the 1960s knows that, quite simply, this view is wrong. Developments in data modelling, generalization, and even spatial analysis, have all been incorporated into standard industrial practice and show that today's research idea often becomes tomorrow's system feature.

The transfers of information that both these flows involve cannot take place without interaction between all of us involved with geographic information and it is this interaction that the Association for Geographic Information (AGI) was, in part at least, set up to facilitate. The Association is the only organization of its kind in the UK that welcomes and represents all who have an interest in geographic information. It was for this reason that it was pleased to help Michael Worboys and his team set up and run the meeting that gave rise to this volume. GISRUK represents a landmark in the development of GIS in UK. For the first time we have not only a forum open to all GIS researchers, we also have its results available to us in book form. It is abundantly clear that not only is GIS research in the UK taking a healthy interest in problems that applications are suggesting, it is also generating solutions, the innovations of this book's title, that sooner or later will become a part of everyday GIS use.

David Unwin
Chair, Information and Education Committee
Association for Geographic Information

Further information on the work of the AGI can be obtained from: The Director, Association for Geographic Information, 12 Great George Street, Parliament Square, London SW1 3AD.

Preface

The chapters of this volume are enhanced and revised versions of papers which were presented and discussed at the conference GIS Research - UK, held at Keele University in the midlands of England in the Spring of 1993. In the early part of 1992, a small group of United Kingdom GIS researchers from several disciplines gathered together at Keele University to discuss the forming of a new conference series. There was a feeling that the needs of GIS researchers in the UK were not being met by conferences at a national level. While established workers found a place on the international conference circuit, there was no real national focus for GIS research. At that preliminary meeting in 1992, the feeling was unanimous that a niche existed for a new event, a steering committee was formed, a strategy was established, and 'GISRUK' , as it soon became known, was born. A call for papers was prepared, which attracted a healthy response. The conference took place in the spring of 1993. It was the first of a series of annual conferences, held in the UK and focusing on research into spatial information handling. GISRUK was attended by delegates from a wide range of backgrounds and subject bases, reflecting the inter-disciplinary nature of the field.

The primary objective of this conference series is to act as a focus in the UK for GIS research in all its diversity, across subject boundaries and with contributions from a wide range of researchers, from students just beginning their research careers to established experts. It was the aim of the steering committee to institute an annual event which would bring together as many as possible of those in the UK and elsewhere who were doing GIS research. The selection process resulted in a collection of high quality papers which covered a broad range of disciplines. Two distinguished international experts were invited to attend and give talks, one of whom (Terry Smith) has contributed an expanded paper to this volume. A session was arranged where leading figures in the applications of GIS in UK local government would present their work. This has resulted in the chapter by Ian Masser and Heather Campbell.

It is difficult to be objective about an event in which I was so closely involved, but the aim of the organizers was that the first conference should be informal, informative, friendly and not too expensive. I hope that these aims were realized and that they will set a tone for the conference series as a whole. With regard to this volume, it is my hope that the reader will get a feel for the general sweep of issues in GIS research at the moment, as well as an appreciation of some of the detailed problems. I remember in late 1992, when having a first skim through the initial papers that were submitted, wondering how any overall structure could be imposed upon such diversity. Now, having had a final read of the chapters in this volume, it surprises me how many connections there are between the ideas of each author. While not claiming that GIS research is an independent scientific discipline, I certainly feel that we have an inter-disciplinary field of study with its own distinctive flavour and identifiable research community.

I am grateful to all those who have helped in the shaping of this conference series, the first meeting of which has led ultimately to this book. The major sponsors were the UK Association for Geographic Information (AGI) and the UK Regional Research Laboratories (RRL) Initiative. The AGI, through its Information and Education Committee, chaired by David Unwin, has been a continuing source of support for GISRUK and for this book. The RRL movement has had an enormous influence upon GIS research in the UK and Ian Masser and Christine Openshaw have contributed significantly to the success of GISRUK. David Parker, a member of the GISRUK Steering Committee, helped greatly with arrangements for this sponsorship. The British Computer Society Special Interest Group in GIS made a generous donation to sponsor a special session on the contributions of new computer technologies to the field. This session was chaired by the Group's secretary, Frank Bennion. Longman GeoInformation, the publishers of GIS Europe helped financially with the conference printed proceedings. The local organizing committee ensured that the conference ran smoothly. Sue Harding made an efficient job of the accounts. Priyantha Jayawardena did a great deal of work, including a mountainous quantity of photocopying. Kendal Allen was a most effective conference administrator. Amongst others who helped were Keith Mason, Phil Story, Chris Yearsley and Petros Bofakos. With regard to the production of the book, Richard Steele of Taylor & Francis has been encouraging throughout.

Mike Worboys
Keele, England, 1993

Contributors

Richard Aspinall
MLURI, Craigiebuckler, Aberdeen AB9 2QJ

Petros Bofakos
Department of Computer Science, Keele University, Keele, Staffordshire ST5 5BG

Chris Brunsdon
Department of Geography, Newcastle University, Newcastle NE1 7RU

Heather Campbell
Department of Town and Regional Planning, University of Sheffield, Sheffield S10 2TN

Hadrian F. Cook
Environment Section, Wye College, Wye, Nr Ashford, Kent

Jason A. Dykes
Midland Regional Research Laboratory, Department of Geography, University of Leicester, Leicester LE1 7RH

Peter F. Fisher
Department of Geography, University of Leicester, Leicester LE1 7RH

Colin Flanagan
Department of Electronic and Computer Engineering, University of Limerick, Ireland

Neil Flanagan
Intaglio Systems, Newtown, Nenagh, Co. Tipperary, Ireland

Mark Gahegan
Department of Geographic Information Systems School of Computing, Curtin University of Technology, GPO Box U 1987, Perth 6001, Western Australia

Richard G. Healey
Regional Research Laboratory of Scotland, Department of Geography, University of Edinburgh, Drummond Street, Edinburgh EH8 9XP

D. Priyantha W. Jayawardena
Department of Computer Science, Keele University, Keele, Staffordshire ST5 5BG

Cullen Jennings
Department of Mathematical Sciences, University of Calgary, Canada

Zarine P. Kemp
Computing Laboratory, University of Kent, Canterbury, Kent CT2 7NF

Arek Kowalczyk
Computing Laboratory, University of Kent, Canterbury, Kent CT2 7NF

Fang Lee
Department of Geography, University of Reading, Whiteknights, Reading RG6 2AB

David Livingstone
Department of Geography, Birkbeck College, University of London, 7–15 Gresse Street, London W1P 1PA

Ian Masser
Department of Town and Regional Planning, University of Sheffield, Sheffield S10 2TN

Tim R. Mayo
Department of Geography, University of Cambridge, Downing Place, Cambridge CB2 3EN

David Miller
Macaulay Land Use Research Institute, Craigiebuckler, Aberdeen AB9 2QJ

Colin Monckton
Midlands Regional Research Laboratory, Department of Computer Studies, Loughborough University of Technology, Loughborough, Leicestershire LE11 3TU

Charlotte Norman
Environment Section, Wye College, Wye, Nr Ashford, Kent

Stan Openshaw
School of Geography, Leeds University, Leeds LS2 9JT

Clive A. Potter
Environment Section, Wye College, Wye, Nr Ashford, Kent

Jonathan Raper
Department of Geography, Birkbeck College, University of London, 7–15 Gresse Street, London W1P 1PA

Gary Robinson
NERC Unit for Thematic Information Systems, Department of Geography, University of Reading, Whiteknights, Reading RG6 2AB

Terence R. Smith
Department of Computer Science and Department of Geography, University of California, Santa Barbara, CA 93106, USA

Philip Story
Department of Computer Science, Keele University, Keele, Staffordshire ST5 5BG

Monica Wachowicz
Regional Research Laboratory of Scotland, Department of Geography, University of Edinburgh, Drummond Street, Edinburgh EH8 9XP

Michael F. Worboys
Department of Computer Science, Keele University, Keele, Staffordshire ST5 5BG

Christopher Yearsley
Department of Computer Science, Keele University, Keele, Staffordshire ST5 5BG

GISRUK Committees

GISRUK National Steering Committee

Peter Fisher	University of Leicester, UK
Bruce Gittings	University of Edinburgh, UK
Roy Haines-Young	University of Nottingham, UK
Zarine Kemp	University of Kent, UK
David Parker	University of Newcastle, UK
Emma Pearson	Perot Systems, Nottingham, UK
Jonathan Raper	Birbeck College, University of London, UK
Michael Worboys (Chair)	Keele University, UK

GISRUK '93 Local Organizing Committee

Kendal Allen (Conference Administrator)
Petros Bofakos
Sue Harding
Keith Mason
Michael Worboys

GISRUK '93 Sponsors

Association for Geographic Information
British Computer Society (GIS SIG)
ESRC Regional Research Laboratories Initiative
Longman GeoInformation

Introduction

This book contains a collection of papers describing research into geographical information systems (GIS). A GIS is a computer-based system which allows the input, management, retrieval, analysis, manipulation and presentation of geo-information. GIS research is a multi-disciplinary enquiry into the nature of such systems and also the role that such systems play in scientific modelling involving geo-referenced information.

In recent years, Michael Goodchild, who is a director of the US National Center for Geographic Information and Analysis, has increasingly been using what he calls the 'S' word in public. He writes that 'there is a pressing need to develop the role of science in GIS' (Goodchild, 1992). He notes two aspects of this role: the science of GIS and the science served by GIS as a tool. There is a parallel here in the editor's own discipline of academic computing, where the notion of computing as science has often been considered problematic. An accepted notion is that computer science is valid when taken as systematic observation and theorizing about a class of artificial processes related to computation, and clearly GIS is a subclass of this collection of general activities. Indeed, the nature and complexity of the underlying information for GIS gives this subclass a special flavour. In the second sense, GIS has many applications, for example, in environmental modelling, where it may be seen in its role as servant of the sciences.

Research into GIS may be seen as falling into the following broad, overlapping and inter-disciplinary categories:

1. development of the scientific foundation for the handling and analysis of computer-based geographically-referenced information;
2. engineering of systems with the capacity to manage and handle geographically-referenced information; and
3. using such systems in real-world applications.

Researchers in the first category include geographers, often influenced by the quantitative geographical tradition and with an interest in the development of statistical and spatio-analytical approaches for the analysis of computer-based geographically-referenced information. However, there are also mathematicians, logicians and computer scientists working on foundational questions for computer-based information systems and with a special interest in spatial information. It is this category which may generate fundamental scientific questions about the nature of geographical information, its handling and analysis. Work in this area is crucial for the provision of a basis for the field as a whole.

The second category is mainly the province of computer technologists, although there are groups with a wider interest who participate in its research problems. Spatial data and in particular geographical data create their own special problems for systems engineers. Such systems are often characterized by huge data volumes, complex data structures and the requirement to manage information that evolves through time. A flagship for research in this area is the Sequoia Project headed by Michael Stonebraker at the University of California at Berkeley, which attempts to extend database technology to solve the problem of managing the enormous volume of global environmental information currently being captured (Stonebaker *et al.*, 1993). There is a wide range of other work, including research into spatial data models, data structures, spatial query languages and new algorithmic approaches. New methodologies are being proposed, including object orientation and logic-based computing. Research in this category has a strong engineering flavour.

The size of the third category reflects the wide variety of current applications of GIS. Such systems have applications in every area where geo-referenced information is useful, and the number of such areas is very large. Applications range from global systems monitoring climate change to local systems for commercial benefit (such as to aid decision making on the siting of a new supermarket). Each application brings its own distinctive requirements and flavour to the subject. GIS are set to play a key role in the computational support for many fields of human endeavour.

Much GIS research is an activity within scientific computational modelling; that is, the use of computers to solve scientific problems. Figure 1 shows in schematic form some of the stages in this modelling process. It is an refinement of the usual modelling schema, where phenomena in the real world are represented as processes in an abstract scientific model. The results of the processing in the scientific model is then re-interpreted in the world of phenomena and conclusions are drawn. When serious use of computation is made, this basic two-tier model may be extended to include several computational models. The conceptual computational model is a representation of the scientific problem domains in a computationally-tractable form, but one which is independent of details relating to any particular computational paradigm. For example, relationships between data domains may be represented as an entity-relationship diagram without getting down to details about specific relational database issues, such as levels of normalization. The logical computational model is a representation which is paradigm dependent. In the case of an eventual implementation involving a relational database, the logic computational model would include details of data tables, indexes, identifiers, integrity constraints, etc. The physical computational model is the eventual representation of the real world phenomena as computational processes inside a computer.

Most GIS research can be situated somewhere within this computer-based scientific modelling scheme. To make a gross generalization, those who are associated with a discipline such as as geography work mainly at the scientific modelling stage, although successful work in this area must take cognizance of later stages in the process. Computer scientists working in the field will generally work with the conceptual, logical or physical computational models, depending whether their interest is primarily analysis, design or implementation. Again, for such work to be successful, notice must be taken of the prior links in the chain, so that the computations can relate meaningfully to the scientific models and thus to the real world phenomena.

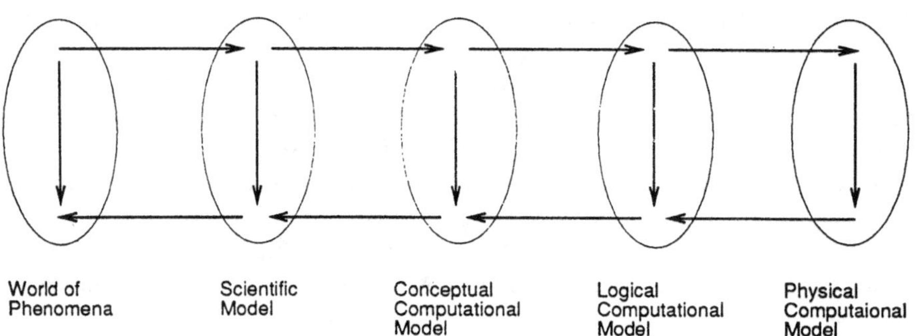

Figure 1: Scientific computational modelling.

The book is divided into three main sections, reflecting three different kinds of activity in the field. The first section on computational support includes papers which focus on issues relating to the computational models in the stages above. As computer science and technology rapidly develop, it is essential to have research which appreciates the new technological developments while being aware of the scientific models which they might support. Issues which are currently subjects of research are the application of new computational paradigms such as object orientation to spatial information management; bringing the computational models closer to the scientific models; and finding ways of supporting not only spatial but also temporal information. The second section on spatial analysis and error is nearer in spirit to the scientific modelling stage. However, all the papers are influenced in their thinking by the computational models that are available. The third section is devoted to applications of this research. The discussion ranges from applications in the local government arena to exploitations of the technology to help solve environmental problems.

Spatial information handling is the province of no one discipline. The papers in this volume, stemming from a wide range of backgrounds provide further evidence of this. It is this multi-disciplinary flavour, coupled with the movement towards a firmer scientific foundation and a strong feeling of the real possibilities of local and global usefulness of results, which make contributing to the GIS research effort a worthwhile human activity.

References

Goodchild, M. F., 1992, Geographical information science. *International Journal of Geographical Information Systems*, **6**(1), 31–45.

Stonebraker, M., Frew, J. and Dozier, J., 1993, The SEQUOIA 2000 project, in Abel, D. and Ooi, B. C. *Advances in Spatial Databases, Proceedings of the 3rd International Symposium, Lecture Notes in Computer* Science, **692**, 397–412, Springer-Verlag.

SECTION I

Computational Support for GIS

Section One: Computational Support for GIS

This section includes chapters which are predominantly concerned with aspects of the computational models underlying spatial information management. Many of the authors in this section have as their main discipline computer science: all have a strong interest in the computational requirements of GIS. The discussion ranges from issues in data collection to data presentation and generalization. There is currently a debate about the best ways in which computer technology can serve the spatial information community, both in terms of the modelling tools provided and the most appropriate paradigms, whether relational databases, object-oriented systems, logic-based approaches or others. Several of the chapters touch on these questions. The section concludes with descriptions of research into the incorporation of temporality into GIS.

From one perspective, a GIS is just a specialized form of computer-based information system. Thus, the issues which dominate information system science and engineering are also likely to be important for GIS. From an engineering point of view, the ultimate goal is the provision of effective methodologies for constructing such systems. While such methodologies are an accepted part of general information systems (studied within the discipline of software engineering) and can be applied to GIS, there seem to be few specialized approaches to spatial information systems engineering. The reasons for this are unclear. It may be that such methods are known but unpublished for reasons of shortage of time or company confidentiality. This seems unlikely as there is no such problem with general software engineering, where the financial stakes are higher[1]. A more likely conclusion is that, due to the relative youth of the field, such a specialized methodology does not exist in any detail. This is clearly a pressing research issue for the field.

Terence Smith, who has contributed the first chapter in this section, is very well placed to address these questions, holding positions in both the departments of geography and computer science at the University of California at Santa Barbara. The chapter that he has contributed to this volume is concerned with some of the large issues in this field. A goal of his research is to provide systems which will 'talk the same language' as scientists involved in applications and not distract users with low-level, internal computational matters. He has been working with earth scientists participating in NASA's Earth Observation System (EOS) programme, and he notes that more than half of their activity is concerned with irrelevant computational issues. The problem here is to provide systems with flexibility and power for scientists, without requiring them to have detailed technical knowledge of the underlying system models. The system which his team is developing allows users to interactively develop their own scientific models in an environment which provides support for high-powered computation and very large database support. The chapter describes an underlying data model, composed of a lattice of structured sets, and a language for interacting with this model. The chapter concludes with some thoughts on the database support required.

The next four chapters are concerned with the collection, conversion and incorporation

[1]However, it is of interest to note that currently an application exists for the first patent for a spatial structure and access method.

of data into GIS. A geographic database contains structured data, where the structure may be topological, feature-based or of other kinds. A major problem is the collection and structuring of the data so that they may be incorporated into a GIS. It has long been predicted that as more and more hardware and software problems become solved, problems with the data themselves will predominate.

Manual digitization is slow, tedious and expensive. In the last few years, there have been significant technological advances which support the capturing of geographical data. **Colin Flanagan, Cullen Jennings and Neil Flanagan** are from the University of Limerick in Ireland, the University of Calgary in Canada and Intaglio Systems in Ireland, respectively. Their chapter addresses the issues of automatic capture and conversion of spatial information from existing paper documents into an electronic form suitable for a CAD system or GIS. They describe the principles of automatic spatial data capture in general and behind their own automatic conversion system. One of the main difficulties here is that such systems cannot rely solely upon syntactic properties of the data. The semantics are often crucial in the decision making process, for example, whether a bump in a line is another line branching out or just a property of the accuracy of the drawing. It is well known that semantics are computationally much harder than syntax. The chapter contains a very useful discussion of the errors that automatic digitization systems make with line data. The authors note that a system which automatically digitizes line drawings must have a very low error rate to be useful, otherwise manual correction time will be larger than the time taken to digitize manually from scratch. For certain classes of data, they claim a higher level of accurate capture for map data than is usual with such systems.

The following chapter by **Tim Mayo** from the Geography Department of the University of Cambridge in England proposes that since fully automatic data capture techniques are problematic, a partially automatic, computer-assisted solution should be sought. The work which he describes forms part of a project, begun in 1988 at Cambridge, which investigates the role that large format raster scanners may play in efficient data capture. He notes the great potential of automatic systems but, as in the previous chapter, recognizes the difficulties. The consideration of partial automation has parallels in computer science with the current work on automatic theorem proving, where semi-automatic approaches are currently popular. With the present state of computer technology, a combination of person and machine appears to be best in many situations where activity is complex and unpredictable, and spatial data capture is just such an area. The chapter concludes with experimental evidence that semi-automatic techniques are useful in this field, both improving the speed and the accuracy.

In the next chapter by **Gary Robinson** and **Fang Lee** of the University of Reading in England, the spotlight is still on the data within a GIS, but the focus shifts from capture to generalization. In cartography, generalization concerns transforming the graphics of a map so that an appropriate amount of semantics is conveyed at varying scales. Here, the semantics are again crucial and the chapter describes methods for handling the semantics which combine those of conventional procedural computation and logic-based systems. Much of the previous work on generalization has concerned small to medium scales. This chapter considers problems associated with large-scale topographic maps, where areal features predominate. The authors note that for areal feature generalization, context and neighbourhoods become important. Thus to properly generalize a spatial object, the relationship of that object to others must be known. One of the techniques used is to create new objects, formed as groupings of basic objects. In detail, the objective of the research is to provide a system for the automatic generalization of complete maps from scales 1:1250 to 1:10 000. The results are embodied in a prototype system called OSGEN.

Another aspect of the data question is the incorporation into GIS of data already in electronic form but requiring transformation so as to fit the system model. An example is the incorporation of remotely sensed image data into a GIS, which forms the subject of the next chapter, written by **Mark Gahegan** of Curtin University, Perth in Australia. The work described was done by Gahegan when he was at Leeds University in England. His idea is to extend the GIS data model so that remotely-sensed data can be incorporated naturally. There is a clear need for such a facility, with more and more satellite sensors in place and the requirement for up-to-date geo-information. A class of problem encountered in this work and described in the chapter stems from the disparity between typical resolutions of remote sensors and geographical feature size. For example, widths of roads often are small compared with the pixel size of the sensor. The solution proposed here is to utilize the semantics of the image by means of scene understanding techniques. The semantics finally constructed are feature-based in nature. Once again, the hidden structures in the data are at the heart of the matter.

The next chapter, written by **Christopher Yearsley, Michael Worboys, Philip Story, Priyantha Jayawardena and Petros Bofakos**, all working at the Department of Computer Science of Keele University in England, moves the discussion forward to more general issues. The chapter considers computational support for geographical data management under the heading of paradigms and algorithms. By paradigm is meant a general model of computation; the chapter focuses on the relational database model, the object-oriented approach and the logic-based approach. Each approach is evaluated in the contribution it has and maybe will make to GIS. Much has already been written about the object-oriented approach and the chapter attempts to clarify some of its foundational concepts. The authors also outline an object-oriented system model for spatio-temporal information. Logic-based approaches have been out of vogue recently, following the partial failure of the Japanese *Fifth Generation* programme. It is felt by many computer scientists that they have thus been undervalued. The authors discuss possible offerings that deductive database technology may have for GIS. The chapter concludes by linking some of the high-level approaches to their algorithmic support.

The theme of spatio-temporality introduced in the preceding chapter is taken up in the last two chapters in this section. Computational support for unified spatial and temporal information systems is currently a pressing issue in GIS research. Computing hardware has now reached the stage where it is beginning to be capable of supporting the large volumes of data and processing speed required of spatio-temporal applications. This raw power must be matched by appropriate conceptual models. The database community have been researching the topic of temporal databases for some years and there is much that can be learned from this research with regard to temporal indexing, query languages and underlying formalism. It is not yet clear how this research will combine with what is now known about handling spatial data to provide support for a unified system supporting spatial and temporal dimensions.

The chapter by **Zarine Kemp** and **Arek Kowalczyk**, who are computer scientists from the University of Kent in England, surveys issues which arise in the design and implementation of spatio-temporal information systems, paying special attention to data modelling and data structures. The authors take as an example a land terrier system. They cover many of the basic concepts in this field, including the distinction between database time and event time, temporal granularity and methods of associating times with events. The chapter goes on to look at possible implementation strategies, concentrating on a temporal enhancement of the geo-relational model. The focus is on the support for real world time. It concludes by looking at other computational models in the context, including the extended relational model and the object-oriented approach.

Finally in this section on computational support is a chapter contributed by **Monica**

Wachowicz and **Richard Healey** from the University of Edinburgh in Scotland. The discussion once more focuses on the handling of spatio-temporal information where the temporality arises from the recording of events in the world (as opposed to transactions with the information system). Most of the temporal database research has assumed the relational model. This chapter considers temporality in the context of object-oriented models. The chapter describes the implementation of such a system using the object-based Smallworld GIS. The authors point out parallels between the definition of spatial object types and temporal object types.

The chapters in this section represent a variety of approaches to computational support for GIS, from the high-level discussion of appropriate computer models to the consideration of specific issues such as data capture, algorithmic support and spatio-temporality. A common thread is the understanding that for GIS to be supported properly, the complex semantics which the information expresses must somehow be incorporated into the system model. With the more advanced computational models being developed, there is hope that more and more of this rich meaning and structure will be explicitly incorporated into new systems.

1

On the integration of database systems and computational support for high-level modelling of spatio-temporal phenomana[1]

Terence R. Smith.

A modelling and data-base system (MDBS) provides computational support that facilitates the iterative development of scientific models in data-intensive and computationally intensive applications in various areas of scientific application. We describe and exemplify core elements of an MDBS that is based both upon a simple characterization of scientific activity and upon the results of joint research with a group of EOS earth scientists. The MDBS that we describe is focused in particular on the representation and manipulation of large numbers of complex, spatio-temporal entities that are modelled in terms of an advanced and comprehensive data model and a high-level database and modelling language (MDBL). The data model is based on a simple characterization of scientific activity that captures the essence of scientific modelling and database activities in terms of four fundamental sets of activities. These activities are defined in relation to large lattices of representational domains (R-domains) that are employed to organize scientific knowledge, data and algorithms. The data model separates the specification and representation of R-domains of elements, and allows multiple representations that are explicitly controlled by the user. Datasets and transformations are organized in terms of an extensible lattice of abstract and concrete R-domains. In particular, the data model allows the easy definition and manipulation of large numbers of domains of spatio-temporal elements. MDBL is a high-level, largely declarative language that is also based on our characterization of science and provides easy expression of the four classes of scientific activities in relation to R-domains. MDBL permits the easy integration of modules that support scientific modelling and database activities. We are implementing a system in which support for the data model and language is integrated with support for a distributed database of large, spatio-temporal datasets.

1.1 Introduction

High levels of abstraction are of increasing significance for many areas of application in which computational methods play an important role. In particular, we believe that there

[1]The Research reported in this paper was supported under NSF IRI91-17094.

[2]The material in this paper is the result of joint research with Divyakant Agrawal, Amr El Abbadi, and Jianwen Su of the CS Department at UCSB.

is an increasing need for such abstraction in languages that support modelling and database activities in many areas of application (Bulletin of the Technical Committee, 1993; Smith *et al.*, 1993). Our justification for this need is that as applications become evermore complex, it is important to hide irrelevant computational details from the attention of individuals whose concern is with the application. Based upon our research interactions with a large number of earth scientists, we estimate that many scientific teams focus up to 50 per cent of their attention on computational issues that are irrelevant to their scientific research. It is therefore clear that significant increases in efficiency are possible for applications in which computation is a major activity. Since these users may then focus their resources on the solution of the domain problems, rather than on 'irrelevant' computational issues, we believe that the efficiency of such individuals, measured in terms of how quickly they find appropriate solutions to problems in their domains of application, could be increased markedly.

In this chapter, we describe a modelling and database system (MDBS) that is intended to provide transparent, high-level computational support for scientific investigators and that, in particular, is designed to support modelling that involves complex, spatio-temporal entities. The system that we describe and exemplify is the result of long-term research collaborations with a number of earth scientists who are involved in NASA's Earth Observing System (EOS) program (NASA, 1990; Dozier, 1990). The general goals of these investigators include the development and application of complex, spatially-distributed models of a variety of earth science phenomena. We have found that these investigators, in common with many other applications scientists, lack adequate computational support for the iterative construction and testing of their models. The generation of research results is heavily and negatively impacted by both the large size and processing requirements of these models, which are frequently coupled, and the need to develop such models using a large and heterogeneous collection of datasets. Such modelling typically requires the definition and manipulation of complex spatio-temporal entities.

The researchers also lack genuine database support for managing their fast growing collection of spatio-temporal and other datasets, which contain information in the form of satellite images and a large collection of digital datasets for elevation, vegetation, geology, soils, meteorology and hydrology. In large part, these data-handling difficulties result from the use of existing file systems as a repository for the datasets. The varied contents, formats, lineage, and the large size of datasets result in an unmanageable collection of files scattered over a network. The absence of database modelling and management further complicates the task of maintaining these datasets, which undergo continued growth and evolution. In addition, efficient access to the contents of these datasets is severely restricted in the current environment. Additional difficulties arise from the fact that while modelling and database activities are closely related from a scientific point of view, they are artificially separated by inadequate computational support. Furthermore, there is no computational support for coordinating the modelling and database activities of researchers involved in joint projects.

An MDBS should provide computational solutions to each of these problems. In particular, it should support the iterative development of scientific models in data-intensive and computationally-intensive applications; the construction, maintenance and access of scientific databases upon which such modelling activities are based; and a variety of important concepts, such as models and multi-investigator projects. Since a great proportion of scientific effort involves the definition and manipulation of spatio-temporal entities, and in particular with the construction, storage, accessing and manipulation of spatio-temporal datasets, it is critically important that much of the support of an MDBS be focused upon such entities.

Our strategy for designing and providing computational support to these researchers involves:

1. basing the design of an MDBS upon a simple model of scientific activity and upon a detailed investigation of the requirements of scientific investigators;
2. adopting a top-down design approach that is based on an appropriate high-level language that permits the simple expression of much of the functionality required by investigators;
3. designing a powerful data model for the organization of data and knowledge, and particularly data and knowledge about spatio-temporal domains, in a uniform manner that is tied, in particular, to our model of scientific activity;
4. providing a simple view of data in terms of datasets and not in terms of a distributed set of files;
5. providing a framework in which many scientific modelling tools can be easily integrated;
6. providing efficient support for:
   (a) exploratory data analysis
   (b) rapid prototyping and testing of models
   (c) production runs of models.

Key aspects of this strategy involve the definition of appropriate, high-level languages in which scientists may express their activities in a manner that is similar to the language in which they express their science and the discovery of efficient computational support for such languages. A final but critical component of our strategy involves the testing of our ideas in terms of applying a prototype MDBS to solve real scientific problems whose solutions lie significantly beyond the computational support currently available to most earth scientists.

1.2 A simple characterization of scientific activities

We believe it essential to base the design of an MDBS upon an appropriate characterization of the goals and activities of scientists, and in particular upon the idea that the fundamental goal of science is the discovery and investigation of symbolic models of phenomena. Both the data model and languages for an MDBS should be closely related to such a characterization. We briefly describe a characterization that we have developed in collaboration with earth science researchers (Smith *et al.*, 1993a; Smith *et al.*, 1993c) and upon which we have based the data model and languages described below. Of central importance for the success of scientific modelling activities, we believe, is the construction of appropriate sets of concepts, and, perhaps even more importantly, appropriate representations of those concepts in some language. We believe that much of the power of scientific investigations derives from the discovery and application of large numbers of clever and powerful representational domains and associated transformations. We therefore conceptualize scientific investigators as organizing much of their knowledge about the world in terms of large numbers of domains of representations of entities and transformations between such domains. In particular, we have found it convenient to differentiate between 'conceptual' domains (C-domains), which relate to abstract views of entities and transformations, and symbolic representations of these abstractions, which we term 'representational' domains (R-domains). In general, we may view a set of R-domains as being necessary for any computations that involve domain elements and transformations. For example, corresponding to the notion of a C-domain of polygons and the associated transformations that provide semantics to the concept of polygons, there are a variety of R-domains whose elements may be interpreted as concrete representations of polygons and associated transformations, including representations based on sequences of points, sequences of line-segments and sets of half-planes.

We may therefore characterize scientific investigations as activities in which scientists:

1. construct, evaluate and employ a large set of R-domains for representing conceptual entities;
2. construct, evaluate and employ representations of many transformations between these R-domains;
3. construct instances of R-domain elements;
4. apply sequences of specific transformations to specific sets of R-domain elements.

In terms of this viewpoint, we may characterize a 'scientific model' as including a set of domains, a set of associated transformations, a set of domain instances and sequences of applications of transformations to domain instances.

We may illustrate some of these ideas in terms of an application problem which we are currently investigating in collaboration with a group of EOS investigators. One of the basic problems with which these earth scientists are faced is to construct models which represent the routing of water, sediments and solutes down the whole of the Amazon watershed. This problem is relatively complicated in terms of the various domains and transformations that are required for an adequate analysis. In order to exemplify the applicability of our characterization of scientific activity, we now briefly describe a highly simplified and schematized view of a subset of the domains, domain elements and transformations that have arisen in this application. We employ this example and its set of domains for illustrative purposes in the remainder of the paper.

In Figure 1.1, we show datasets and operations that provide a simplified view of the problem. Each of the datasets corresponds to an instance in some concrete R-domain and the operations correspond to transformations on R-domain elements. In Table 1.1, we exemplify a sequence of the four classes of actions described above that accomplish the computation represented in Figure 1.1. In this figure, we represent these commands in natural language, while in section 3.3, we provide their translation into our high-level modeling and database language MDBL. It is to be stressed that expressing these commands in MDBL requires little more than an appropriate set of domains, domain instances and

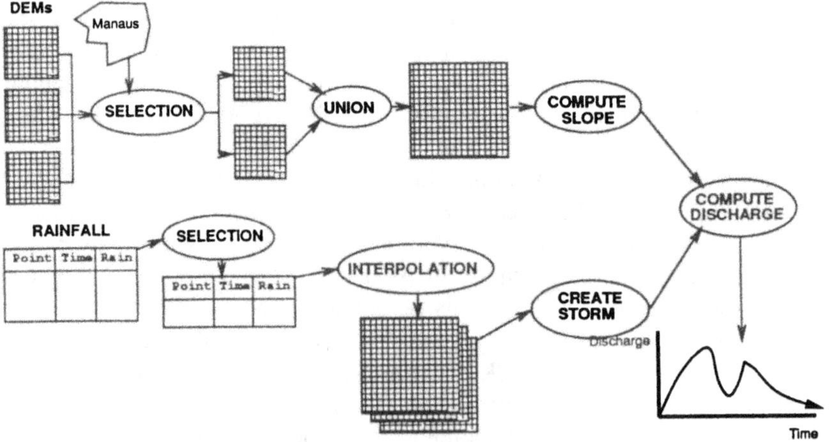

Figure 1.1 Computing discharge: A simplified set of operations.

transformations. For example, one requires R-domains for digital elevation models (DEMs) and for hydrographs (Hydrographs) and their associated transformations.

In brief, the sequence of the computations involves first choosing DEMs of interest by intersecting them with an appropriate area and then combining them into one single DEM. Rainfall data from points inside this area and for certain specific time periods are retrieved and interpolated over the DEM. A transformation, find-storms is employed to extract all Storms from the rainfall data for each hour. The DEM is used to generate an isochrone 'map' that models the flow of water from each point in the DEM to the mouth of the river. Hydrographs is a domain of representations of the discharge of water at the mouth as a function of time. It should be noted that the sequence of commands involves the creation of new domains, as well as the application of transformations such as display to domain elements.

It is clear from this simple example that the data model and languages of the MDBS should support the definition and application of domains of elements and their associated transformations, and in particular of domains of entities that incorporate space and time in some essential manner.

We are designing an MDBS that adheres to this view of scientific activity and, in particular, that supports the four general classes of activity listed above. Central to the MDBS is a high-level, largely declarative modelling and database language (MDBL). In relation to the iterative development of scientific models of phenomena and the associated activities of database creation and access, MDBL may be employed by investigators to construct and apply an extensible and potentially very large collection of R-domains of elements and associated transformations. Such a lattice of domains may be viewed as representing a relatively complete set of concepts that a scientist needs for modelling some set of phenomena, from low-level datasets to high-level models. MDBL supports the declarative specification and efficient construction of transformations associated with the domains; the creation of domain instances; and the application of specific domain transformations to instances, or sets of instances, of domain elements.

MDBL also supports the representation of database activities in a manner that is in

Table 1.1 Computing discharge: a simplified sequence of actions

1. Collect the DEMs that intersect the Manaus area.
2. Mosaic this set of DEMs and produce a single resulting DEM.
3. Show a 3D display of this DEM.
4. Create a general class of maps that represent the slope function defined over a DEM.
5. Create the appropriate slope map from the new DEM constructed for the Manaus area and store it as a member of the general class just constructed.
6. Create a general class of maps that represent the equal travel distance for water flow over a given slope map.
7. Create the appropriate equal travel distance map from the slope entity just created for the Manaus DEMs and store it as a member of the general class of maps just constructed.
8. Find time-slices of rainfall from hour 1 to hour 12 on January 21 1989 for rainfall records within the Manaus area.
9. Create a rainfall raster for each of the previous datasets using a rainfall interpolation routine.
10. Create a general class of raster representations of rainstorms.
11. Find and store all the rainstorm rasters that can be found in the rainfall raster just created.
12. Compute the hydrograph that would occur if this storm were to fall on the equal distance travel surface constructed above.
13. Display a plot of this hydrograph.

accordance with our model of scientific activity. For example, scientists may view any dataset as an element of some domain, rather than as a collection of files that may be distributed in the current computational environment, and may also access datasets by content. In general, the database support underlying MDBL incorporates:

1. a uniform structuring and organization of the data that is independent of both the physical implementation of the system and the distributed nature of the database. In particular, users see data in terms of named 'virtual' datasets rather than in terms of files.
2. simple access to datasets that avoids problems related to I/O. For example, access to portions of datasets may be made in terms of domain names and declaratively expressed constraints on the values of the elements sought.
3. the construction of, and access to, transformations on the domains.
4. support for concepts such as model and project including support for concurrent access to datasets within a project, that are independent of the system implementation.

1.3 The modelling and database system

1.3.1 A data model for R-domains

The data model is formulated in terms of R-domains. Each C-domain is represented in terms of a single abstract R-domain with a set of associated transformations, and a set of concrete, equivalent R-domains. Abstract R-domains serve as the specification of the domains, including the domain names and transformation names. Each concrete R-domain implements its abstract R-domain in terms of the structure of the elements and the transformations applicable to these elements with respect to a specific representation. In computational terms, this permits us to define new concrete R-domains independently of the representation of the domain elements employed in its definition and facilitates the construction and use of an inheritance structure over the set of domains. We allow several equivalent concrete R-domains to represent a single abstract R-domain (as in the case of Polygons mentioned above), since it is frequently advantageous for scientific investigators to employ different representations of a C-domain. Different concrete R-domains of the same abstract R-domain may have distinct representations but they should be equivalent. A general illustration of this structure of is shown in Figure 1.2.

The purpose of abstract R-domains is then to define a 'uniform' interface for all concrete R-domains. The defining characteristics of any concrete Rdomain include:

1. the domain name;
2. the 'structure' (or type, or representation) of the domain elements, which is described using tuple, set, sequence, array, finite function, etc. constructors;
3. constraints on the values of domain elements, which are expressed declaratively and naturally in a logic for the data structures;
4. implementation of transformations on the domain elements that provide, in part, the semantics of the domains;
5. sets of transformations relating both to equivalence classes of elements within a concrete R-domain and to homomorphisms between equivalent concrete R-domains.

In general, abstract R-domains form a lattice on which inheritance of both structures and transformations occurs naturally.

An important mechanism for the inductive construction of new R-domains involves the application of constructors (e.g. set, tuple, sequence, function) to elements from previously

REPRESENTATIONAL DOMAINS

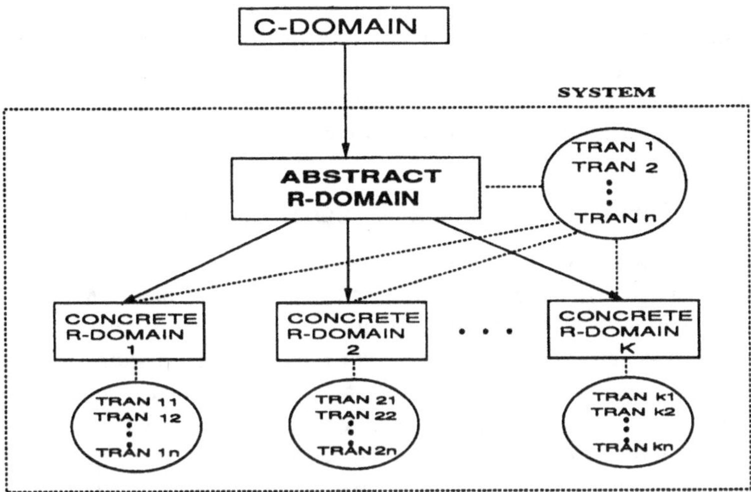

Figure 1.2 An abstract R-domain and its n *concrete R-domains.*

defined domains, which may also have constraints placed on their values. The placing of such constraints induces an 'inheritance structure' on the R-domains with respect to both the representation of domain elements and the transformations applicable to elements. We have focused attention on defining domains of elements that possess non-trivial 'spatial' or 'spatio-temporal projections' and particularly on domains of pointsets and domains of mappings from points and pointsets into other domains. For such domains, the concept of 'finite representability' is of central importance, permitting the definition and application of domains having elements of infinite extent. We have a complete theory of such domains. Examples of typical R-domains are Bools, Computable_Reals (primitive domains); Points, Finite_lattice_pointsets, Polygonsets (purely spatial domains); Elevations, Rainfalls (non-spatial domains); Digital_elevation_models, Drainage_basins (geographic domains); Hydrographs, Rainfall_sequences (temporal domains).

A special subset of abstract and concrete R-domains correspond to the recently-established Federal Standard (US Geological Survey, 1992). In this case, specific instances of concrete R-domains correspond to data structures for the electronic exchange of datasets.

1.3.2 A model of metadata

The metadata of a database is also represented in the same database and allows simple queries. The model of metadata possesses the following characteristics:

1. it is based upon our concepts of concrete and abstract R-domains;
2. it has a simple representation using relations;
3. it permits simple queries concerning the existence and properties of abstract and concrete R-domains; the various concrete R-domains associated with a given abstract R-domain; the existence and properties of the transformations associated with R-domains and elements of the various R-domains; and the interrelations between the domains.

More specifically, the metadata is represented by four sets of relations:

- the **abstract relation**, which contains all abstract R-domains;
- the **concrete relation**, which describes all concrete R-domains for each abstract domain;
- the **homomorphisms relation**, which has the transformations for converting elements between different concrete R-domains;
- the set of **element relations** (one for each abstract R-domain), each of which includes the element identifiers (or addresses) and the descriptions of the elements.

1.3.3 MDBL: A high-level modelling and database language

Based on our conceptual data model of domains and transformations, we have designed a simple, declarative, high-level modelling and database language, 'MDBL', to allow investigators to express easily and naturally most of the operations that are employed in the iterative development of their models, while hiding computational issues that are irrelevant to the scientific goals of the investigators. It is designed to provide support for about 80–90 per cent of scientific activities in terms of a very simple functionality, and to support a translation into a powerful visual interface. As noted in our characterization of scientific activity, the most important activities in scientific investigations may be characterized as:

1. defining new domains, new domain elements, new transformations and storing them in some database;
2. searching for datasets that satisfy certain constraints;
3. searching for transformations that satisfy certain constraints;
4. applying transformations domain elements in general and to datasets in particular.

Hence MDBL has been designed to provide scientists with straightforward support for creating, retrieving, storing, and deleting domains, transformations, and elements. Queries and function expressions defined in the previous section are used for specifying what a scientist wishes to access in a database and how computations should be carried out.

As one may note from Table 1.2 the language supports four sets of high-level commands that are intended to support the four major classes of scientific activity. We briefly describe these commands and then provide examples to illustrate the use of the language. The first set of commands (access) permits accessing of information on domains, transformations and domain elements from public and private databases. The second set of commands (create) permits the creation of new domains, new transformations and new domain elements. In particular new transformations may be created by a variety of methods, including function expressions that involve existing transformations, while new elements may be constructed by applying transformations to existing or constructed elements. The third set of commands (store) permits the storage of newly constructed domains, transformations and elements in the public database. The final set of commands (delete) permits deletion of domains, transformations and elements. The where conditions in each

Table 1.2 MDBL commands overview

ACCESS CREATE STORE DELETE	DOMAIN *name* TRANSFORMATION *name* ELEMENT *name* IN *domain-name*	WHERE conditions

of the commands are specified declaratively in terms of a simple syntax. MDBL combines the conventional ideas of data definition and data manipulation languages.

MDBL permits system access at a variety of levels appropriate for different classes of users and tasks, and is designed to facilitate the translation of its commands in terms of an easy-to-use visual language that may be used as the basis for constructing a graphical interface. A user commences an MDBL session by specifying a 'public' database (shared by many users) and a 'private', persistent workspace. The workspace provides a database and computational modelling environment for scientists, and in particular permits the creation of new domains, elements, transformations as well as persistent variables that can hold objects (identifiers) and structured values. Both the public database and the private database may be accessed, and domains, transformations, and elements may be moved to the public database.

To exemplify the simplicity and the use of MDBL, we present the MDBL translation of the queries shown in Table 1.1, Suppose we already have domains DEMs and RainFalls defined and Manaus is a variable holding a polygon object which represent the Manaus area. The operations can simply be expressed as the following MDBL commands.

1. access X = {D in DEM
   where spatial-projection(D) intersect spatial-projection(Manaus)}
   The function *spatial-projection* projects objects to their spatial components and *intersect* is a spatial predicate.
2. Manaus-dem = apply DEM.union to X
   where spatial-projection(Manaus-dem) = spatial-projection(Manaus)
   DEM.union is a transformation defined on the domain *DEM*, which combines many DEMs into one.
3. apply 3dplot to Manaus-dem
   The operation *3dplot* displays the DEM as a three-dimensional mesh.
   Such display functions are provided from other tools (such as MAT-LAB) and 'registered' in an MDBS as transformations. The abstract/concrete domains in our model makes such integration of tools relatively easy.
4. create concrete domain Slope
   where abstract domain = Slope
   structure = {[Point, [1..4]]}
   transformation = dem2slope: DEM → Slope is <fun-exp>
   The function expression <func-exp> defines the body of the transformation.
5. create element Manaus-slope in Slope
   where value = apply dem2slope to Manaus-dem
6. create concrete domain Binary
   where abstract domain = ETDM
   structure = raster
   transformation = dem2etdm:DEM → ETDM is dem2etdm
   transformation = compute-discharge:
   (ETDM, {Storm}) → Hydrograph is discharge
   Here the concrete domain is represented in binary format as 'black boxes'. The transformations are thus provided as executable programs.
7. create element Manaus-etdm in ETDM
   where value = apply dem2etdm to Manaus-dem
8. create element Y in rain_tpr where value = <query-expression>
9. create element Z in RainFall.RF-PxR
   where value = apply rain-interpolate to Y

10. create concrete domain S-PxR
    where abstract domain = Storm
    super-domain = RainFall.RF-PxR
    transformation = create-storms: RainFall {Storm}
    is <fun-exp>

    Since the type for the new domain is the same as its super-domain, the type specification can be omitted. The function expression can be easily written.
11. create elements in Storm where values = apply create-storms to Z
12. Manaus-discharge = apply compute-discharge to (X, W) where W in Storm
13. apply 2dplot to Manaus-discharge

1.3.4 Database support

Storage support for MDBL

In order to cope with the size of individual data objects and the volume of scientific data, it becomes necessary to design and develop operating systems support for efficient retrieval, storage, and organization of scientific data. The attribute-based naming scheme used in conventional databases is not suitable for scientific data since it assumes a flat name-space. Hierarchical naming, which is widely prevalent for file-systems, is useful for efficient management of large numbers of data objects. Clearly, due to the large volume of data in scientific databases and due to the need for attribute-based browsing in such an environment, a fusion of the above two approaches becomes necessary. Furthermore, the naming scheme for scientific databases must allow customization of the name-space for individual scientists. This is because, in general, a scientific database will be shared by a variety of scientists and each of them will have different needs resulting in conflicting requirements for name organization. Recently, such a naming scheme, which provides customization, has been developed for the Prospero project (Neuman, 1992). Although this scheme is designed for filesystems, it provides mechanisms to integrate attribute-based naming scheme necessary for databases.

In order to implement our ideas concerning database support, we are currently investigating the integration of several available database tools, including the the data management system POSTGRES (Stonebraker *et al.*, 1990) and the virtual file system PROSPERO, as well as software packages that support scientific modelling and data processing operations. Some of the main aspects of our current design include:

1. the integration of the Prospero file system with MDBL and POSTGRES;
2. a file-based interface to access database objects from POSTGRES;
3. the integration of database objects that are distributed over a network via a virtual file-system based on Prospero;
4. the tailoring of the virtual system according to a user's needs.

In the context of storage support, we are also investigating the use of POSTGRES to provide database storage support for very large objects in MDBS and extensions to the notion of Prospero filters, which are used with file directories, in the context of databases. The select filter mechanism allows the users to selectively view data objects in a given directory. The derived filter mechanism permits the users to organize objects in a directory based on the selected attributes of the objects (i.e., it creates virtual levels in the name space). Finally, the union filter mechanism allows the users to seamlessly combine multiple directories stored at different places into a single directory. Of course, these mechanisms can be combined to provide a rich and powerful naming mechanism for scientific data. In order to extend these notions in the context of databases, we need to apply the filters to the files instead of restricting them to directories. These filters can be used to implement efficient indexing for scientific data.

Database support for geographic data

Geographic Information Systems (GIS), or database technology applied to geographic data, have several characteristics that distinguish them from traditional database systems. In particular, geographic data is often encountered in many different formats (paper, text files, raster files, vector files, relations, numbers, etc.), under many different representations (graphic, numeric, textual, etc.) and in large quantities (exceeding traditional database applications by orders of magnitude). Furthermore, geographic data is not only used for retrieval and consultation purposes but also as input to complex transformations, as part of the ongoing scientific research in the area. In many cases, geographic information is highly interrelated and isolated pieces of data may not be easily interpreted without external references to other data. The fact that the meaning of the data is exogenous, i.e. not contained in the data itself, adds another layer of complexity to the design of these systems.

Since a database cannot be properly designed without a clear understanding of the data it will contain, a first goal in our project (Alonso and Abbadi, 1993) was to formalize the contents and meaning of geographic data. This was done in the three steps of:

1. incorporating into our model the recently adopted Spatial Data Transfer Standard (SDTS), which defines the additional information required to provide geographic data with meaning;
2. providing means to associate the additional information specified by the standard to the actual data by means of an object-oriented approach in the representation of geographic data;
3. providing accounting mechanisms for the transformations performed on the data.

When geographic data is not the product of direct observation (field measures, radar images, satellite photos, etc.) it is derived data, i.e. the product of a series of transformations applied to one or more sets of data. Most of the research in this field is based on the interpretation of this derived data, hence it is crucial to provide the means to recreate those transformations and store the relations created between the input and the output data. One of the most interesting results of our work has been to show that several problems encountered in these databases, usually regarded as independent and treated as such, are in fact only different aspects of the same problem, to keep track of object interdependencies. The object-oriented model we propose provides the data structures necessary to store the information relevant to those interrelations. A geographic object that contains its meaning and the information necessary to recreate its genesis has proven to be a very powerful abstraction. To prove the flexibility of the model we have developed several concise and efficient algorithms for:

1. **lineage tracking:** describing the sources and methods used to create the data;
2. **coherency checking:** in a database with multiple versions of each object the system must provide support to detect inconsistencies in the retrieval of data;
3. **change propagation:** the updating of derived data when the sources are replaced by newer versions;
4. **version maintenance and storage optimization.**

A model of collaboration and cooperation

We have also developed a more advanced model of collaboration and cooperation in scientific databases (Agrawal *et al.*, 1993). Several scientists may wish to collaborate together to correctly execute their algorithms concurrently. In particular, a scientist may start working on some raw data and then start using derived data that resulted from another scientist's computations. Hence, conflicting accesses to the same set of data may occur, and the need for concurrency and data flow control is necessary. The traditional database approach for concurrency control requires the complete isolation of the various users as long as their computation (transaction) has not terminated. We have recently developed a relaxed correctness criterion for such an environment referred to as 'relative atomicity'. Using this model, a scientist explicitly states when and where other scientists may observe data resulting from incomplete computations. Hence, the traditional 'isolation' criterion for transactions is relaxed in an organized manner. In particular, we have proposed a model that can efficiently recognize the correct execution of a set of concurrently executing processes, and have proposed protocols for ensuring such correct executions. In the course of this project, we plan to further relax atomicity requirements for scientific environments where teams of researchers access a common scientific database. Often, in such cases, groups of scientists may be divided into teams, where within each team, the scientists observe each others' results, while across teams, results are only made visible at the end of well defined computations. In such cases, a team of researchers collectively execute a 'collaborative transaction', while between teams of scientists, either standard or relative atomicity criteria are used to control and restrict information flow.

Detection of corrupted data

In large geographic databases, files are often replicated at several sites in a distributed system for purposes of efficient and fast access. If a copy of the file is available locally, the cost of read operations is negligible. This is especially important since many data files are very large and are often used as a source for deriving data, i.e., most operations are read operations. One prominent example of such systems involves image databases where many large (e.g. gigabyte-sized) images, often referred to as raw data, are transmitted to a central repository of data, and scientists across the globe acquire copies of this raw data in order to derive more structured images. Due to failures, some pages may get corrupted. Since the files are quite large and the portion of corrupted pages is expected to be small, exchanging all copies and comparing them is unacceptable. We have developed an approach, based on coding theoretic techniques, which provides an efficient and optimal solution for a coordinator-based model, in which a single site acts as the coordinator and identifies all corrupted pages in all the copies (Abdel-Ghaffar and Abbadi, 1993). In particular, the solution is a decentralized solution with less communication overhead. The coordinator model is especially appropriate in the satellite-based image database where the scientists with copies of the raw data may only be known to the central coordinator site.

1.4 Prototyping efforts

We are currently working with a group of EOS earth science investigators to jointly implement and apply a high-performance version of an MDBS that incorporates the data model and high-level languages described above. This system is intended to support large-scale earth science investigations by complementing the EOS information system (EOS-DIS) (Dozier, 1990) in terms of support for high-level modelling and data management. As noted above, MDBS is intended to facilitate the iterative development of models in data-intensive and computationally intensive applications. The project represents a major extension of a 3-year, interdisciplinary NSF-supported research project on scientific databases, which has resulted in the design of the data model and languages (Smith *et al.*, 1993b, 1993c). MDBS is being developed with the specific goal of solving several major EOS-scale problems in earth science, including the routing of water and sediment down the entire Amazon basin, the integration of GCM's and regional hydrological models, and the modelling of snowmelt and the routing of runoff from regional snowpacks. It is also designed to be generally applicable to a large class of scientific modelling problems that involve both data-intensive and computationally intensive aspects. Based upon our data model, we are constructing a large lattice of representational domains in order to organize the data, algorithms and computational modelling activities of the applications scientists. A simple prototype of the data model is currently being built. It implements the data model and MDBL based on the POSTGRES database management system (Rowe and Stonebraker, 1987; Stonebraker *et al.*, 1990). The database contains the datasets from applications mentioned above and is distributed over a network. A virtual file system is employed to provide network transparency.

Based on this model and the set of R-domains, the system will support a visual interface that implements, in part, the MDBL languages. The system will provide easy access to distributed databases of datasets and algorithms, and will permit the efficient reuse of code for building new algorithms. Support will also be provided for lineage and error tracking, for the concepts of model and project, and for 'model debugging'. In order for systems such as MDBS to be acceptable to scientists it is critical that they possess a high degree of performance. Such performance will stem in part from parallelization and also from the provision of high-level modelling tools that are based on the data model and permit scientists to construct creative solutions to apparently intensive computational problems.

References

Abdel-Ghaffar, K. A. S. and El Abbadi, A., 1993, Efficient detection of corrupted pages in a replicated file, in *Proceedings of the Twelfth ACM Symposium on Principles of Distributed Computing*, August. To appear.

Agrawal, D., Bruno, J., El Abbadi, A. and Krishnaswamy, V., 1993, Relative serializability: A theory for relaxing the atomicity of transactions. Technical report.

Alonso, G. and El Abbadi, A., 1993, A versatile modelling approach for geographic databases. Technical report.

Bulletin of the Technical Committee, 1993, on *Data Engineering*, **16**(1) (special issue on scientific databases).

Dozier, J., 1990, Looking ahead to EOS: The earth observing system. *Computer in Physics*, May/June.

NASA, 1990, *EOS: A Mission to Planet Earth*, Washington, DC: NASA.

Neuman, B. Clifford, 1992, Prospero: A tool for organizing Internet resources. *Electronic Networking: Research, Applications, and Policy*, **2**(1).

Rowe, L. A. and Stonebraker, M. R.. 1987, The POSTGRES data model, in *Proc. Int. Conf. on Very Large Data Bases*, pp. 83–96.

Smith, T. R., Su, J., Agrawal, D. and El Abbadi, A., 1993a, Database and modelling systems for the earth sciences. In IEEE [6]. (Special Issue on Scientific Databases).

Smith, T. R., Su, J., Agrawal, D. and El Abbadi, A., 1993b, MDBS: A modelling and database system to support research in the earth sciences. Technical report, Dept. of Computer Science, UCSB.

Smith, T. R., Su, J., Dunne, T., El Abbadi, A. and Agrawal, D., 1993c, Requirements for modelling and database systems supporting investigations of large scale flows in the amazon watershed. (In preparation).

Stonebraker, M., Rowe, L. and Hirohama, M., 1990, The implementation of POSTGRES. *IEEE Transactions on Knowledge and Data Engineering*, **2**(1):125–142.

US Geological Survey, 1992, Spatial Data Transfer Standards.

2

Automatic GIS data capture and conversion

Neil Flanagan, Cullen Jennings and Colin Flanagan

2.1 Introduction

The majority of recent engineering diagrams and maps are produced by CAD or GIS systems. Processing the data in these documents in electronic form provides substantially more flexibility than paper affords, allowing updates to a document to be easily reproduced, or documents to be combined or have additions made to them. For instance, from a utility company's GIS system a user could determine the location of the house nearest to a river where a chemical spill occurred. From CAD electrical diagrams a computer simulation of the circuit being designed might be created. Because electronic forms are easy to archive and update, they are valuable media for document storage. Although new documents are often drafted and stored in an electronic format, many existing documents survive only on deteriorating paper. During the early 1980s many companies attempted to scan these documents and convert them automatically into an electronic format. Scanning and converting these documents has turned out to be a very challenging problem, so that currently most conversion is done by using digitizing boards to re-draft the documents into a CAD or GIS system.

2.1.1 Conversion Systems

Presently a great deal of conversion is done through manual re-drafting on a digitizing table. Re-drafting is time-consuming, however, making automatic conversion systems an attractive idea. It seems unlikely that any automatic system will ever perform 100 per cent correct conversion on complex documents, but even a system that was 90 per cent correct and helped the user to fix the remaining 10 per cent would considerably reduce the time required to convert paper documents into electronic forms. Many commercial systems for solving this problem have been designed. Usually these systems are claimed to be much faster than manual conversion, operating generally around seven times as quickly and at one tenth of the cost. Surveys suggest that systems frequently fail to perform at these levels (Bono, 1988; Anderson Report, 1986). In fact, many of these systems fail to solve the conversion problem and end up slower than re-drafting on a digitizing board. Researchers who have applied themselves to this problem include J.P. Bixler (Bixler and Sanford, 1985; Bixler *et al.*, 1985), T. Donnelly and W.N. Martin (1987), D. Dori (1989), S.H. Joseph (1989), S. Suzuki and T. Yamada (1990), T. Taxt (1989), T. Wakayama (1982), and H. Yamada *et al.* (1991). The work described in this chapter illustrates the progress to date in an ongoing program to successfully and efficiently capture documents that exist solely on paper. Many commercial and research systems were evaluated during the course of the research. Drawing on the examples of these a suite of software tools has been created that takes a scanned image and produces vector output. Like most commercial and research systems, this system adopts ideas suggested by (Musavi *et al.*, 1988). Our system produces comparatively clean output vectors, regardless of whether the original diagram was a town plan or a contour map, lacking many of the artefacts commercial systems produce. This in turn means that less user time is needed to process the scan. Only 0.3 per cent of the vectors from a contour map used to benchmark the system needed retouching. This research and the system created demonstrate, however, that although the successful digital capture of maps and drawings is possible, 100 per cent capture is not yet feasible, due to fundamental problems in current approaches to thinning.

2.1.2 Maps

Certain conversion problems are most common with maps. The simplest maps are composed solely of contour lines. Conversion of these maps involves following the lines. Often a user identifies the elevation associated with each contour line before it is stored in a GIS system. More complex maps have symbols like houses or mines that must be located and recognized. Road and boundary lines must also be identified. One of the most complex problems faced is *colinearity*, in which a single feature on a map represents more than one aspect of the physical world. For example, on American forestry maps, if a road runs along the boundary of two counties, only the road will be drawn on the map – the boundary will simply be implied. Because of such problems and because accuracy must be very high (typically above 1/100th of an inch) the conversion of maps is a difficult problem. Maps formed the majority of the data set used in evaluating the system developed.

2.1.3 Engineering diagrams

Engineering diagrams consist mainly of lines, arcs, and symbols, including text. One

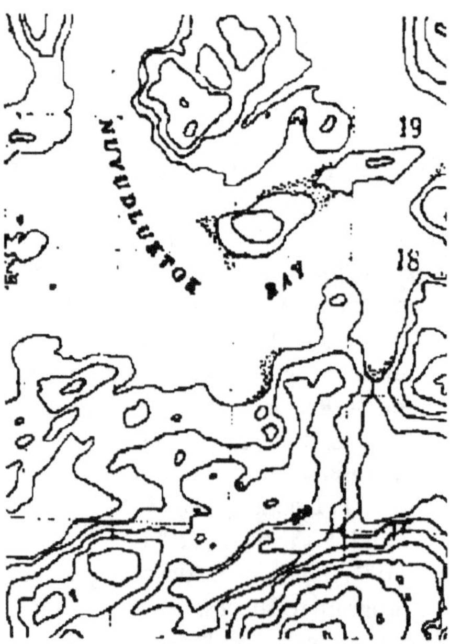

Figure 2.1 An example image, a section of coastal contour map.

problem that arises is that the lines and arcs are often dashed: the recognition system must be able to represent all the individual dashes as one line with a particular dash pattern rather than as miscellaneous short lines. The lines often run on precise angles, and converting, for example, 30.0–30.1 degrees is not acceptable. In addition, lines running through symbols often make both harder to detect.

2.2 Conversion methods

The steps taken in the conversion of images are drawn from the suggestions of Musavi (Musavi *et al.*, 1988).

1. **Scan**: Scan the paper image to produce a grey scale raster image.
2. **Threshold**: Form a binary image from the grey scale image.
3. **Salt and pepper filter**: Remove isolated black and white pixels that are noise.
4. **Thin**: Thin the lines on the image to one pixel wide.
5. **Chain code**: Follow the thinned lines and produce chain codes representing them.
6. **Vector reduction**: Reduce straight segments of the chain code into long lines that represent the chain codes.

Most commercial and research systems use schemes similar to Musavi's.

2.2.1 Scanning

Currently most scanners use CCD camera technology. Most of these produce 8-bit gray scale images. A resolution of 300 dpi is standard for the low-end scanners, while 800 dpi is common in commercial-grade scanners.

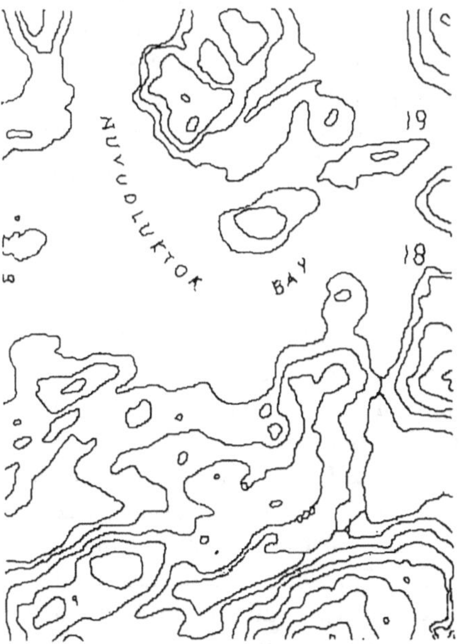

Figure 2.2 The vectors extracted.

2.3 *Preprocessing*

The procedures outlined here are applied to the image that comes from a scanner before this image is passed to the vectorization software. These techniques simply clean up the scan and put it in a form acceptable for the vectorization processing.

2.3.1 Thresholding

Thresholding converts an image from gray scale to black and white. In the simplest thresholding technique, a fixed gray level is chosen. Pixels darker than this level are interpreted as black, and lighter ones as white. Because this technique works well for many drawings, many scanners implement it in hardware. Another common thresholding algorithm, which dithers the image to simulate gray scale, is unsuitable for vectorization purposes, because it turns a grey line into a series of black and white dots, which give poor results with vectorizing algorithms. The strengths and weaknesses of different thresholding algorithms are covered well in the literature. (Taxt *et al.*, 1989) discusses several algorithms that produce good results but require interactive input. Parker's (1991) algorithm produces excellent results without interactive input. Because most scanners produce high quality scans with little noise, thresholding is often easy, one fixed threshold working for the whole image.

2.3.2 Adaptive thresholding

The major problem with the fixed threshold method is that it thresholds the whole image with the same value. Imagine an image with a large coffee stain in one place and another area that is very faint. If the lines in the faint area fail to be as dark as the background of the coffee stain, there is no value that will give a good thresholding result. The two regions require different thresholding levels, or adaptive thresholding. Adaptive methods usually first build a local histogram for each pixel based on its neighbouring pixels and then threshold the pixel. Domain information – such as the fact that engineering drawings are mostly white with a few black lines and that there are no large black areas – can easily be incorporated into these schemes and used as *a posteriori* probabilities for constructing a Bayesian classifier for separating the black and white. Taxt describes some of these methods (Taxt *et al.*, 1989). They provide a way to use knowledge about the class of image in thresholding. The results are excellent even with very poor input, but considerable computation time is often required.

2.3.3 Noise removal

CCD scanners often leave noise - usually black and white speckles – on the image. These speckles occur because when monochromatic light scatters from a surface whose roughness is of the order of the wavelength of the light, interference among the waves produces nodal patterns that leave bright and dark spots (Jain, 1989). A simplistic method is to find the area of all black and white regions. If the area is too small then the area is filled with appropriate colour (black or white). The area is calculated by counting the pixels that would be filled if the area were flood filled. The area is filled using the same flood fill algorithm. The flood fill algorithm used to fill in white areas is described as: 1. find a starting pixel; 2. set this pixel to black; 3. recursively set all this pixel's white pixel neighbours to black.

The algorithm used was modified slightly so that if the area exceeds a threshold, then the algorithm starts backtracking without filling the rest of the pixels. If only an area is being calculated and not filled, then the algorithm still sets the pixels to black but as it backtracks, it clears them to white so that the image is not modified when the algorithm terminates. Better noise removal algorithms that use up less stack space exist, but since the areas in question are small, using one of these would not be much of an advantage. Some of the more sophisticated speckle removal algorithms were developed for Synthetic Aperture Radar (SAR). Lim presents methods for complex speckle removal (Lim and Nawab, 1980).

2.4 *Erosion vectorization*

Most current systems thin the preprocessed image until all its lines are only one pixel wide and then chain code these lines, understanding them as the centre of the original thicker lines. The chain codes are then reduced into straight line segments that can be identified as vectors and stored.

2.4.1 Thinning

Many thinning algorithms approximate the Medial Axis Transform (MAT), which was originally defined by Montanari (1969). A common algorithm of this type is presented by Zhang (Zhang and Suen, 1984). All MAT techniques result in artefacts that are highly undesirable for the vectorization of maps and engineering diagrams. Several thinning methods that do not generate MAT-type image have been developed (Baruch, 1988). These techniques have often produced other artefacts just as undesirable for the vectorization of maps and engineering diagrams as those produced through MAT. One method that tries to eliminate some of the artefacts that happen at intersections is described by Govindan (Govindan and Shivaprasad, 1987). This method fails to prevent artefacts in some other cases, however. Li describes another method for thinning, which uses knowledge about where traditional thinning fails (Li and Suen, 1991). A final method works directly from the gray scale image and produces a binary skeleton (Yu and Tsai, 1990); it does an excellent job, but it works only on images whose only lines run at angles that are multiples of 45°. Many different thinning algorithms - which produce different types of results – exist. A reasonable survey of thinning methods is presented in Smith (1987) and Kwok (1988). These methods all vary in the exact type of results they produce but regardless of the method used, the results of the final vectorization are greatly effected. The problems and artefacts created by thinning are discussed below.

2.4.2 Implementation

The erosion method devised by Zhang and Suen (1984) was chosen for the project, as it is more commonly used than any other in conversion systems, making it easier to compare results obtained using the system developed. Understanding the algorithm requires the definition of some terminology at the outset. The pixels that are neighbours of a particular pixel p_1 are labeled $p_2, p_3, \ldots, p_8$, as shown in Figure 1.3.

Each pixel is either black or white. The convention used here is that we want to erode black regions where black regions are represented as having a pixel value of 1. Two functions of a pixel are defined:

$$N(p_1)=p_2+p_3+\ldots+p_8+p_9 \tag{2.1}$$

$$S(p_1)=\text{number of 0 to 1 transitions in } (p_2,\ldots,p_9) \tag{2.2}$$

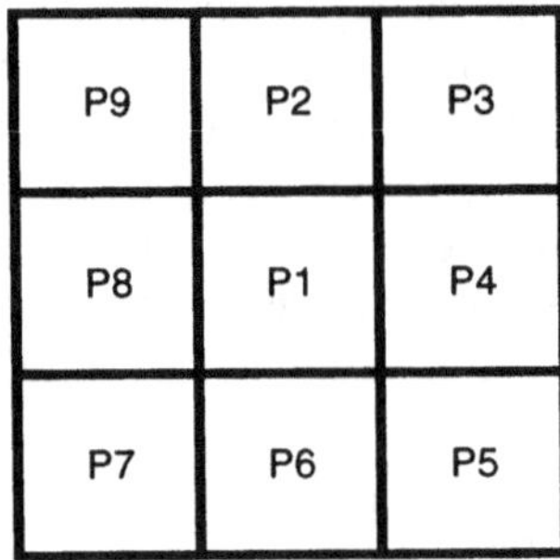

Figure 2.3 Labelling of pixel neighbours.

From these, two conditions C_1 and C_2 are defined for a given pixel. Condition one of a pixel $C_1(p_1)$ is true only if all of the following are true:

$$2 \leq N(p_1) \leq 6 \tag{2.3}$$

$$S(p_1) = 1 \tag{2.4}$$

$$p_2 \cdot p_4 \cdot p_6 = 0 \tag{2.5}$$

$$p_4 \cdot p_6 \cdot p_8 = 0 \tag{2.6}$$

Condition two of a pixel $C_2(p_1)$ is true only if all of the following are true:

$$2 \leq N(p_1) \leq 6 \tag{2.7}$$

$$S(p_1) = 1 \tag{2.8}$$

$$p_2 \cdot p_4 \cdot p_8 = 0 \tag{2.9}$$

$$p_2 \cdot p_6 \cdot p_8 = 0 \tag{2.10}$$

The algorithm can now be easily defined as a two-step algorithm. Step one involves checking all pixels on the image and, if $C1$ is true for the pixel, marking the pixel. The end of step one consists of deleting all the marked pixels by setting their value to 0 after all the pixels on the image have been checked. Step two is the same as step one, except that C_2 is checked instead of C_1. Steps one and two are repeated in order until there is an iteration of either step one or two in which no pixels are marked. When the algorithm terminates, a thinned skeleton is all that remains of the image. The image of the letter H in Figure 2.4 has been thinned using this algorithm, and the result is shown in Figure 2.5.

A small problem with this thinning algorithm that has also been noted by others (Abdulla *et al.*, 1988) is that it does not always give unitary skeletons. On some of the skeletons, points that are not intersections have $N(p_1)=3$. A more complex image is shown in Figure

Figure 2.4 Image of an H.

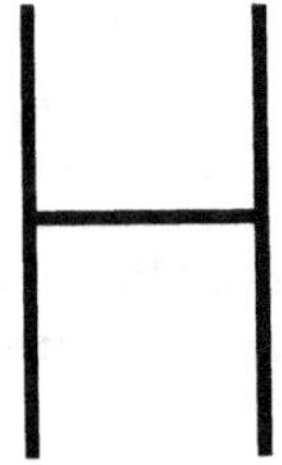

Figure 2.5 Image of thinned H.

2.6. This is thinned into the image shown in in Figure 2.7. To make it easier to compare the images, a section from the top centre of Figure 2.6 is overlayed in Figure 2.8 with the thinned segment. In this example a small section along one of the lines is magnified in Figure 2.9 to show that several of the pixels along the centre line are not unitary. They are connected to 3 other pixels but they are not intersection points.

2.5 Chain coding

Representing a line by chain coding was first described by Freeman (1961) and has since been described by others, including Gonzalez (Gonzalez and Wintz, 1987). Pixels on the

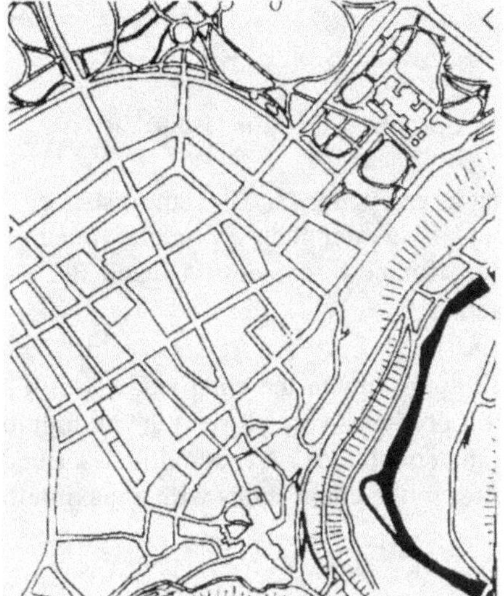

Figure 2.6 Image to be thinned.

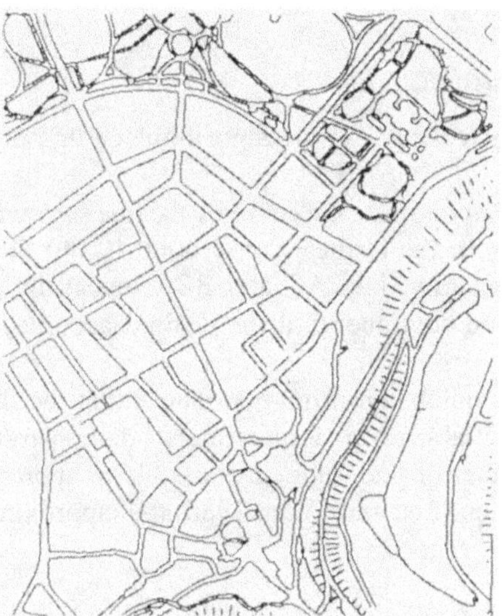

Figure 2.7 Thinned Image.

Figure 2.8 Thinned overlayed on original.

Figure 2.9 The problem of non-unitary pixels.

thinned image are followed to chain code the image. The only problem at this stage is in determining when a pixel is part of a line, when it is an end point and when it is an intersection. This problem is solved by finding out how many of the pixel's 8-connected neighbours are black. One black neighbour means the pixel is an end point, two mean it is part of a line, and more than two means it is an intersection point. Zhang's thinning algorithm (Zhang and Suen, 1984) occasionally leaves points that should not be intersection points with 8-neighbour counts larger than 2, but a slight modification of the algorithm fixes this (Abdulla *et al.*, 1988).

2.5.1 Vector reduction

Several systems are described in (Bixter *et al.* (1988) and Jain (1989).

Reduction to lines

An excellent, simple algorithm for this process is described by Jain (1989). It is:

> Approximate the curve by the line segment joining its end points (A, B). If the distance from the farthest curve point (C) to the segment is greater than a predetermined quantity, join AC and BC. Repeat the procedure for new segments AC and BC, and continue until the desired accuracy is reached.

This quick algorithm guarantees that the lines will approximate the chain code within a specifiable error. Repeated tests demonstrated that it produces very close to the minimum number of line segments possible to approximate the chain code. A method is described by Leu (Leu and Chen, 1988) that approximates lines to the chain codes with a maximum error criterion.

Reduction to arcs and splines

Methods for fitting arcs to a set of points are given by Rosin and West (1989). Methods for fitting splines to a set of points are given by Bixler *et al.* (1988) and Jain (1989). Chain coding is following a line one pixel wide by starting at an end pixel and then finding the next pixel by using a single number that describes the direction in which to move. The coding of the directions is shown in Figure 2.10. To chain code an image, the algorithm

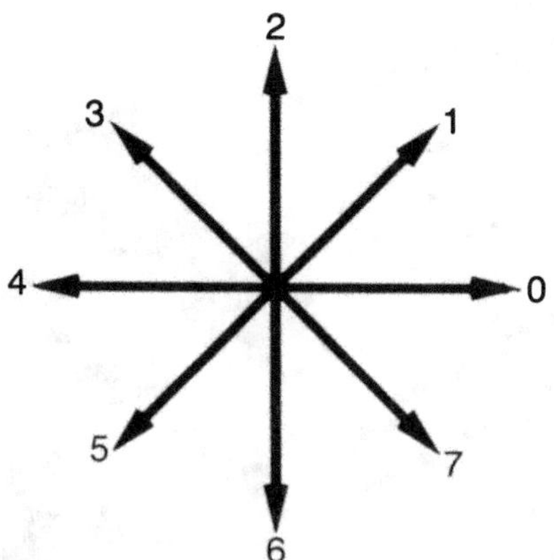

Figure 2.10 Chain code directions.

finds a pixel that is an end point of a line segment and chain codes along the line until it reaches a pixel that is an end point or an intersection. Pixels that represent end points and points of intersection are simply defined such that $N(p_1) \neq 2$. If $N(p_1) > 2$ the pixel is an intersection point, and if $N(p_1) = 1$ it is an end point. If $N(p_l) = 0$ it is an isolated bit of noise that should be deleted. Once the algorithm finds an end point or an intersection of the line, the chain code for this line is output, and all the pixels that are part of it are deleted. This chain coding process is repeated until all the black pixels on the image are deleted. A problem with this system is that it never chain codes figures, like the letter O, that have been thinned down to one contour and have neither a starting point where the chain coding could begin nor intersection nor end points where the system would stop. A solution to this problem lies in first chain coding the image just as described above but then doing an extra pass over the whole image and, if any black pixels have not been deleted, assuming they are part of a closed loop and starting chain coding at one of these points.

2.5.2 Chain reduction

In this stage, the chain codes are examined and long vectors that closely represent the chain codes are formed. The algorithm employed first takes a chain code and approximates the whole sequence by a single line going from one end to the other. This line is shown as AB in Figure 2.11. The farthest point on the chain code is found and labelled as point C

If the distance from the line AB to the point C is less than the maximum deviation the user has specified, the vector is a valid approximation of the chain code. If the distance is greater than the maximum allowable deviation, then the algorithm tries to approximate the chain codes from A to C with the line AC and the chain codes from C to B with the line CB. This approximation continues recursively until the distances are less than the maximum deviation. This second approximation is shown in Figure 2.12. If the user had specified a maximum deviation of 2 pixels, the line CB would be close enough, but AC would need to be further subdivided into the the configuration shown in Figure 2.13. At this point all the vectors are within the specified deviation, so the algorithm terminates.

2.5.3 Results

A section of an image was taken through all the steps to produce a vectored image. The image, see Figure 2.14, was vectorized and reduced with an error value of 2. The output

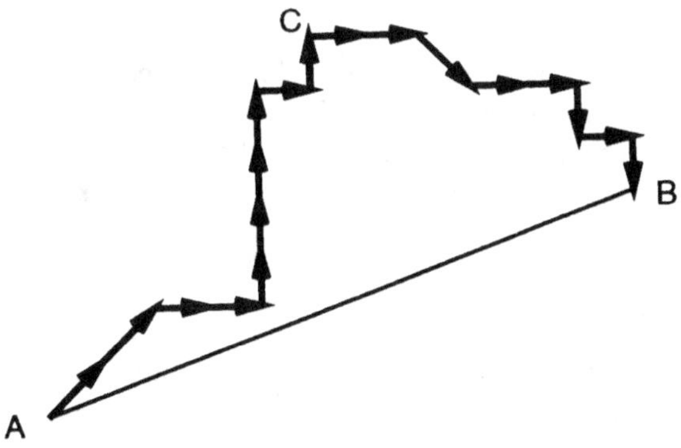

Figure 2.11 Vector approximation.

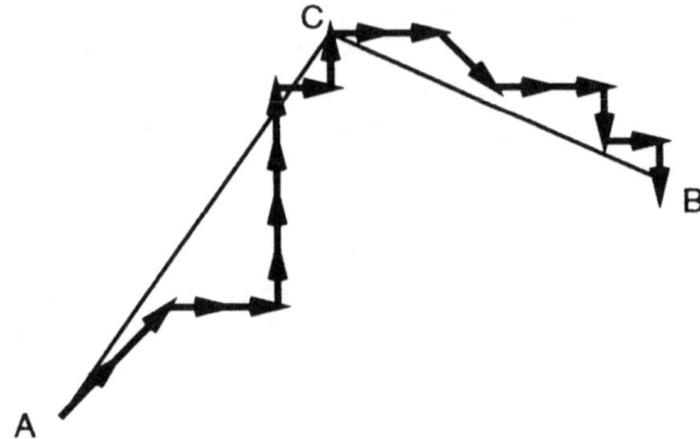

Figure 2.12 Second vector approximation.

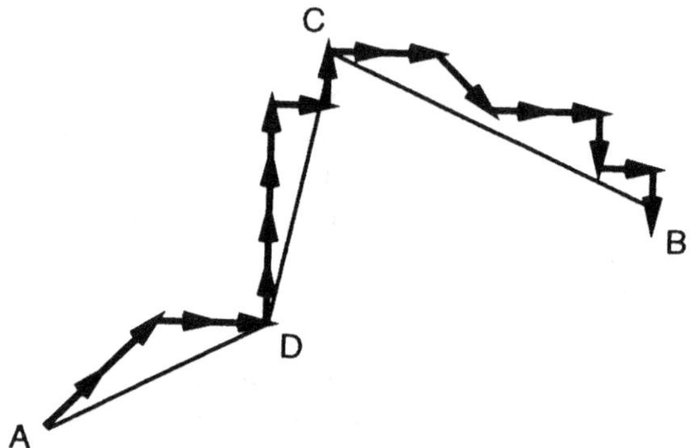

Figure 2.13 Third vector approximation.

vectors are shown in Figure 2.15. Trials with a reduction error values greater than 5 proved too large and resulted in a large misrepresentation of the image. In the case where the reduction error was set at 2 pixels a much better vectorization was achieved. A section of the vectors produced was overlayed on the corresponding section of the contemporary city plan shown in Figure 2.6 using the software developed.

2.6 Problems

2.6.1 Line fuzz

If a line on the scan has a broken edge terrible things happen. The thinning algorithm thinks the points of the rough edge are lines themselves and leaves little lines that run from the centre of the true line to where a peak on the broken edge was. This gives the final line many short lines that diverge off the true line and look like fuzz as seen in Figure 2.17.

2.6.2 Problem of scale

The scale problem is a fundamental problem in vectorization. Consider a line with a bump in it as shown in Figure 2.18. If this bump is very small, then this should be vectorized as just one line. If the bump is very large, then it is not a bump but another line that touches the first line. If the bump is somewhat in between, then it is undetermined. Given a line

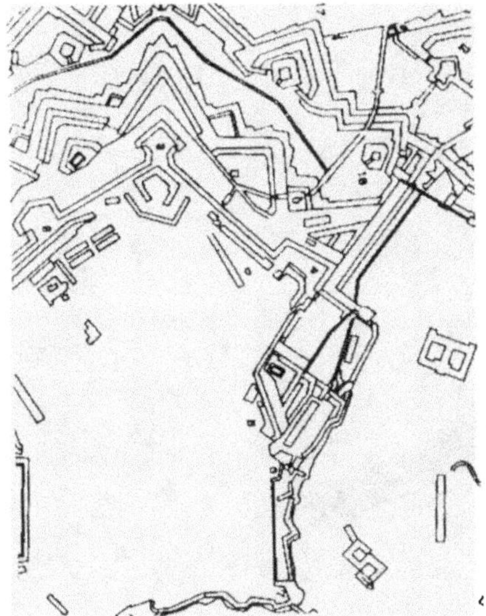

Figure 2.14 Original image.

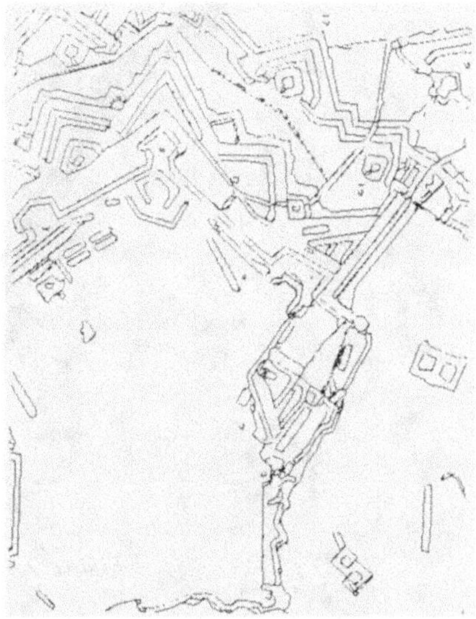

Figure 2.15 Reduction with error 2.

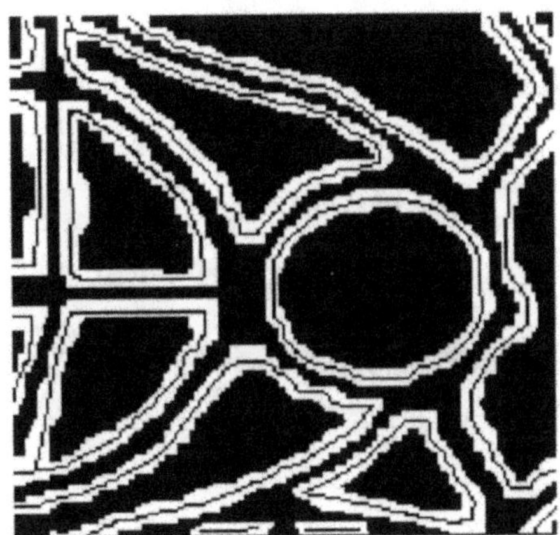

Figure 2.16 Thinned overlayed on city plan (detail).

with a bump, it might be a silhouette of a submarine or it might be a plain line. A human can tell what it is by using knowledge about the rest of the drawing and what it might represent.

2.6.3 Intersection artefacts

Just about all thinning processes cause a distortion of the image near intersections. This falls into two major catagories of T displacement and X destroying. The T displacement distortion is shown in Figure 2.19. The straight line is pulled down towards another line that intersects it. This causes a T to look more like a Y. The second major category of distortion is the X destruction shown in Figure 2.20. In this case two lines that intersect at low angles are incorrectly vectorized as two lines merging together into one line then diverging apart again into two lines.

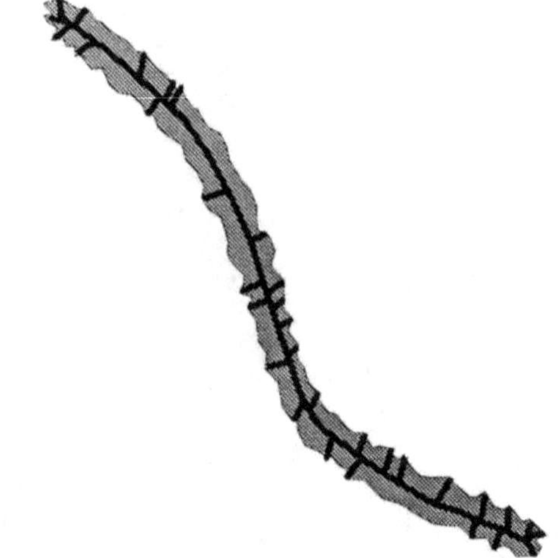

Figure 2.17 Fuzz on line.

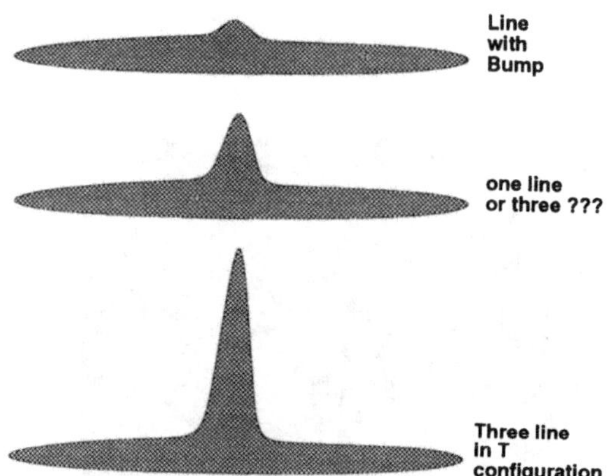

Figure 2.18 Problem of scale.

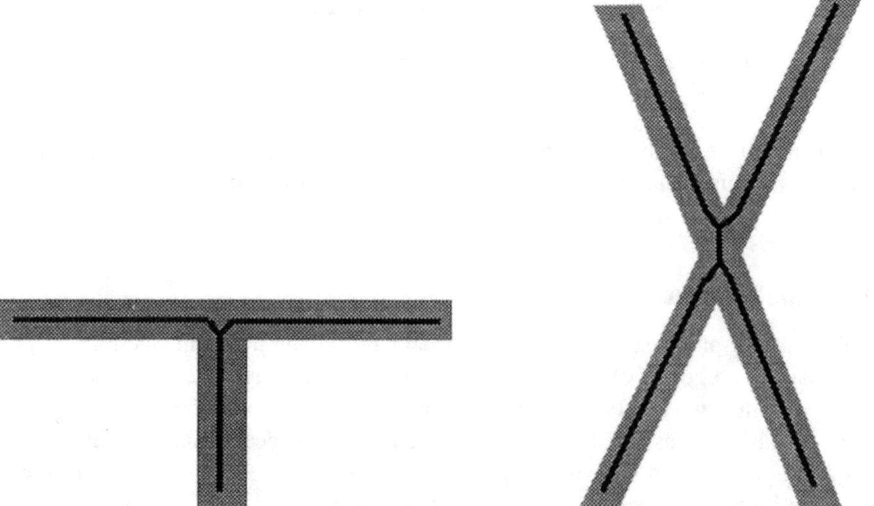

Figure 2.19 Intersection T displacement.

Figure 2.20 X intersection destruction.

2.6.4 Inside bias

One of the major problems with the algorithm employed is that it has inside bias meaning that the vectors approximating the curve are heavily biased towards the inside of the curve. A dotted line is shown in Figure 2.21 that approximates a solid curve. Because of the way the algorithm breaks the vectors, every vector is on the inside of the curve. This is the inside bias.

2.6.5 End point artefacts

The ends of lines often have noise introduced by the thinning process. In Figure 2.22 the right end has been bent into one corner and the left end of the line has split and diverged to both corners. The reason that this type of noise is a particular problem is that the direction of the end of a line is often used by higher level processes. For example, a system that was finding dashed lines would like to find dashes that point at each other and group them together into one dashed line. To do this it would use the end point direction information. The split ends and end deviation corrupt the direction information.

2.7 Conclusions

Currently there is a considerable need for computer vision systems that can recognize 2D line drawings of engineering diagrams and maps. Significant research and development effort has been put into solving this problem. Automatic conversion in a way that takes an order of magnitude less human operator time than just re-drafting the drawing is not currently possible for most drawings. In a manual system a long straight line can be digitized by two keys pushed on the digitizing table cursor. In an automatic system, if the user has to correct one end point or delete a piece of fuzz on the line or move the line, it will take longer than digitizing the line manually. Because of this, systems that try to automatically digitize line drawings must have an extremely low error rate to require less human operator time than manual digitization. The type of artefacts caused largely by the thinning process have a lot to do with why current automatic system are not much faster than manual digitization. However, results from our work are encouraging. The coastal contour map in particular showed very few areas that require user correction. Designing efficient and accurate thinning and vectorization processes leads to output from our system that is accurate with low artefact counts.

The system described is the subject of ongoing refinement with current emphasis on the perfection of the thinning stage. The system holds promise that the low error rates necessary to make the process commercially viable can be attained.

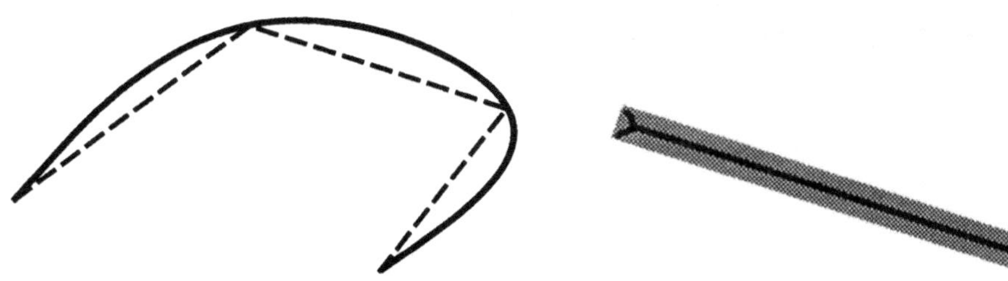

Figure 2.21 Inside bias.

Figure 2.22 Line end noise.

References

Abdulla, W.H., Saleh, A.O.M. and Morad, A.H., 1988, A preprocessing algorithm for hand-written character recognition. *Pattern Recognition Letters*, **7**, 13–18.

Baruch, O., 1988, Line thinning by line following. *Pattern Recognition Letters*, **8**, 271–6.

Bixler, J.P. and Sanford, J.P., 1985, A technique for encoding lines and regions in engineering drawings. *Pattern Recognition*, **18**(5), 367–77.

Bixler, J. P. , Waton, L.T. and Sanford, J.P., 1988, Spline-based recognition of straight lines and curves in engineering line drawings. *Image and Vision Comput.*, **6**(4), 262–9.

Bono, P.R., 1988, Uses for CGM in Raster-to-Vector Conversion, in Mumford, A. and Skall, M., (Eds) *CGM in the Real World*, 113–43. Springer, Berlin.

Donnelly, T. and Martin, W.N., 1987, Dracap: Drawing capture for electronic schematics, in *Proc. IEEE Computer Society Workshop on Computer Vision*, 274–6. IEEE, New York.

Dori, D., 1989, A syntactic/geometric approach to recognition of dimensions in engineering machine drawings. *Computer Vision, Graphics, and Image Processing*, **47**, 271–91.

Freeman, H., 1961, On the encoding of arbitrary geometric configurations. *IEEE Trans. Elec. Computers.*, **10**, 260–8.

Gonzalez, R.C. and Wintz, P., 1987, *Digital Image Processing*. Addison-Wesley, Reading, Mass., 2nd edn.

Govindan, V.K. and Shivaprasad, A.P., 1987, A pattern adaptive thinning algorithm. Pattern Recognition, **20**(6), 623–37.

Jain, A.K., 1989, *Fundamentals of Digital Image Processing*. Prentice Hall, Englewood Cliffs, NJ.

Joseph, S.H., 1989, Processing of engineering line drawings for automatic input to CAD. *Pattern Recognition*, **22**(1), 1–11.

Kwok, P.C.K., 1988, A thinning algorithm by contour generation. *Commun. ACM*, **31**(11), 1314–24.

Leu, J-G. and Chen, L., 1988, Polygonal approximation of 2-d shapes through boundary merging. *Pattern Recognition Letters*, **7**, 231–8.

Li, B. and Suen, C.Y., 1991, A knowledge-based thinning algorithm. *Pattern Recognition*, **24**(12), 1211–21.

Lim, J.S. and Nawab, H., 1980, Techniques for speckle noise removal. *Proc. SPIE*, **234**, 35–44.

Montanari, U., 1969, Continuous skeletons from digitized images. *Journal of the ACM*, **16**(4), 534–49.

Musavi, M.T., Shirvaikar, M.V., Ramanathan, E. and Nekovei, A.R., 1988, A vision based method to automate map processing. *Pattern Recognition*, **21**(4), 319–26.

Parker, J.R., 1991, Gray level thresholding in badly illuminated images. *IEEE Trans. Pattern Analysis and Machine Intelligence*, **13**(8), 813–9.

Rosin, P.L. and West, G.A.W., 1989, Segmentation of edges into lines and arcs. *Image and Vision Comput.*, **7**(2), 109–14.

Smith, R.W., 1987, Computer processing of line images: A survey. *Pattern Recognition*, **20**(1), 7–15.

Suzuki, S. and Yamada, T., 1990, Maris: Map recognition input system. *Pattern Recognition*, **23**(8), 919–33.

Taxt, T., Flynn, P.J. and Jain, A.K., 1989, Segmentation of document images. *IEEE Trans. Pattern Analysis and Machine Intelligence*, **11**(12), 1322–9.

The Anderson Report. Automatic Digitizers - Special Report, October 1986.

Wakayama, T., 1982, A core-line tracing algorithm based on maximal square moving. *IEEE Trans. Pattern Analysis and Machine Intelligence*, **4**(1), 68–74.

Yamada,, H., Yamamoto, K., Saito, T. and Matsui, S., 1991, Map: Multi-angled parallelism for feature extraction from topographical maps. *Pattern Recognition*, **24**(6), 479–88.

Yu, S.S. and Tsai, W-H., 1990, A new thinning algorithm for gray-scale images by the relaxation technique. *Pattern Recognition*, **23**(10), 1067–76.

Zhang, T.Y. and Suen, C.Y., 1984,. A fast parallel algorithm for thinning digital patterns. *Commun. ACM*, **27**(3), 236–9.

3

Computer-assisted tools for cartographic data capture

T.R. Mayo

The cost of cartographic data capture has long been recognized as a major problem in GIS. Conventional methods of data entry rely on manual digitizing and are both slow and expensive. Fully automatic conversion systems hold the promise of large savings in cost and time but have so far failed to deliver a general purpose capture solution. This paper investigates a proposed new class of capture techniques, computer-assisted systems, which offer smaller but still useful benefits and can be implemented using existing technology.

3.1 The cartographic data capture problem

Cartographic data capture forms a major cost in the implementation of most GIS. Most organizations need to convert a vast archive of map-based records into digital form before the benefits of GIS can be enjoyed. Conventional data capture techniques based on digitizing tablet technology are slow and expensive but currently form the only practical method of capturing this foreground data in the required structured form. A survey by Dickinson and Calkins (1988) suggests that data capture costs can constitute up to 70 per cent of total implementation costs of GIS systems. There can be little doubt that the development of cheaper and faster techniques for foreground and background data capture would yield significant benefits to the GIS user community.

The effort involved in cartographic data capture should not be underestimated. Digitizing of the Ordnance Survey basic scales series involves the conversion of some 220 000 map sheets and when complete will have cost at least £176 million (Chorley, 1987). Government organizations and public utilities need to convert archives consisting of tens or hundreds of thousands of sheets. Electricity companies, for example, typically maintain records of their distribution infrastructure at scales of up to 1:500 whilst HM Land Registry holds data on over 16 million land parcels at 1:1250 scale. Even relatively small improvements in capture efficiency can lead to useful cost savings when such large volumes of data are being processed.

Cartographic data is generally captured by manual digitizing or raster scanning. Manual digitizing allows the capture of the attributed, structured data required in most applications by employing a human operator to interpret the content of the drawing during conversion. Raster scanning dramatically reduces capture costs by eliminating the human element but the resulting data lacks structure and is generally only suitable for use as a map backcloth. There has been a long history of attempts to improve capture efficiency using advanced techniques, either by modifying the manual capture process or improving the raster product, but none has lead to a general purpose replacement for conventional techniques. The vast majority of cartographic data capture is still performed by manual, tablet-based digitizing.

In 1988 a project commenced at Cambridge University Geography Department aimed at developing more efficient capture techniques using data from the large format raster scanners that were starting to become widely available at that time. The project concentrated particularly on the processing of the low quality foreground documents often encountered in GIS applications. This paper describes a new class of advanced technique identified during the project, the computer-assisted system. Such systems may hold the potential to improve data capture efficiency over a wide range of problems.

3.2 Advanced techniques for cartographic data capture

Previous attempts to improve capture efficiency can be divided into four approaches; novel hardware, vector structuring systems, raster-to-vector conversion and digital overlay systems.

Early work focused on the development of novel hardware to improve operator performance during digitizing. In the groovy digitizing system described by Rhind (1974) an etched copy of the target document was used to constrain the movement of a sharp stylus during manual digitizing. Cameron (1984) describes a system based on an array of light-sensitive diodes built into the digitizing puck. Most such systems failed to achieve widespread popularity, mainly due to their strict data requirements. An important exception was the Fastrak digitizer (Fulford, 1981) which has been used in the capture of contour data for many years. The Fastrak uses laser deflection technology to track linework on a photographic image of the target document under guidance from an operator. Unfortunately the high cost of the equipment restricts its use mainly to national mapping agencies. In the last few years several line following systems based on similar principles but using conventional hardware have become available; see Woodsford (1991) and Shapiro (1992).

Vector structuring systems attempt to reduce digitizing costs by automatically structuring and feature coding manually digitized vectors. Most systems make use of an expert systems approach, processing the vector data by applying a variety of geometric and topological rules. A typical application is the automatic association of buildings and text in Ordnance Survey data described by Bridger and Fukushima (1989).

Automatic raster-to-vector conversion systems attempt to remove the human element entirely by extracting features directly from a conventional raster scan, yielding a result similar to that achieved by manual digitizing. Algorithms to convert a binary scan into unstructured spaghetti vectors are well known but the result is of limited use in the GIS context since it contains no more information than the original raster data. Most systems therefore attempt to structure the results of the vectorization procedure using a rule base similar to that found in vector structuring systems. A classic example is MARIS (map recognition input system) developed for the capture of Japanese 1:2500 scale maps (Suzuki and Yamada, 1990). As yet such systems provide only limited structuring and attribute association facilities and reported performance is poor. MARIS, for example, provides only 90 per cent correct feature recognition on a relatively simple source document despite at least four years of research effort.

Digital overlay (or heads-up) systems attempt to make use of raster imagery in a much less ambitious manner. Such systems replace the paper document and digitizing tablet with a scanned image displayed on the computer screen. An operator traces off features from the image in much the same manner as from a conventional tablet although the system does provide some benefits in the form of increased feedback and variable display scale. These benefits must be set against the problems introduced by limitations in screen size and resolution.

Of these advanced approaches automatic systems offer greatest potential savings in capture costs and it is in this area that most research has been concentrated; see, for instance, Lichtner (1985), Musavi *et al.* (1988) and Bjerch and Taxt (1988). Unfortunately the difficulty of specifying a rule base that copes correctly with every conceivable configuration of map entities makes fully automatic systems prone to error even on high quality topographic data. There is little evidence to suggest that they will provide an effective solution to the more complex problem of foreground data capture within the near future. This research investigates a proposed new class of advanced capture technique, the computer-assisted system, which offers smaller potential cost savings but is inherently robust and can be implemented using existing technology.

3.3 Computer-assisted techniques

Assisted systems form a natural development of digital overlay techniques. They still rely on an operator to initiate and guide the capture process but take advantage of the power of modern computers to remove some of the drudgery involved in digitizing. The operator examines each feature as it is captured, correcting mistakes and resolving ambiguities as they are encountered. The human element reduces sensitivity to minor imperfections present in the original document or introduced during scanning. As in digital overlay, data structuring and attribute entry can be combined with the digitizing activity if desired.

The line-following systems described above illustrate the application of an assisted technique. In such systems the computer assists in the selection and accurate positioning of data points leaving the operator free to guide the tracking process. Line-following systems are well suited to the capture of sinuous linework where the choice of appropriate control points requires skill and judgement but may be less efficient in other situations. This work investigates the possibility that a range of assisted techniques can be developed covering all the basic cartographic data types, providing a general purpose capture system. Evidence that such techniques might exist is provided by work in the field of image analysis, where researchers have used interactive techniques to improve operator performance in data capture from remotely sensed imagery. The systems described by Kass *et al.* (1988), Buchhem and Lillesand (1988) and Schowengerdt and Pries (1988) all rely on computer-assisted principles. The SAFE system for operator guided structuring of cartographic vector data (Goodson and Lewis, 1988) could also be considered an assisted technique.

Assisted systems are never likely to provide the improvements in capture efficiency that fully automatic systems aspire to, since the human element will always limit the capture rates that can be achieved. They may, however, be able to deliver significant benefits to the user community within a short timescale, a factor that is of great relevance to the many large organizations currently embarking on cartographic data capture programmes. The following sections outline a theoretical framework for the design and analysis of computer-assisted systems, describe some systems developed during this research project and present a quantitative analysis of their performance.

3.4 A framework for the analysis of interactive capture systems

To facilitate the design of assisted capture systems a framework for the analysis of the interactive digitizing process is required. Such a framework can be developed from an examination of conventional tablet-based digitizing. Analysis suggests that this process consists of nine key tasks:

1. *Point selection:* the selection of the next point in the current feature to capture.
2. *Point location:* the accurate alignment of the digitizer puck crosshairs on part of a feature and the recording of a location.
3. *Orientation specification:* the accurate positioning of the puck in one direction to indicate text or symbol orientation.
4. *Text entry:* the use of the keyboard to enter text strings and attribute data.
5. *Marking:* the marking of a feature as processed on the original to prevent duplicate entry.
6. *Search:* the selection of the next feature in the document to capture.
7. *Review:* the examination of the document for areas or features not yet processed. This differs from search in that features detected may be flagged rather than digitized immediately and because it is generally performed in a systematic manner.
8. *Option selection:* the modification of the current digitizing environment (for example, feature code or snap tolerance) by menu item or puck button selection.
9. *Pick:* selection of an existing digital feature (in snapping or attribute entry).

The order in which the tasks are performed depends on the type of data being captured and the digitizing software used to control the process. Figure 3.1 illustrates a typical sequence of operations for two common capture tasks; the entry of topographic linework and capture of attributed network link data.

Each task has an associated time penalty and error characteristic. The contribution of the task to overall digitizing costs depends on the magnitude of the time penalty and the frequency with which the operation is performed. The capture of sinuous linework, for example, might be expected to be dominated by point selection and point placement costs, whereas network data capture might well contain a large text entry component. Some experimental investigations to measure these costs directly have been conducted during this project. The results will be presented in future publications.

Analysis of the digital overlay technique using the 9-task framework identifies both strengths and weaknesses compared to conventional systems. Marking costs are removed since captured vectors are automatically displayed over the raster image and point placement times might also be expected to be reduced since digitizing can proceed at enlarged scales. The time taken to perform point selection and search will however rise dramatically if frequent panning of the display is required (that is, if the average feature size is large compared to the image display area). Review is also likely to be more difficult since it is usually impossible to display the complete document on a workstation screen at a useful resolution. The balance between costs and benefits will therefore be determined by the properties of the dataset being captured and in particular its average feature size and density.

Computer-assisted capture techniques aim to improve the performance of digital overlay systems by increasing operator speed or reducing error rates in one or more of the capture tasks. Many strategies for achieving this can be identified, for instance assisting in point placement by local analysis of the raster image or automating text entry using character recognition techniques. In addition, efficient systems must provide a quick and simple mechanism for correcting the errors that will inevitably occur and be reasonably insensitive to minor faults in the scanned data. Systems that exhibit these characteristics can be termed

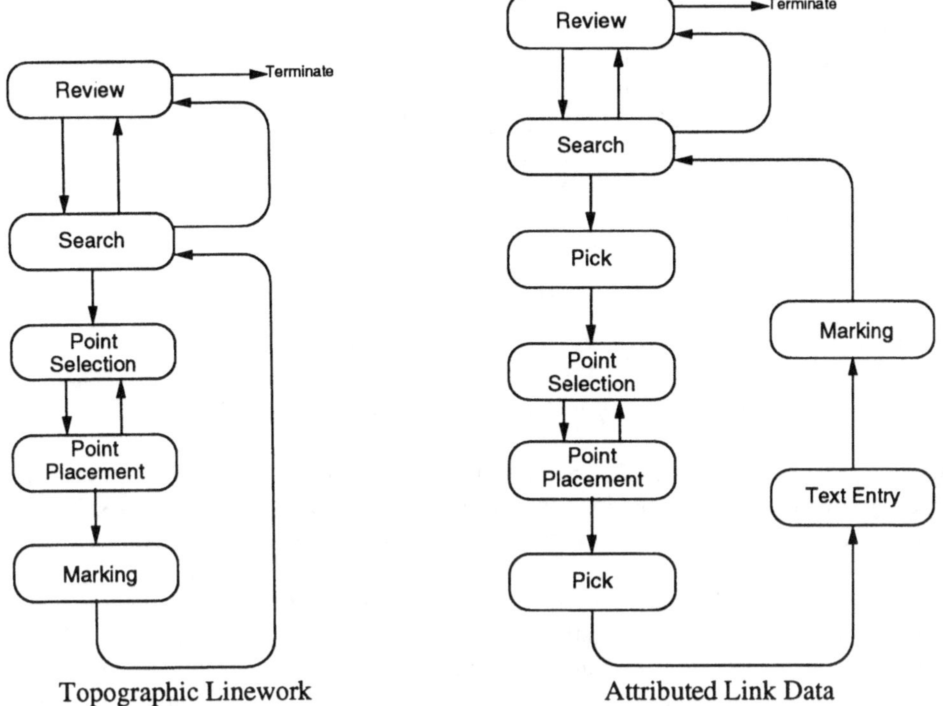

Figure 3.1 Typical sequences for two common capture tasks.

'robust'. The next section describes four computer-assisted tools designed to address some basic cartographic problems in an efficient and robust manner.

3.5 Some assisted tools for basic cartographic data types

The research project has investigated a number of possible designs for assisted data capture tools. Four of these have been developed into a fairly advanced state for use in experimental evaluations. Each of the four is aimed at a basic capture problem, either straight linework, sinuous linework, polygon data or printed text.

The **assisted digitizing tool** attempts to aid the operator in point placement during linework capture. To capture a line the operator points to its approximate start and end points using the screen cursor. The system searches a strip of raster between the hits, selects the pixels that are judged to form the line and replaces them by a vector approximation (Figures 3.2a and 3.2b). Further segments can be added by selecting additional points with the system computing knot locations by line intersection (Figure 3.2c). The size of the strip searched is indicated by a circular disk or locator and can be reduced for detailed work. Since the vector approximation is based on the raster image, high placement accuracy can be achieved from quite poor operator hits. The system is aimed primarily at the capture of straight linework although it can be used with any type of linear data.

The **line-following tool** assists the operator in point selection and point placement during sinuous line capture. It presents a similar interface to other line-following systems although details of the implementation differ significantly. To capture a line the operator points to its approximate start position and indicates the rough direction of tracking required

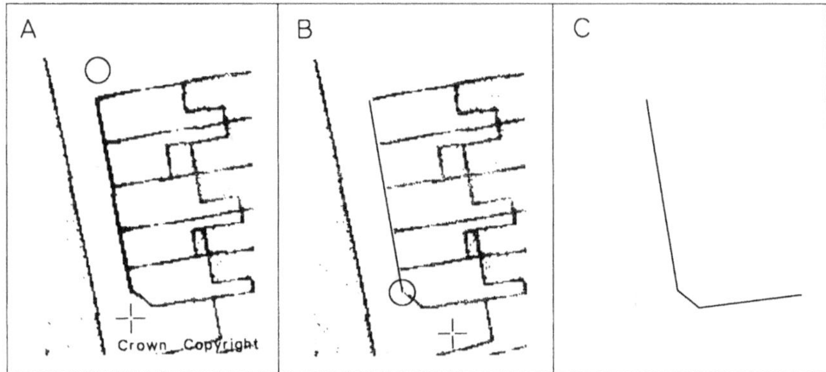

Figure 3.2 Operation of the assisted digitizing tool.

(Figure 3.3a). Depressing a mouse button causes the system to attempt to track the line, with progress indicated by a moving screen locator. The locator attempts to pass through occlusions and junctions by reconstructing the line stoke from its local orientation (Figure 3.3b).

The operator can back up and correct the tracking process or force the locator to move to a desired location using a combination of mouse strokes and keyboard operations. When a satisfactory raster representation of the line has been identified it is replaced by a vector approximation using another mouse click (Figure 3.3c). The locator automatically jumps small gaps during tracking and, by selecting appropriate control parameters, dashed lines may also be handled. Although primarily designed for sinuous data the procedure can handle all types of linework.

The **polygon capture tool** assists in point selection and placement during polygon capture. To capture a polygon the operator simply clicks the mouse on a point inside it. The system searches the image for the nearest enclosing boundary, smooths the result and displays it on the screen (Figure 3.4a). For composite polygons such as land parcels the boundary can be extended by merging in adjacent polygons using additional hits (Figure 3.4b).

When the required boundary has been achieved another button click is used to convert the raster outline into a vector feature (Figure 3.4c). The tool can cope with polygons containing stipple and text but cannot, as yet, handle regions delimited by dashed linework. It has been used successfully on both polygons with simple boundaries, such as building outlines, and complex regions, such as water bodies and closed contours.

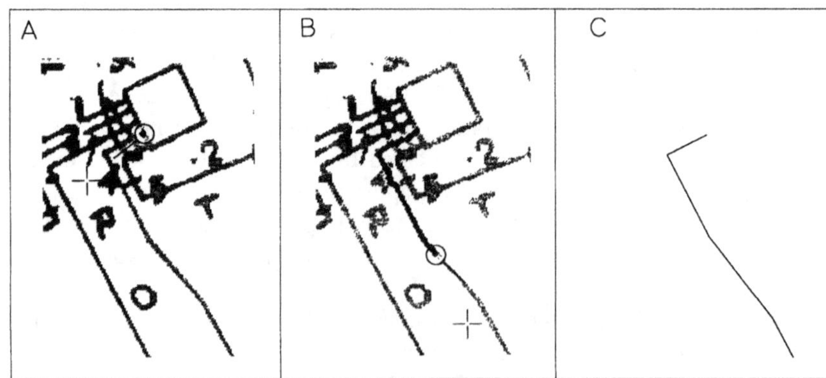

Figure 3.3 Operation of the line-following tool.

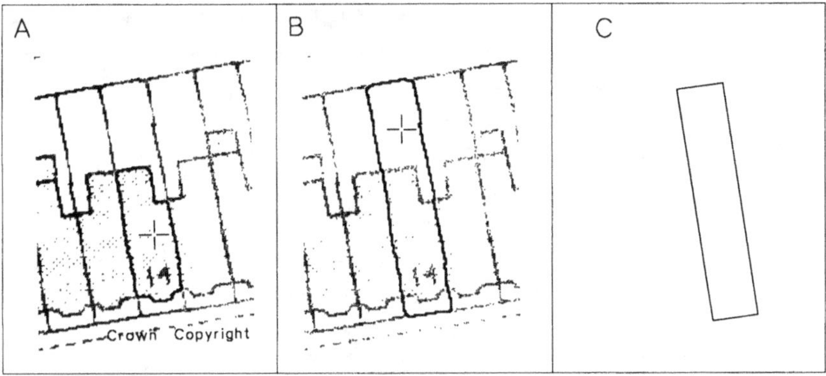

Figure 3.4 Operation of the polygon capture tool.

The **text capture tool** is designed to assist in entry and placement of printed and stencilled text. To capture a text string the operator points to its approximate start and end points in the image. The system searches the region between the hits for probable character objects and highlights them (Figure 3.5a). Subsequent hits can be used to extend the selected set (Figure 3.5b). When the required pixels have been identified the operator instructs the system to classify each object against a user supplied training set and replace successful matches with a vector representation (Figure 3.5c). The system can handle text at any scale and orientation, on a straight or curved baseline and containing more than one font. The current implementation does not cope reliably with broken or merged characters and provides only crude error correction facilities. Recognition rates are high compared to other reported work but could be further improved by the use of syntax analysis or data dictionary approaches.

The tools are implemented as separate programs under the VMS DecWindows environment but are based on a common user interface. The interface supports both monochrome and indexed colour raster imagery, permits zooming and panning of the data and provides facilities for the storage and retrieval of raster and vector coverages (Figure 3.6). Each tool is designed on the direct manipulation principle (Took, 1990) with the concept of a locator having been found to be particularly useful in this respect. All the algorithms attempt to segment out a raster image of the target feature before it is vectorized, thereby avoiding many of the problems encountered using conventional vectorization approaches (Smith, 1987; Jennings and Flanagan, 1993).

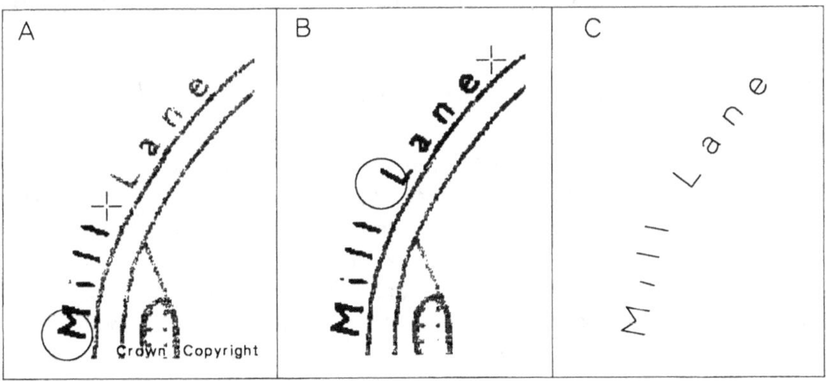

Figure 3.5 Operation of the text recognition tool.

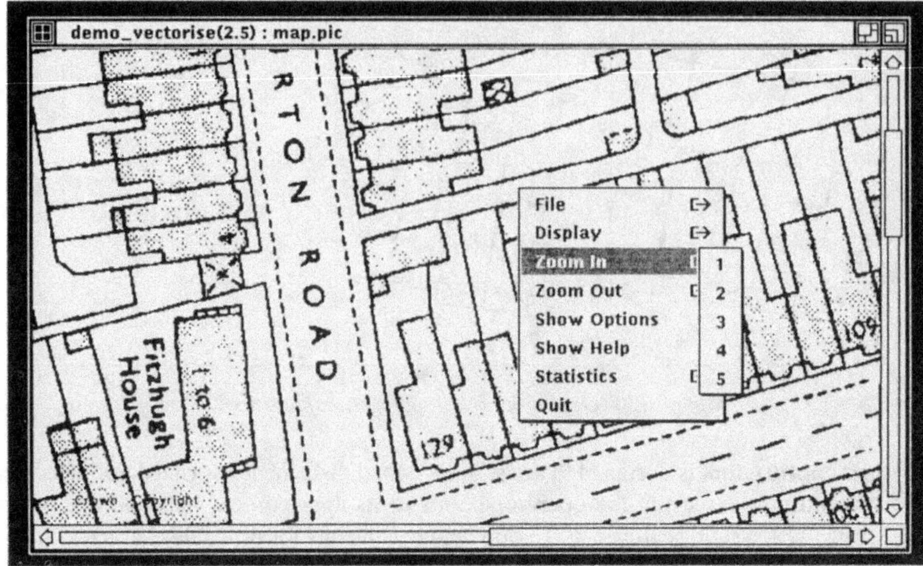

Figure 3.6 The common user interface.

Apart from its specific aim each tool attempts to assist in the search and review activities by 'greying out' the pixels that have already contributed to vectorization. These pixels and the associated vector features can then be removed from the display using a program option, facilitating the location of unprocessed features. All the programs attempt to locate features to sub-pixel accuracy and are generally able to perform satisfactorily on fairly low resolution data (200dpi for most topographic documents). They also provide 'undo' facilities to correct mistakes within the current feature but do not permit editing operations as these were not required for the experimental work. A conventional digital overlay system based on the same user interface has also been constructed for use in comparative evaluations.

The assisted tools have been used to capture data for a number of environmental research projects within the Department of Geography. Applications have included capture of a topographic and meteorological database for southern Spain, the extraction of parcels from land use survey maps of the UK and the entry of elevation data for a terrain modelling study in the Swiss Alps. They have also been the subject of a number of comparative experiments designed to assess their performance on various kinds of cartographic data.

3.6 Experimental evaluation of the assisted tools

In order to examine the efficiency of the tools developed and to draw implications for the design of other assisted techniques a series of timed digitizing experiments have been undertaken. The experiments compared operator performance using conventional tablet-based digitizing, digital overlay and assisted techniques. A range of source documents have been considered including examples containing topographic, network and polygon data. Evaluations performed so far have concentrated on the linework capture tools which were developed earlier in the research programme and are the most extensively tested. The results provide an indication of increases in efficiency achievable using assisted tools under a range of conditions and also give a comparison of the efficiency of conventional tablet-based and digital overlay systems on various documents.

Twenty comparative experiments have been conducted to date. Each experiment involves an operator digitizing the same document four times using the conventional, digital overlay, assisted digitizing and line-following techniques. In all, six documents and five operators have been examined, with each document processed by at least two operators. The documents were taken from published sources or from samples collected during a small-scale survey. They represent a wide variety of capture problems from dense rural topographic data consisting mainly of sinuous linework to sparse network schematics which are dominated by text. Table 3.1 lists the documents used and provides an indication of their data content based on statistics recorded for the author using a tablet-based capture system.

During each test the operator was asked to record the time spent on the capture of each feature type. The results, which exclude bookkeeping operations such as document registration and file creation, were subsequently aggregated to line, symbol and text classes. Operators were not expected to edit or structure the data they captured although foreground data was entered in a form suitable for structuring. Conventional capture was performed using a TDS A0 digitizing tablet controlled by the ArcEdit module of the ArcInfo package. The screen-based techniques used a DEC VaxStation model 38 computer fitted with a 16 ins colour monitor and a 3-button mechanical mouse. Table 3.2 shows the average linework capture times recorded for each technique. Figure 3.7 expresses the results as effective point capture rates based on the data volumes set out in Table 3.1. Examination of these results leads to several conclusions:

1. Digital overlay (DO) does not appear to provide a significant increase in point capture rates over conventional (CV) methods. Recorded capture rates nowhere exceed those for conventional methods by more than 30 per cent and in two cases overlay is actually slower than manual data capture.
2. The assisted digitizing (AD) technique shows a consistent improvement in speed over digital overlay, suggesting that assisted point placement does indeed increase capture

Table 3.1 Document extracts used in the comparative evaluations

		Features					Average
Document series	Description	Lines (cm)	(pnt)	Text (str)	(char)	Symb (n)	line length (cm)
TopoR	Topographic (rural)	960	(3301)	102	(535)	182	0.29
TopoU	Topographic (urban)	487	(1377)	48	(221)	116	0.35
Poly	Polygon (urban)	473	(214)	157	(327)	0	2.21
Net	Network	860	(1071)	185	(859)	159	0.80
SchemD	Network schematic (dense)	594	(309)	191	(1370)	61	1.92
SchemS	Network schematic (sparse)	1183	(163)	132	(1410)	19	7.26

Table 3.2 Capture times for linework

	Average Time (Min)					
Technique	TopoR	TopoU	Poly	Net	SchemD	SchemS
Tablet-based (CV)	108	78	16	71	16	16
Digital overlay (DO)	137	70	17	64	15	12
Assisted digitizing (AD)	75	53	16	41	14	11
Line following (LF)	76	69	28	56	18	14

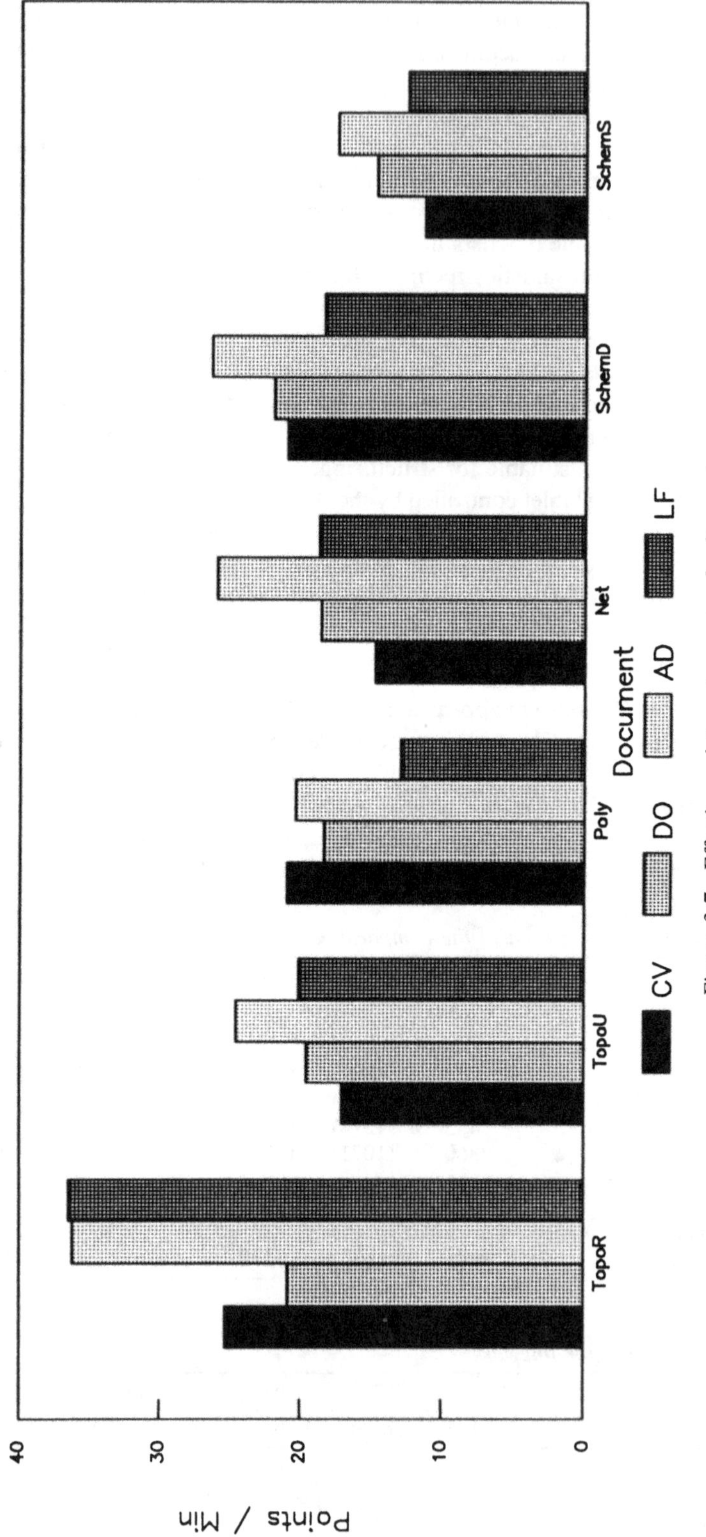

Figure 3.7 Effective point capture rates for linework.

rates. The increase is greatest for the dense rural topographic data and smallest for the polygon data (a rather poor quality colour image derived by density slicing greyscale data). The technique gives an average saving in digitizing time of 26 per cent compared to conventional methods.

3. The performance of the line-following tool is much more variable. On the rural topographic data capture rates are similar to, although surprisingly not higher than, those achieved using assisted digitizing. On other documents placement rates are similar to or worse than those for unassisted overlay with particularly poor performance on the low quality polygon data scan. In these documents the benefit obtained from assistance in point selection is more than cancelled out by the time taken to control and correct line tracking, at least using the algorithms developed in this project.

The results therefore suggest that assisted techniques can significantly improve capture rates provided an appropriate tool is chosen for the job. Selection of an inappropriate tool can lead to much more limited improvements or even reductions in performance.

Error rates also have a strong impact on overall capture efficiency since error correction is a time-consuming task. The digital datasets captured during the experimental work were therefore analysed to assess levels of omission and commission. Figure 3.8 presents the results of this analysis, expressed as the average proportion of points that were in error (missing, grossly misplaced, incorrectly feature coded or duplicated).

The screen-based systems generally give lower rates of error than tablet-based digitizing, although in both cases performance is strongly related to the type of data being captured. Digital overlay gives better performance than manual capture on all the test documents (60 per cent less errors on average) as one might expect. Assisted digitizing, however, exhibits consistently worse results than digital overlay which is surprising given the similarity of the interfaces of the two techniques. Examination of the raw data suggests that the increase is mainly due to extra errors of omission, perhaps indicating that assisted interaction encourages a less rigourous approach from the operator. Results for the line-following tool show a similar pattern and appear to be due to the same causes. There would therefore appear to be a case for the implementation of a simple facility to assist the operator in review activity, perhaps involving a systematic guided search of the raster image.

Point placement accuracy is currently under investigation. Previous work using a prototype of the assisted digitizing tool showed a 37 per cent improvement in placement consistency between operators compared with tablet-based digitizing (Devereux and Mayo, 1990).

The polygon and text capture techniques have not yet been subjected to rigourous comparative evaluation. Initial tests with the polygon tool indicate that effective capture rates of 31 points per minute (on buildings) to 148 points per minute (on water bodies) can be achieved with the rural topographic document. The text recognition program provides entry rates similar to those for manual capture due to the high computational costs incurred, although performance increases dramatically on faster RISC-based hardware.

3.7 Conclusions

The experimental results presented above indicate that assisted techniques can indeed improve data capture efficiency, both in terms of increased capture speed and reduced errors rates. Relatively simple techniques can increase effective point placement rates by around 50 per cent compared to conventional capture methods provided that the tool used is well matched to the task. Choice of an inappropriate tool can lead to much smaller benefits

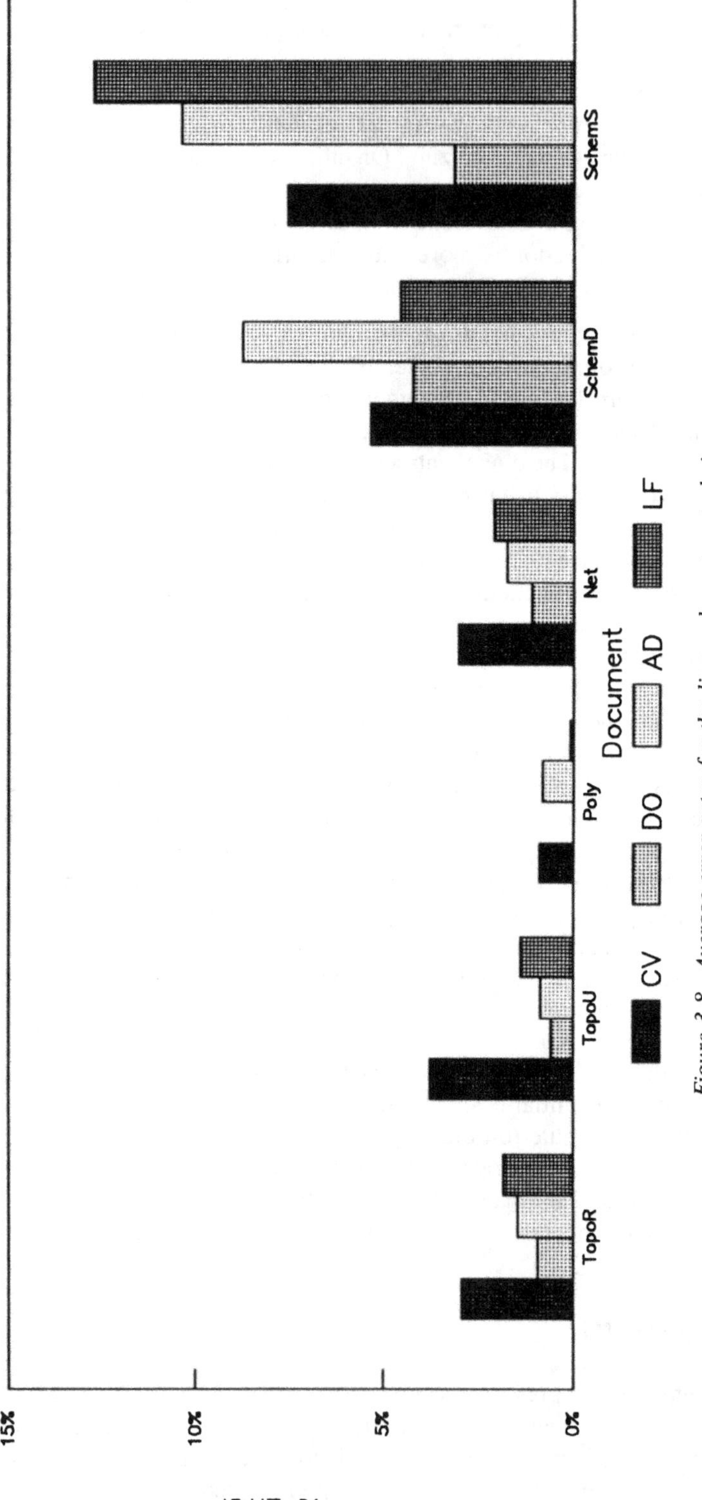

Figure 3.8 Average error rates for the linework capture techniques.

or even a reduction in capture rates. In an ideal system the operator would be able to swap between capture techniques at will, perhaps even during the entry of a single feature. A major aim of the next phase of this project will be to develop such an integrated environment for experimentation and analysis.

The initial tests of the polygon capture tool indicate that more specialized techniques are likely to lead to further improvements in capture rates. The techniques developed in this project are currently being used as the basis for the investigation of composite tools or supertools that address application specific capture tasks (Devereux and Mayo, 1992). The potential of the text recognition in this regard is particularly exciting.

In conclusion, this paper has investigated a proposed new class of cartographic data capture techniques termed computer-assisted systems. It has provided a theoretical framework for their design and analysis and described some examples constructed for research purposes. Experimental results indicate that modest but useful increases in capture efficiency can be achieved using simple assisted tools aimed at basic cartographic capture problems. More specialized tools based on these techniques hold out the possibility of substantially increasing the savings that can be achieved.

Acknowledgements

The work described in this paper was conducted as part of a research projectg funded by Alper Systems Ltd, Cambridge.

References

Bjerch, T., and Taxt, T., 1988, Syntax analysis in automatic digitizing of maps, in *Procs. 4th International Conference On Pattern Recognition*, British Pattern Recognition Society, Cambridge, 222–9.

Bridger, I. and Fukushima, Y., 1989, Towards intelligent data, in *Procs. AGI 89*, Birmingham, UK, sec B.2, 1–8.

Buchhem, M.P., and Lillesand, T.M., 1988, Semi-automated training field extraction and analysis for efficient digital image classification, in *Photogrametric Engineering and Remote Sensing*, Sept.

Cameron, E.A., 1984, Manual digitizing systems, in Marble, D., Calkins, H. and Peuquet, D. (Eds), *Basic Readings in GIS*, Spad Systems Ltd, sec 3, 11–8.

Chorley, Lord, 1987, *Handling Geographic Information*, HMSO.

Devereux, B.J., and Mayo, T.R., 1990, In the foreground: intelligent techniques for cartographic data capture, in *Procs. AGI '90 (2nd National Conf. of the AGI)*, Brighton, UK, Oct. 1990, sec.6.2, 1–9.

Devereux, B.J., and Mayo, T.R., 1992, Task oriented tools for cartographic data capture, in *Procs. AGI 92*, Birmingham, UK, sec 2.14, 1–7.

Dickinson, H.J. and Calkins, H.W., 1988, The economic evaluation of implementing a GIS, in *Int. Jnl. of GIS*, **2**(4), 307–28.

Fulford, M.C., 1981, The Fastrak automatic digitizing system for line drawings, *Int. Jnl. of Pattern Recognition*, **14**, 65–73.

Goodson, K.J., and Lewis, P.H., 1988, Feature extraction from line drawn images, in *Procs. 4th International .Conference On Pattern Recognition*, British Pattern Recognition Society, Cambridge, 216–21.

Jennings, C., Flanagan, N., 1993, Automatic GIS data capture and conversion, in *Procs. GIS Research UK 1993*, Keele, UK, 26–35.

Kass, M., Watkin, A., and Terzopoulos, D., 1988, Snakes: active contour models, in *Int. Jnl. of Computer Vision*, **1**(4), 321–31.

Lichtner, W., 1985, Pattern recognition procedures for automatic digitizing of cadastral maps, in *Procs. Auto-Carto 7*, Washington DC, 337–41.

Musavi, M.T., Shirvaikar, M.V., Ramanathan, E. and Nekorei, A.R., 1988, A vision based method to automate map processing, in *Int. Jnl. of Pattern Recognition*, **21**(4), 319–26.

Rhind, D.W., 1974, An introduction to the digitizing and editing of mapped data, in *Automated Cartography*, The British Cartographic Society, 51–66.

Schowengerdt, R.A., and Pries, R.A., 1988, Interactive image feature compilation for geographic information systems, in *Journal of International Society for Optical Engineering* (SPIE), **924**, 305–11.

Shapiro, J., 1992, GIS data conversion for complex documents, in *Procs. AGI 92*, Birmingham, UK, sec 2.11, 1–5.

Smith, R.W., 1987, Computer processing of line images: a survey, *Int. Jnl. Pattern Recognition*, **20**(1), 7–15.

Suzuki, S. and Yamada, T., 1990, MARIS: map recognition input system, in *Int. Jnl. of Pattern Recognition*, **23**(8), 919–33.

Took, R., 1990, Putting design into practice: Formal specification and the user interface, in Harrison, M. and Thimbleby, H. (Eds), *Formal Methods in Human-computer Interaction*, CUP, 63–96.

Woodsford, P.A., 1991, Cartographic and GIS development at Laserscan 1987 to 1991, in *The Cartographic Journal* (Journal of the British Cartographic Society), **28**, 43–9.

4

An automated generalization system for large scale topographic maps

Gary J. Robinson and Fang Lee

This chapter describes a practical approach to the automated generalization of Ordnance Survey large scale (1:1250) topographic maps, with a target scale of 1:10 000. The method involves creating complex meta-objects from the basic elements on the map, determining appropriate spatial relationships amongst them, and then performing generalization operations using rules based on those employed by Ordnance Survey in manual generalization. To speed up these procedures the spatial data are partitioned using the natural networks formed by roads and other linear features. The system is written in Prolog, C and embedded SQL and has elements of conventional procedural programming and expert systems.

4.1 Introduction

Map generalization is the process of reducing the amount of detail in a map so that the character or essence of the original features is retained at successively smaller scales. In the context of topographic map generalization this usually involves omission, aggregation, simplification, displacement, exaggeration and symbolization of either individual features or groups of features. The application of these operators involves considerable skill and expertise, in order to obtain an acceptable balance between retention of information and legibility. Cartographers in many countries have built up many rules and guidelines to undertake these processes in manual map production. For example, Ordnance Survey GB (OS) has incorporated the guidelines used internally into one module of a series of formal drawing instructions used by surveyors, draughtsmen and cartographers in the production of derived scale products, including the 1:10 000 series.

The translation and application of manual generalization processes to digital maps has been an active area of computer-assisted cartography for many years. Theoretical work by Perkal (1986) and Tobler (1966) established the foundations of this lively topic (Shea, 1991). Subsequent workers have extended these ideas, although they have tended to focus on the simplification of linear features (Lang, 1969; Douglas and Peucker, 1973; Jenks, 1981; Devean, 1985; McMaster, 1987). In contrast, relatively few methods for displacing cartographic features and generalizing areal features have been developed (Lichtner, 1979; Monmonier, 1983; Meyer, 1986; Nickerson and Freeman, 1986). While this early work concentrated on the generalization of individual features, there has been a noticeable shift towards the investigation of the relationships between multiple features and objects. As a result some rule-based or knowledge-based models have been proposed in recent years (Nickerson, 1986; Müller, 1991).

Most research into map generalization has been at small or medium scales. One reason for this may be that at these scales maps are relatively simple in terms of the variety of features on them. Another reason arises from the different dominant types of features at each scale. Large scale maps, especially in urban areas, mainly contain areal objects such as buildings, land parcels and road surfaces, whereas small scale maps mainly contain linear and point features such as contours, rivers, (stylized) roads and point symbols (Vicars and Robinson, 1989). The form of features and their spatial relationships with one another, also tend to be simpler at smaller scales. This often allows features to be treated independently during generalization. The considerably more complex spatial interactions at large scales prevents this, and forces the adoption of more sophisticated approaches. Another complicating factor is that generalization, at any scale, can be a highly subjective process: even experienced cartographers can and will come up with different results. This highlights the fact that generalization guidelines, where they exist, are not rigid. Any system designed to generalize a digital map automatically must therefore take this into account.

Automated generalization is becoming of increasing interest within national mapping agencies, possibly due to the growing market for digital base products and recognition of the benefits of deriving smaller scales products from these. This paper describes a project that addresses the problem of large scale topographic map generalization, using a more pragmatic approach than those employed in previous research, and which is specifically targeted at the derivation of 1:10 000 scale from 1:1250 scale OS digital data.

4.2 Topographic map generalization

The general nature of the problems posed by large scale topographic map generalization may be shown by considering the treatment of a building, using the guidelines used by the OS.

If the building is below a certain size it may be omitted. However, if it is also situated in an open area it could be topologically significant, that is, it may be a landmark in its own right, or may need to be retained to indicate the nature of the surrounding area. In this event it will have to be retained and expanded to the minimum size of building at the target scale. If the building is not isolated but is attached or close to one or more larger buildings it may be merged with them. The presence of nearby roads and other features may also influence how the building is handled. A common example of this is the need to displace certain building types away from road edges, keeping the intervening space to indicate the nature of the building, i.e., it may possess a front garden or forecourt. The generalization of a building depends therefore not only on simple geometric properties but also on the spatial context in which it occurs and its type. Clearly the more traditional and conventional algorithms used at smaller scales cannot handle this type of problem.

This example also illustrates one of the main problems with generalization, and in particular, in attempting to automate the process. This is that some generalization rules are extremely subjective. For example, what is meant by the term 'topographically significant'? It would be extremely difficult to express this concept in hard computer code given the enormous number of possible spatial configurations a building may possess with respect to neighbouring features. This raises the more general question of whether a truly fully automated generalization system can ever be developed. It may be that such a system can only go so far in producing useable results, at least one capable of closely reproducing those created by manual methods. This suggests that either compromises in terms of differing graphic styles and non-optimal solutions have to be accepted, or it is recognized that occasionally some situations will require manual intervention. In the case of the latter option the effectiveness of automated techniques will depend on how much of the process can be carried out successfully by the computer, and more importantly, whether those features that have to be processed manually can be completely and correctly identified.

4.3 The strategy

The starting point for our approach in developing a computer based system for the automated generalization of OS large scale topographic data is to consider the definitions of the various features or 'objects' found on the map, and in particular the nature of the spatial relationships amongst them. This is an essential step because the actions applied to an object during generalization strongly depend upon the spatial context within which the object occurs, which in turn is governed by the type and distribution of neighbouring objects. Object definitions, and in particular the associated spatial relationships, therefore need to be designed to allow the rules and guidelines use in manual generalization, with appropriate modifications where necessary, to operate in as natural a way as possible. The form of these definitions and the interactions involved, and the fact that generalization processes are already expressed in rule form, suggests that expert system techniques coupled with an object-based approach might provide appropriate solutions.

At first sight the complicated interactions amongst objects appears to make the task of defining, let alone applying, generalization operations extremely difficult. However, in practice certain properties of topographic maps simplify the problem. The first is that objects separated by certain types of linear features, such as roads rivers and railways, interact very little, if at all. Linear features, especially roads, thus provide a convenient way of dividing the map into natural spatial units, or blocks, so that the amount of processing necessary is substantially reduced. The second property is that certain categories of objects dominate others in terms of the order in which generalization actions are applied. In other words, a set of priorities for generalization exists. Roads, for example, can nearly always be generalized without regard to other features, whereas buildings must have their relationships with other buildings and proximity to road edges taken into account.

Another relevant feature of generalization rules, such as those employed by the OS, is that they sometimes refer to groups of objects, which may be connected, such as terraced houses, or not, as in a factory complex. Many rules are also context dependent, and have qualifiers that depend on the overall character of the region in which they are being applied. These properties suggest that a hierarchical scheme for defining objects and groups of related objects (meta-objects) would be appropriate, on which such generalization rules can be applied.

Using these observations the strategy we have adopted can be outlined: road centre lines, with the exception of those from cul-de-sacs, are used to create a network. The spaces, or blocks within the network, are then used to index the features on the map. Features are also placed into layers to assist with prioritizing the order in which generalization is performed. For example, roads, building and fences are put in one layer, and street furniture into another. Next, basic objects such as buildings are formed from the spatial data, followed by meta-objects, which correspond to groups of associated basic objects. Finally, generalization of the entire data set is then carried out. The implementation of these steps in a working system is now described in more detail.

4.4 OSGEN: A prototype generalization system

The ultimate objective of this project is to develop a computer-based system to generalize complete maps from 1:1250 to 1:10 000 scale automatically as a batch process. The system, called OSGEN (*O*rdnance *S*urvey *gen*eralization), currently only handles roads and buildings because these features comprise the overwhelming majority of the features on a typical OS large scale urban area map. Other features such as fences, walls and street furniture will be incorporated into the system later. OSGEN is being developed using purpose-written software within a VAX/VMS environment. Standard software components are used throughout to facilitate further development: Prolog is used in building the expert systems, embedded SQL is used to communicate with the database, and C, as a host language, is used both in linking the parts and for developing the generalization algorithms. One of the key features of OSGEN is that facts and other data held in the Prolog databases can be dynamically updated as generalization proceeds.

The OS topological data model is used as the basis for storing and manipulating the basic elements of the map and simple objects or polygon features, e.g., buildings. However, because it was designed for other purposes (e.g., customized map production), the OS data model is inappropriate as it stands for our purposes. It was therefore necessary to introduce more complex data structures to allow adjacency and other spatial relationships to be handled. At the simplest level these additional structures define groups of similar connected aligned objects, for example, a row of terraced houses. At the other extreme they are used to represent dissimilar unconnected unaligned groups of objects, such as a factory complex. Because of the natural hierarchy present in many object types, the model also incorporates a mechanism to deal with this. Each level within the hierarchy has appropriate information about that level, e.g., the lowest level handles proximity to nearby objects such as roads and other buildings, whilst higher levels cope with more abstract information, such as dominant building character. This higher level information can also play an important part in how individual features are handled during the generalization process. An example of this would be the slightly different treatment of houses in an open-plan housing estate compared to one in which the houses are aligned in a more regimented fashion.

In addition to controlling the generalization processes, an expert system is also used in creating the basic level objects and the hierarchy of meta-objects. The main reason for using this is the flexibility it provides: new object descriptions can easily be added, or existing ones modified, to cope with changes in any modifications in the generalization rules and procedures in other parts of OSGEN. Figure 4.1 shows schematically the component parts of OSGEN.

4.4.1 Partitioning: The street block

The polygons formed by the road network, termed blocks, provide a way of partitioning the elements on an urban area map. The main reason for doing this is that it allows features that are unlikely to interact during the generalization process to be kept apart, for example, buildings lying on opposite sides of a road. However, it also means processing can operate more conveniently, on a block-by-block basis. Despite the reliance on complete blocks for processing, situations where blocks are intersected by sheet boundaries do not cause problems because the data-base design allows data to be extracted independently of sheet boundaries. Another important advantage of using a block-oriented approach is that update becomes more controllable – new or changed data only affects the appearance of the other features in the same block.

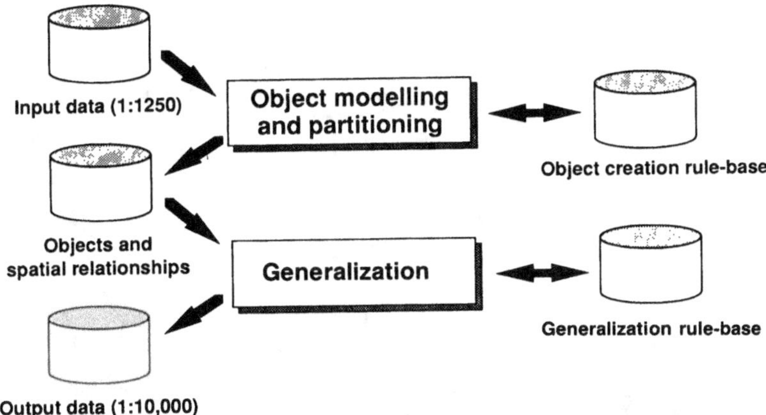

Figure 4.1 Schematic view of OSGEN generalization system.

The first problem is how to define a street block. One method is to use road centre lines extracted from the database. This has the benefit of directly using the existing data without much additional effort. (The road centre line network is also used in subsequent stages of the generalization process, for example in expanding road casings). However, from Figure 4.2a we can see that typical road centre lines do not form suitable polygons because of many cul-de-sacs, whilst 'sliver' polygons are also found along the centres of dual-carriageways. Both of these unwanted artifacts have to be removed. In theory this could be done automatically using appropriate algorithms, but to save time the road centre lines were simply manually edited on an interactive workstation. The resulting line work (Figure 4.2b) is held in the Oracle data-base as a link-node structure, which a Prolog searching algorithm accesses to generate the street blocks. Bounding rectangles are also created in order to speed up the process of assigning objects to the appropriate containing blocks.

4.4.2 Higher level data structure: meta-objects

At the target scale of 1:10 000 building objects are categorized into the following main classes: important building; isolated building; detached house; semi-detached house and terraced house. These are further grouped together where necessary into meta-objects classes: aggregated building (e.g., a factory or school); group of detached houses; group of semi-detached houses and row of terraced houses. Similarly, individual sections of road edges are assembled to form complete road meta-objects.

In order to identify the correct components of building meta-objects, the characteristics of the basic objects and their relationships are determined. These include feature code, area, number of vertices, dominant orientation, shape and adjacency. A set of rules is used to decide which objects are likely to go together, for example, the two houses forming a pair of semi-detached houses. This process is repeated for each street block until the complete hierarchy of meta-object is complete.

The operation of this stage of OSGEN can be illustrated using the relatively simple example of group of connected buildings (Figure 4.3a). These can be represented as an abstract (Figure 4.3b) and expressed in the following relation:

$$G=\text{graph }([a, b, c, d,], [e(a,b), e(b, d), e(b, c), e(c, d])$$

The second argument in the right hand side of this expression represents a series of known implicit facts, namely that the distance between adjacent objects is zero. (A more general

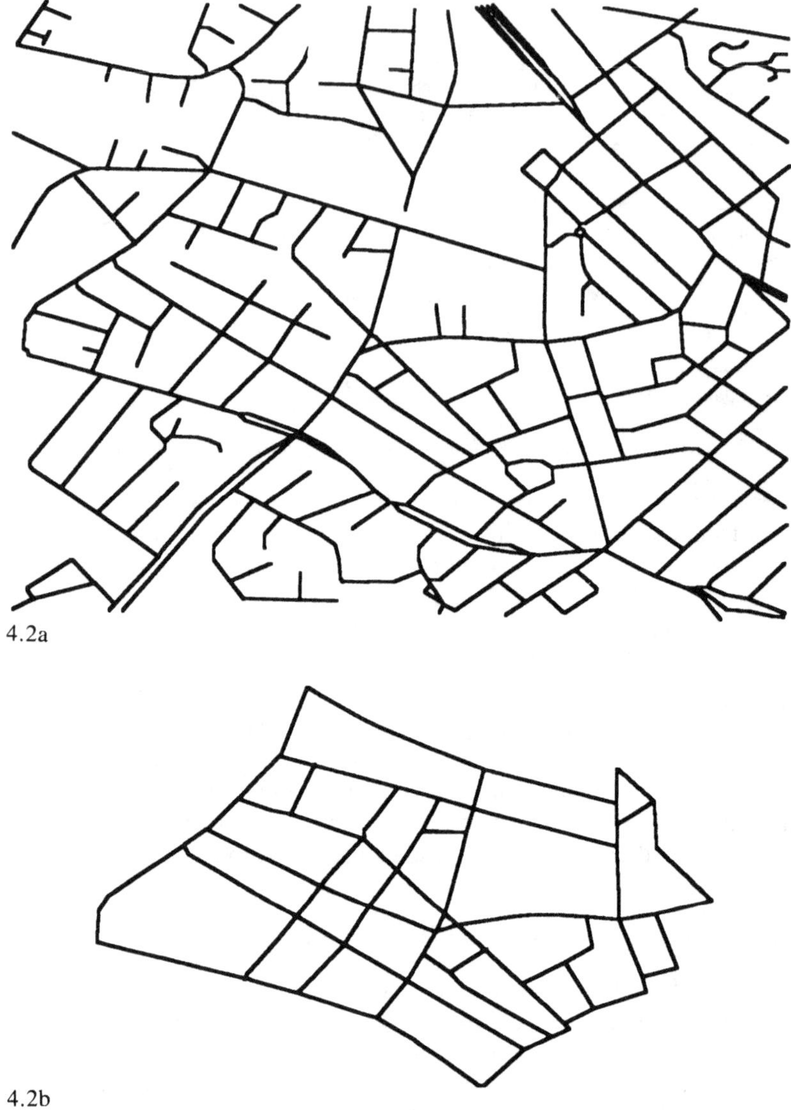

Figure 4.2 Road centre lines: (a) from a typical 1:1250 scale OS dataset; (b) used to create polygon 'blocks' for partioning data.

case would clearly involve a series of explicit non-zero distance relationships between pairs of buildings). Using these relationships and the other spatial parameters that have been calculated beforehand the expert system uses the following rules from the rule-base to determine higher level information about the meta-object (these are expressed in pseudo IF–THEN form here, rather than in Prolog clause form as used in the rule-base):

Rule 1. IF number of potential meta-objects = 1
AND important building flag = 1
THEN meta-object is an important building

Rule 2. IF number of potential meta-objects = 1
AND important building flag = 0

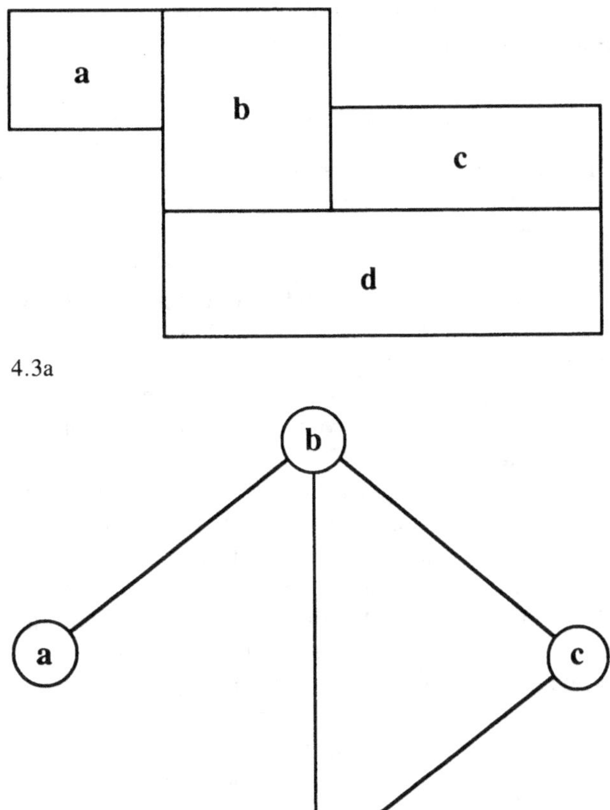

4.3b

Figure 4.3 (a) A group of connected buildings and (b) the same group abstracted as a graph.

AND it has no neighbour
THEN meta-object is an isolated building

Rule 3. IF number of potential meta-objects=2
AND important building flag=0
AND their area difference<tolerance1
AND their number of vertices differ by<tolerance2
AND their dominant orientation difference<tolerance3
THEN meta-object is a semi-detached house

Rule 4. IF number of potential meta-objects>2
AND important building flag=0
AND their area difference <=tolerance1
AND their vertex difference <=tolerance2
AND their main orient difference <=tolerance3
THEN meta-object is a terraced house

Rule 5. IF number of potential meta-object>2
AND important building flag=0
AND their area difference>tolerance1
AND their number of vertices differ by>tolerance2

AND their main orientation difference>tolerance3
THEN meta-object is an aggregated building

4.4.3 Rule-based expert system for map generalization

Once the necessary information, in terms of the meta-object hierarchy has been generated, the actual generalization process can begin.

The first step is to generalize roads as these have the highest priority in terms of their effect on other features. The separate components of each road edge meta-object are joined to eliminate small gaps caused by drives and other access points. The road edge is then simplified using a 'tram-line' algorithm, before the casings are expanded to the minimum named road width at 1:10 000 scale. Finally, road junctions and cul-de-sacs are tidied up and the opposing road edges are referenced to the road centre line network. Figure 4.4a and 4.4b shows an example of this procedure applied to a complete 1:1250 scale map of part of central Birmingham.

The generalization of buildings is handled in two stages. Firstly, individual buildings are checked to see if they need to be omitted where they are smaller than a certain area threshold. However, the action taken is also dependent upon the spatial context of the building. If it is sufficiently close to another larger building then it is combined with it, with common internal walls dissolved. Otherwise, if it is topographically important it is enlarged to the minimum size at the target scale of 1:10 000. Building outlines, particularly those of combined buildings, are then simplified using a purpose-written algorithm that attempts to follow the practice of manual cartographers as much as possible. This means that projections and recesses are removed where appropriate, but retained and enlarged if they reflect the character of the building. The second stage involves the adjustment of building positions, which may also have had significant changes in their geometrical limits, relative to the new position of the expanded road casings. This is where the spatial relationships between buildings and roads come in, because these simplify the task of working out which buildings have to be moved (and where). The majority of movements involve shifting buildings onto road edges, or away from road edges, retaining a minimum gap if necessary. Figures 4.5a, 4.5b and 4.6a, 4.6b show respectively how a variety of

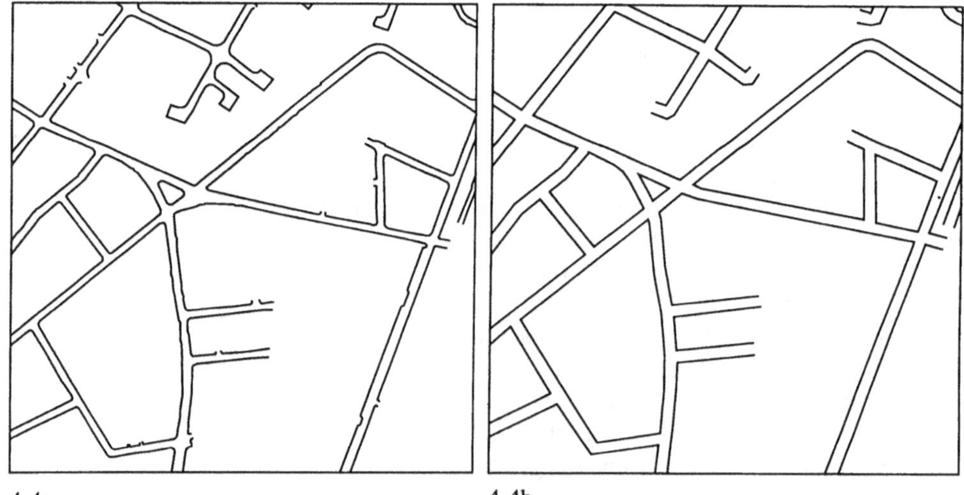

4.4a 4.4b

Figure 4.4 (a) Original road casings from 1:1250 scale data. (b) Generalized road casings to suit 1:10 000 scale.

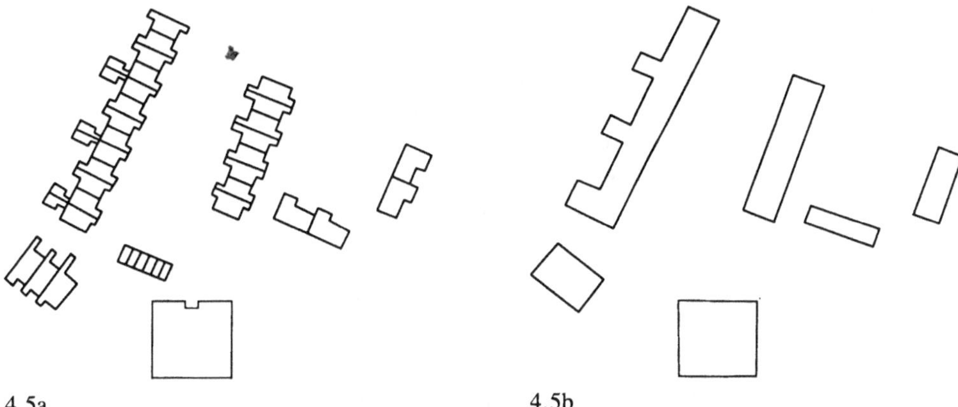

Figure 4.5 (a) Original houses at 1:1250 scale. (b) Generalized houses to suit 1:10 000 scale.

residential dwellings and commercial buildings are generalized in OSGEN. Note particularly how small outbuildings and garage blocks have been removed automatically, while other small buildings have been merged with others. Comparison with the manually generalized version of these objects is difficult because of the higher degree of generalisation of the latter to suit the current production requirements of the 1:25 000 scale series. However, the general reaction of OS and others is that the results are quite acceptable, with a good balance between removal of enough detail and the retention of the general character.

All these activities are implemented in an expert system and associated rule-base. The geometric aspects of the generalizing processes are performed by procedural subroutine algorithms. Here are some examples of generalization rules for the case of an isolated building:

Rule 1. IF number of meta-objects in list=1
AND its size < =tolerance1
AND the distance to the closest meta-object< =tolerance2
THEN meta-object is omitted

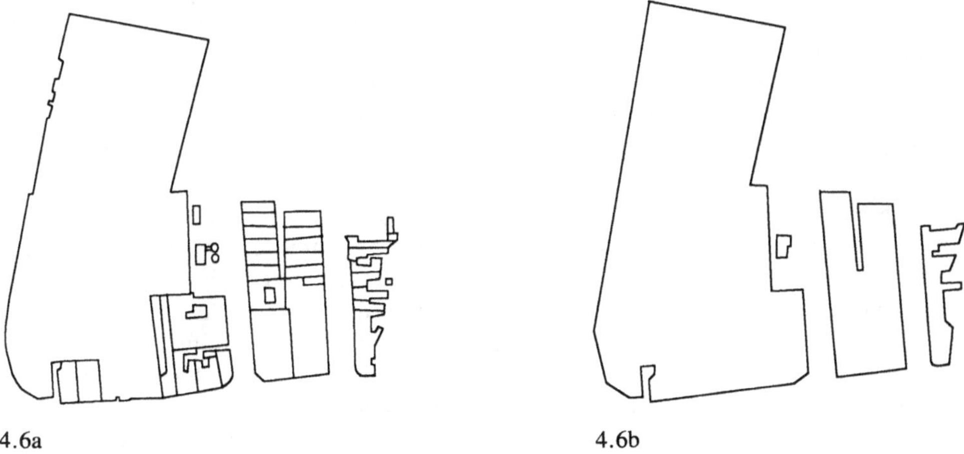

Figure 4.6 (a) Original commercial buildings at 1:1250 scale. (b) Generalized buildings to suit 1:10 000 scale.

Rule 2. IF number of meta-objects in list=1
AND its size< =tolerance1
AND the distance to the closest meta-object>tolerance2
THEN meta-object is retained and exaggerated to minimum size
Rule 3. IF number of meta-objects in list=1
AND its size< =tolerance1
AND the distance to the closest meta-object< =tolerance3
THEN meta-object is merged with that meta-object

In the case of meta-object displacement after road enlargement, the rules look like this:

Rule 1. IF meta-object faces one road segment
AND distance of meta-object behind it>tolerance1
THEN meta-object is displaced away from the road segment
Rule 2. IF meta-object faces one road segment
AND distance of meta-object behind it< =tolerance1
THEN the side of the meta-object is moved away from the road segment
Rule 3. IF meta-object faces two road segments
AND distance of meta-object behind it >tolerance1
THEN meta-object is displaced according to the direction of
average angle of the two road segments
Rule 4. IF meta-object faces two road segments
AND distance of meta-object behind it< =tolerance1
THEN two sides of the meta-object are moved according to
the direction of average angle of two road segments

4.5 Conclusions

The OSGEN rule-based system described here has been shown to be an effective approach for automated map generalization at large scales. In addition, OSGEN offers the benefits of being an open model: it can be adapted to other kinds of feature generalization and to different scales. There is, however, still a long way to go in solving some of the problems that have cropped up in the development of the prototype, for example, in processing times. The remaining feature categories found on OS large scale maps also have to be incorporated into the system.

Apart from achieving the objective of successfully generalizing features on large scale maps the development of OSGEN has also resulted in two potentially useful spin-offs. The first is the capability of improving the quality control of the original data, particularly in feature coding and geometric fidelity. This capability was realized quite early on in the project after loading the data into the system, when it was noticed that certain objects such as buildings could not be closed because some of the features forming their boundaries were mis-coded. The second spin-off is that the system can add additional information to the original data, such as building type, by using the derived geometric characteristics and spatial context of objects as they are created. Whether this is useful, for example, in GIS applications such as market research, is debatable. Even the sophisticated method presented here may not identify correctly all features on the map: some would argue a success rate of less than 70 per cent. However, the alternative of manually coding the information necessary for generalization is even less attractive. Either way, for generalization to operate successfully, requires that this additional information is available.

4.6 Acknowledgements

This research project is jointly funded by the Ordnance Survey and a Departmental Scholarship from Reading University. The cooperation of staff at the OS and Reading University is greatly appreciated.

References

Deveau, T.J., 1985, Reducing the number of points in a plane curve representation, *Proceedings, AUTO-CARTO 7, International Symposium on Computer Assisted Cartography: Digital Representations of Knowledge*, 152–60.

Douglas, D.H. and Peucker, T.K., 1973, Algorithms for the reduction of the number of points required to represent a digitized line or its caricature. *The Canadian Cartographer*, **10**(2), 112–22.

Jenks, G.F., 1981, Lines, computers and human frailties. *Annals of the Association of American Geographers*, **71**(1), 1-10.

Lang, T., 1969, Rules for robot draughtsmen. *Geographical Magazine*, **62**(1), 50–1.

Lichtner, W., 1979, Computer-assisted processes of cartographic generalization in topographic maps. *Geo-Processing*, **1**, 183–99.

McMaster, R.B., 1987, Automated line generalization, *Cartographica*, **24**(2), 74–111.

Meyer, U., 1986, Software-developments for computer-assisted generalization, in *Proceedings Auto-Carto London*, **2**, 247–56.

Monmonier, M., 1983, Raster-mode area generalization for land use and land cover maps. *Cartographica*, **20**(4), 65–91.

Müller, J.C., 1991, Building knowledge tanks for rule-based generalization, in *Proceedings of the 15th Conference of the International Cartographic Association*, 257–66.

Nickerson, B.G., 1986, Development of a rule-based system for automatic map generalization, in *Proceedings of the Second International Symposium on Spatial Data Handling*, 537–56.

Nickerson, B. and Freeman, H., 1986, Development of a rule-based system for automated map generalization, in *Proceedings of the 2nd International Symposium on Spatial Data Handling*, Washington, 537–56.

Perkal, J., 1986, An attempt at objective generalization. Michigan Inter-University Community of Mathematical Geographers, Ann Arbor, Michigan, Discussion paper No. 10.

Shea, K.S., 1991, *Design Considerations for an Artificial Intelligent System. Map Generalization: Making Rules for Knowledge Representation*, London: Longman, 3-20.

Tobler, W.R., 1966, Automation in the preparation of thematic maps. *The Cartographic Journal*, **2**(1), 32–8.

Vicars, D.W. and Robinson, G.J., 1989, Generalization from large-scale to medium and small scale Ordnance Survey maps using expert systems techniques, in *Proceedings Auto-Carto London*, **2**, 267–75.

5

A consistent user-model for a GIS incorporating remotely sensed data

Mark Gahegan

The integration into GIS of remotely sensed imagery is problematic for two reasons. First, the sophisticated scene interpretation tools and operations introduced are difficult to understand and apply, and second, the user interface can become very complex. This chapter discusses problems arising from the use of many different types of data within the same GIS model including image data, and attempts to show how these problems can be overcome by improvements to the data model and the inclusion of expert knowledge.

5.1 Introduction

The motivation for this work is as follows. The use of image data directly within GIS is not currently widespread because of logistic, economic and technical factors. However, there are several reasons why it is desirable to include such data directly. Examples are the need for up-to-date information, the need for a flexible feature recognition strategy, or the need to keep track of the errors and uncertainties inherent in the capture and manipulation of data. The logistic factors of low coverage rates and patchy availability of data are slowly being improved by the launch of new satellites, and the widening of accessibility to existing image libraries. The economic factors are still important, but at the time of writing, the US government is dramatically reducing the cost of LANDSAT imagery to encourage take-up. The technical problems (of interpretation and integration) associated with the inclusion of image data form the subject of this chapter.

Throughout this chapter the following naming convention is used: a feature is a single geographic entity or object that has a unique identity, for example a certain lake or road. A feature-type is a set or class of geographic entities of a given type, such as all lakes or all roads.

5.1.1 The need for consistency in the user interface

Spatial data for use within GIS is widely available from many different sources. These sources not only differ by the format in which they supply data, but also by the meaning that the data conveys. For example, a road network, digitized from a paper map, is conceptually different from a LANDSAT or SPOT image of the same scene. In the former there is a theme to the map and an interpretation process has already been applied at some stage to extract the information of interest, by a process that may be manual, automated, or a combination of both. In the latter, no interpretation has been applied, but features of interest may be readily available by classifying the image according to ranges of values in one or more reflectance bands.

Essentially, the remotely sensed data is a raw image, and exactly how it should be interpreted is a matter of some debate. It is, however, true to say that the interpretation techniques applied can greatly influence the result. In short, digitized overlays (and vector data in general) are usually feature-based, whereas imagery is not. A typical GIS is highly biased in favour of just one type of spatial data (either image or feature, which is often synonymous with raster or vector) and hence supports only a limited degree of interaction between the various data types. Usually this involves a conversion from one form to another. The complexities arising from the combination of such data are reflected in the user model presented by many of the software products available, to the detriment of the user. To allow GIS to freely use image data it is highly desirable to remove these complexities from the interface and this requires the introduction of expert knowledge.

5.1.2 Data used in GIS

Before proceeding further it is appropriate to outline the types of spatial data used in GIS. Some applications, such as the siting of a new road or reservoir, may require the combination of data from more than one of the following sources.

Image data

Image data is captured from remote sensing platforms such as aircraft or satellites, or even from the scanning of paper maps. An image consists of an ordered set of pixels that can be logically grouped in various ways in order to describe different feature-types (Ton *et al.*, 1991). Typically such images capture data in more than one spectral window and hence are, in fact, several layers or bands deep. For example, LANDSAT thematic mapper (TM) scenes consist of seven bands, though not all of these may be used in a particular exercise. Aircraft-borne sensors may capture even more layers.

Digitized overlays or coverages

Much spatial data is in the form of overlays; that is a single, non-overlapping classification of an area. This type of data has already been processed to provide descriptions of various features. Typically, this data is in one of two forms. The first is the result of a digitizing process, described with vectors, and containing identified features. Overlays of this type are often made for a specific purpose, so may contain only one type of feature, an example would be a road network. The second is the result of a land use classification operation performed on image data using an image processing system, in which case it is likely to contain identified feature-types, but will not discriminate between individual feature instances. For example, a land use overlay may contain information on water bodies, urban areas, roads, woodlands, agriculture and so forth. When using such overlays in GIS it is common to first extract from them the features or feature-types of interest before using them to perform specific tasks.

Feature data

Descriptions of individual features may be supplied to supplement existing data, or may represent planned changes, such as a new section of road.

5.1.3 Problems caused by different data sources at the user level

Operations carried out on overlays and images have a different behaviour to those carried out on specific single features (Roberts and Gahegan, 1993) in that they involve a set of features, or in the case of image data a set of pixels that may have been classified as belonging to a particular feature-type. The outcome of an operation such as intersection can therefore be either a feature, a collection of features, a feature-type, a collection of feature-types, or an image. The type of the result (feature, feature-type, image) when intersecting two images will be an image, but the type of the result of intersecting a feature with an image, or a feature with a feature-type is not clear. The result could be considered by default to be of the lowest type, but this is certainly a cause of confusion, and possibly contention. In many cases the user's perception of what is produced will depend not on the types of data used, but on the appearance of the result; for example the number of uniquely identifiable regions produced.

Another problem caused by presenting so many different representations of spatial data to the user is that the user is then required to know the location of the data within the system. That is, which images, overlays or coverages a particular feature or feature-type occurs within. Of course this information may be provided by the GIS, but it still introduces a layer of unnecessary indirection into the model, and makes the structure of the system more difficult to understand. The grouping of feature-types into overlays can also appear very arbitrary, since it tends to be data driven (from the capturing process) and not problem driven; that is not organized to reflect how it might be used. Further problems may occur when features are present in more than one overlay. This can happen if a land use overlay (derived from a remotely sensed image) is being used with a digitized road map. Portions of the roads may appear in the land use overlay as well. If, for example, the user wishes to operate on the roads in some way, it is unclear which representation the system should use.

5.1.4 More on image data

Thematic remotely sensed data from LANDSAT and SPOT satellites has a resolution of 10–80 (square) metres per pixel. This factor, coupled with a general blurring of the image (caused by factors such as the point spread function of the sensing apparatus, sensor position, ground illumination and atmospheric conditions) combine to produce images that are difficult to work with. Features such as roads and rivers often have a width that scales to a (sometimes small) fraction of a pixel, and so appear as no more than an aliasing effect, with unreliable connectivity between the pixels that contain the feature. Hence, even a good classification of the image will not produce data of the same type as (say) a vector overlay that has been digitized by hand. As a general principle, the geometric connectivity or continuity of features is not guaranteed in raster imagery. Much of the discussion that follows is given in terms of data captured by satellite. However, the problem of insufficient resolution is a general one and so still persists in data obtained from aircraft-borne sensors, even though their resolution is usually higher.

As a result of inadequate resolution, it is often necessary to employ sophisticated scene understanding techniques to aid in the interpretation of images, with the end results being very dependent on the techniques chosen; for example see Taylor *et al.* (1986) and Pavilidis and Liow (1990). It is clear that there is no single best interpretation that will suit all subsequent user requests, that is we cannot produce a single classified overlay that will be useful for solving all problems. Consequently the interpretation can only be constructed accurately when the use that it will be put to is known beforehand (Marr, 1982). To continue with the theme from the above example, if it is known that a user is interested only in road networks, then the remotely sensed scene can be analysed using a line following algorithm, to emphasize the connectivity of all appropriate linear features. A filter to remove areas with inappropriate reflectance characteristics may also be of use initially. On the other hand, if the user is interested in forestry then a scheme based on pixel classification may give the best results, but may degrade the appearance of the roads still further. What is proposed then is a GIS that uses a dynamic approach to the interpretation of imagery, and that interpretation proceeds only when the user's needs are known.

5.2 Model overview

The model proposed essentially offers a late binding between an image and its interpretation and subsequent use in solving a particular query. In order to support this notion, the inter-relationships between images and features of interest must be carefully constructed. Figure 5.1 shows these relationships in the form of an E-R diagram.

5.2.1 Using different scene interpretations

An image can have more than one valid interpretation at a given time, each interpretation (here called a view) is a different representation of (some of) the features present. A set of such views is constructed so that each one emphasizes a particular feature-type of interest. Since data at different conceptual levels is included within the model it is possible that a feature may be described more than once (as mentioned in section 5.1.3). A feature then has a set of spatial descriptions, in the form of collections of pixels or vectors, and these are not necessarily identical. Since the descriptions attempt to denote accurately the shape of a feature they are known as *faithful*.

A special case is made for *iconic representations*, which are used to indicate the presence of a feature when it would not normally be visible at the current scale, given the resolution of the output device. Iconic representations fall into two categories; linear icons are line representations of features that are in reality long thin regions where connectivity must be emphasized (such as rivers and roads), point icons are symbolic representations of other region or point features such as buildings, towns, public telephones and so forth. A linear

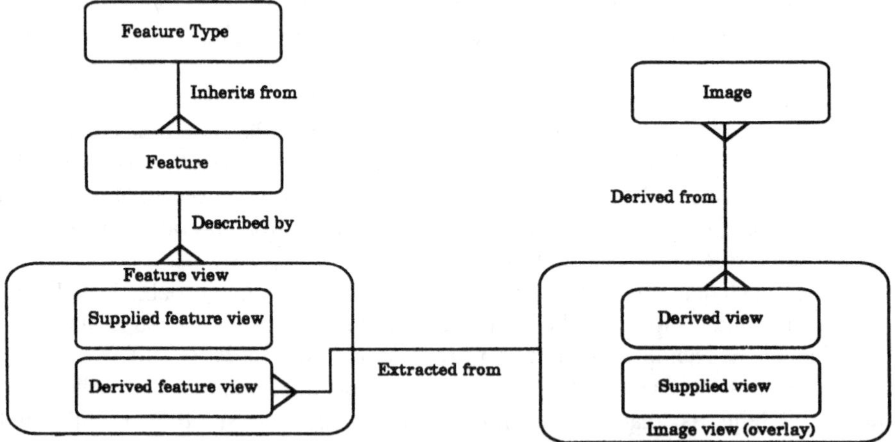

Figure 5.1 E-R diagram of spatial data sources. Spatial and temporal properties of a feature are inherited from the currently selected view. Feature-types form part of an object hierarchy.

icon can be constructed from a regional representation, and will differ from feature to feature, whereas a point icon may be defined for a feature-type. A more detailed account of how faithful and iconic representations are used when displaying features is given by Roberts *et al.* (1991).

5.2.2 View formation

A view is formed as the result of applying scene understanding or image processing technique(s) to image data. Existing feature or feature-type representations in the form of overlays can also be included here. To facilitate view formation, a library of image processing and scene understanding tools form a part of the GIS, and contains functionality that is largely subsumed from existing image processing systems. Generally, one appropriate faithful view may be sufficient for each feature-type. However, there are applications where more than one may be required, such as when analysing changes over time, or when a particular feature is indistinct and so could have two or more possible spatial descriptions, perhaps representing minimum and maximum extents.

Views must be selected or constructed at query time, and this is achieved by utilizing stored knowledge about the behaviour of feature-types within a scene. This knowledge is associated with the class description for each feature-type within an object-oriented framework (Worboys *et al.*, 1990), and contains information about spectral, spatial and scale behaviour. The mechanism used for view formation also depends on the feature-types under investigation and is also stored here. The approach taken is to combine probablistically the evidence derived from the application of various scene understanding techniques. For example, the extraction of forested regions involves the combination of edge evidence, edge enhancement and probablistic classification. A full account of this process and some example results is given by Gahegan and Flack (1993a).

In keeping with the object-oriented model, the existence of a particular feature is not dependent on its spatial description, rather it has an independent existence. As a consequence of this, spatial descriptions of a particular feature and any derived spatial properties such as area and perimeter length, can be inherited from the currently employed view. A natural extension to this is to allow temporal behaviour to be modelled by including a time value or window with each feature representation. The time value associated with a feature is inherited from the data from which the feature is formed. This would normally be the date that an image was captured or a survey completed. With further modification, this allows for temporal studies to be carried out where the spatial properties of a feature are subject to change over time (Corr *et al.*, 1989; Langran, 1992). The derived spatial properties such as length, area, etc. are also inherited, but from the currently employed view. Hence both of the above are not necessarily fixed.

5.2.3 A feature-based conceptual model

From the user's perspective, the model proposed is based entirely on features. Overlays and images are not normally presented to the user, although they may be reconstructed to their original form from the query language. The user can manipulate features by grouping or selecting on any combination of properties, spatial or aspatial. The mapping between features and their original source (overlays or images) is handled by the system. This way the user operates with entities that have both a consistent level of abstraction, and a consistent behaviour.

The system is constructed as a series of layers, each one being a further abstraction of the raw data, in much the same way as is found with conventional databases. Figure

5.2 shows how the various layers interact. In fact the design is an extension of the three layer architecture used for many relational databases; see for example the description given by Korth and Silberschatz (1991). However in this case, the mappings between the physical data and the conceptual model of the domain are necessarily more complicated. An extra layer of indirection is introduced to handle data that is not in feature form.

5.2.4 A feature-based query language

The user builds new maps by combining features and feature-types selected from a list maintained by the system of all feature-types about which knowledge is held. Individual features can be isolated by selection on the name, location, or other unique property. This can be supported interactively via a mouse by highlighting features in the display, or in a further selection list. This user interaction has the effect of constructing automatically the appropriate selection criteria.

Where a selected feature appears in more than one view then rules are applied to ensure that the most appropriate view is used. This will of course depend on the task at hand, and under the following general areas: choosing a representation that best emphasizes the feature, choosing a representation whose format (raster, vector, etc.) is best suited to the operation to be performed, and choosing a representation that best suits the map being produced.

5.3 Rules governing the user interface

With the inclusion of some simple rules, much of the work involved in choosing suitable representations of features and ensuring the consistency of the resulting map can now be handled by the system, and not by the user. The user interface utilizes four sets of rules to govern: feature-view selection; display consistency; temporal consistency and cartographic accuracy. These are used to guide the user and constrain the potentially explosive number of meaningless or inconsistent maps that could be created.

In the description given below, conflicts are discussed and possibilities for their resolution

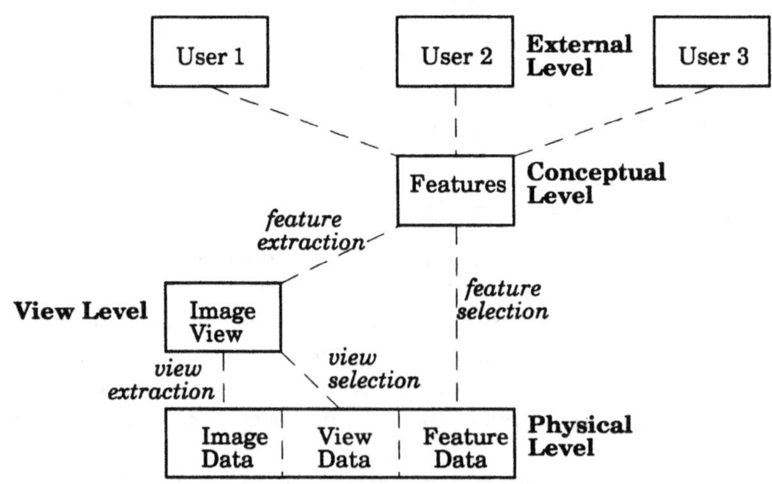

Figure 5.2 Overview of layers and bindings.

are presented. In practice the system is set up to take a default action for each conflict without prompting the user for a decision. These defaults can of course be changed to suit the nature of each problem studied.

5.3.1 Feature-view selection

For each selected feature, the system must decide which of the views held should be provided for the user. This depends on the current scale and resolution of the output media; if the feature would not be visible when drawn using a faithful representation, then an iconic representation is used, if one is available. In practice this is not so straightforward, since features may become partly visible before they are totally lost. A simple cut-off point for a region is to use the number of pixels (or area) occupied by the feature as it appears on the current output device, and to not let this fall below a certain (user defined) threshold.

For each selected feature the view chosen should fall within the current time window (see temporal constraints), and if no time window is defined, then priority is given to the most up-to-date view available. This may however be superseded by other more pressing constraints.

When using or combining features in some spatial operation, then the representation chosen is one deemed appropriate for the task in hand. This usually means a choice between raster and vector based views, with raster views being more efficient for area based calculation and overlay operations, and vector views more useful where connectivity is important. If the feature of interest is an agricultural area, then the most suitable representation may be one in which linear features such as roads and rivers are not emphasized, and perhaps even removed, since although they occupy a relatively small area, their effect is exaggerated when quantized onto a grid of coarse resolution and this can lead to inaccurate measures of area. If no suitable representation exists then one is constructed automatically, and added to those already available for future use.

5.3.2 Display consistency

A displayed map has amongst its properties a spatial window that governs its extent, and a scale. The spatial window may be derived by selecting the smallest window to exactly surround all selected features, in which case the scale may change. Alternatively, the window may be fixed, which allows for the possibility that some selected features may not fall wholly within it. Here there is a potential conflict between the displayed scale of the map and the extents of the spatial window, and this may be resolved by either increasing the scale (to enlarge the window) or by ignoring the outlying selected features.

It is possible that a selected feature may be entirely obscured from view by other coincident features. This may occur since overlapping segmentations are allowed, or because of a limited resolution. The problem is handled by the following rules. Features are displayed in an order that depends on their spatial type. Regions are drawn first, followed by lines and finally points. Within these groupings, features can be drawn by size, starting with the largest and proceeding to the smallest. If conflict still occurs (the most likely examples being due to several point icons occurring at the same location on the output media) then the conflict exists with the resolution of the output media. This conflict may be resolved in one of two ways, either by increasing the resolution (by zooming in and hence changing the displayed scale) or by using cartographic placement rules to position the coincident features as near to their true location as possible, whilst still maintaining legibility.

5.3.3 Temporal consistency

These constraints are only required when the problem being investigated has a temporal dimension, such as the detection of changes in land use over time, and hence for many applications these constraints can be relaxed entirely. A map has a time window or value that describes its temporal extents. Since time is part of the underlying model, then it forms a parameter upon which queries can be built, and constraints expressed. There are two necessary temporal constraints.

1. A selected feature does not exist within the map's temporal window. The system can choose either to relax the temporal window, or to sacrifice the feature. Note that for planning or modelling purposes it is often necessary to add features that do not yet exist into a map, such as a proposed building or road. For this purpose temporal windows are allowed to proceed beyond the current time.
2. Selected features are derived from raw data (images, overlays) with differing temporal windows. This is quite common since many forms of spatial data are not regularly updated, or there may be difficulties involved in obtaining up-to-date data. Within a study, the dates attached to individual source data may vary widely. However, applying this constraint can force the system to supply data from a particular time, allowing a map to be built which accurately represents that time. Upon this basis, changes over time can be detected and analysed.

5.3.4 Cartographic quality

Since features may be derived from many different data sources, we must take measures to ensure that cartographic consistency is maintained where it is judged to be important. Selected features may be drawn from data with different projections, scales and accuracies. It is often not sensible to change all data used in a study to have the same projection, since this usually introduces further errors. If cartographic accuracy is important, then the representation of the features used should all share a common projection, or at the very least the user should be informed of the possibility of errors arising from mixing projections. In practice this can be resolved by selecting representations from a common source if possible, or by reprojecting some features. The same argument holds for the accuracies of data sources. The effects on accuracy of combining multi-source data are discussed fully by Burrough (1987).

5.3.5 The role of the database administrator (DBA)

This type of system is designed for use by the non-expert, and hence attempts to provide expert judgement in the form of rules to govern its behaviour. The upshot of this approach is that the system must initially be configured by (or in cooperation with) an expert user or DBA in the following ways.

- For the domain of study, descriptions of all feature-types of interest must be provided and this will include the knowledge and methods used to extract feature-types from image data, by utilizing the scene understanding and image processing tools provided. This is of course a highly involved task. Various mechanisms are under investigation to automate the formation of appropriate rules.
- The rules governing feature-view selection must be set up to suit the type of problems or analysis to be conducted. This includes the provision of icon symbols, and cut-off points for iconic feature display, as well as the default settings for the resolution of

conflicts. The latter is a simple matter of prioritizing the different rules. Rules with a lower priority are relaxed first.

- Default settings for spatial (and temporal) extents must be supplied, along with other cartographic information, such as preferred scale and projection settings.

5.4 Examples

Some constraints are directly translated into qualifying sub-clauses that are then appended to the original query or operation. Recall that the spatial and temporal properties are inherited from the view of the feature currently employed, hence they cannot be used in an operation until a view is selected, so the following is used initially.

```
//INSTANTIATE VIEW
x=count_views (woodland, faithful) //[faithful/iconic/all]
if x=0
    v=make_view (woodland, faithful)
else
    v=select_view (woodland, faithful)
woodland.view=set_view (v)
```

The function set_view() instantiates the feature-type with an appropriate view from which to select features. The function select-view() uses criteria from the default settings to select an appropriate view by examining the source of each view and comparing it with the criteria for extraction held with the feature-type definition, or by appropriateness to the operation to be carried out. The function make_view() utilizes the same knowledge to construct an appropriate view if none is currently available. The new view is added to the set of views already held by the system for future reference.

If spatial and temporal constraints are being applied, then the query

```
select all W
```

is augmented by

```
where W.time in time_window [t1, t2]
and W.extents in spatial_window[x1, y1, x2, y2]
```

(W is used here to differentiate instances from the object type woodland)

When a view is selected, the query or operation can proceed. Note that before each individual feature can be displayed, the following test is required if iconic displaying is selected.

```
if displaysize (W, output_media_resolution, scale)<min_display_size
    draw_iconic (W)
else
    draw_true (W)
```

Results showing the adaptive behaviour of the view creation mechanism in response to different user requests are given in Gahegan and Flack (1993a,b).

Conclusion

Two significant advantages are clear from the work so far. The first is that the integration of scene understanding tools in a GIS can provide a great deal of flexibility in interpretation, without placing the burden of understanding on the user. The second is that a consistent and simple user model has been defined, where difficult concepts are hidden from the user and managed by the system. In both cases, expert knowledge is used to support this increase in functionality. A further advantage, as yet unexplored, is likely to accrue from keeping all processing within a single environment. The passing of classified overlays from image processing systems into GIS does not allow for the management and control of error and uncertainty measures. Placing all of the functionality within the same environment provides an opportunity to manage this effectively across each stage of the processing.

The model described is currently still under development, with various sections needing more work. However, the initial results are very encouraging. Two areas currently being pursued are the acquisition of knowledge by automated means and improving the results of feature detection by the inclusion of other feature attributes in the classification process. Measures of size, shape and location can be good descriptors of some feature-types, and have been shown to improve the quality of interpretation (Taylor *et al.*, 1986).

References

Burrough, P.A., 1987, *Principles of Geographical Information Systems for Land Resources Assessment*. Clarendon Press, Oxford, chapter 6.

Corr, D.G., Tailor, A.M., Cross, A., Hogg, D.C., Lawrence, D.H., Mason, D.C. and Petrou, M., 1989, Progress in automatic analysis of multi-temporal remotely-sensed data. *International Journal of Remote Sensing*. **10**(7), 1175–95.

Gahegan, M.N. and Flack, J.C., 1993a, A model to support the integration of image understanding techniques within a GIS, in *Proc. 25th International Symposium on Remote Sensing and Global Environmental Change*, Graz, Austria, 4–8th April. Published by Environmental Research Institute of Michigan (ERIM), PO Box 134001, Ann Arbor, MI 48113-4001, USA.

Gahegan, M.N. and Flack, J.C., 1993b, Query-centred interpretation of remotely sensed images within a GIS, *Proc. European conference on Geographic Information Systems*, Genoa, Italy, March, 942–9. Published by EGIS foundation, PO Box 80.115, 3508 TC Utrecht, The Netherlands.

Korth, H.F. and Silberschatz, A., 1991, *Database system concepts* (2nd Edn), McGraw-Hill, Ch. 1.

Langran, G., 1992, *Time in geographic information systems. Technical issues in geographic information systems*, London, Taylor & Francis.

Marr, D., 1982, *Vision: A Computational Investigation into the Human Representation and Processing of Visual Information*, E.H. Freeman & Co., 270.

Pavilidis, T. and Liow, Y.-T., 1990, Integrating region growing and edge detection. *IEEE Transactions on Pattern Analysis and Machine Intelligence*, **12**(3), 225–33.

Roberts, S.A. and Gahegan, M.N., 1993, An object-oriented goegraphic information system shell. *Information and Software Technology*, **35**(10), 561–72.

Roberts, S.A., Gahegan, M.N. and Hoyle, B.S., 1991, Application of object-oriented databases to geographic information systems. *Information and Software Technology*, **33**(1), 38–46.

Taylor, A., Cross, A., Hogg, D.C. and Mason, D.C., 1986, Knowledge-based interpretation of remotely sensed images. *Image and Vision Computing*, **4**(2), 67–83.

Ton, J., Sticklen, J. and Jain, A.K., 1991, Knowledge-based segmentation of LANDSAT images. *IEEE Transactions on Geoscience and Remote Sensing*, **29**(2), 222–31.

Worboys, M.F., Hearnshaw, H.M. and Maguire, D.J., 1990, Object-oriented data modelling for spatial databases. *International Journal of Geographical Information Systems*, **4**(4).

6

Computational support for spatial information handling: models and algorithms

C. Yearsley, M.F. Worboys, P. Story, D.P.W. Jayawardena and P. Bofakos

6.1 Outline of the Chapter

This chapter discusses some of the contributions that the discipline of computer science can make to spatial information management. There are many areas involving spatial data in which computer scientists are active. This chapter does not aim to make a complete survey, but concentrates on some areas of interest to the authors. The chapter begins with some general considerations about computational frameworks in which spatial information management can occur. It then discusses some of the issues surrounding the support of such systems with appropriate structures and algorithms.

It is now recognized (Healey, 1991) that the pure relational approach has not solved all the problems for the wide range of information systems now required. Some newer approaches, especially object-oriented and logic-based systems, offer further possibilities. The problem of how these new approaches can be applied to spatially and temporally referenced information is not yet fully solved, but a substantial amount of progress has been made. The discussion of these issues forms section 6.2 of this chapter.

Any high-level computational model must be well-supported at ground level by efficient data structures, access methods and algorithms. An argument that has been around for some time arises over the appropriateness of triangulated data structures for general purpose spatial systems. Clearly such preprocessing (i.e. triangulation) will make many geometric operations work faster. However, there is the substantial computational overhead of increased storage. The pros and cons in this debate are considered later in section 6.3.

6.2 Computational frameworks for spatial information management

There are two important modelling issues that lie at the foundations of spatial information management. One of these concerns the underlying scientific models in which the analysis of spatial data takes place. The other relates to tl e underlying computational models, with respect to which such systems are constructed. These areas are not mutually exclusive and each has an effect on the other. In general, the former is the concern of the application scientists (geographers, geologists, planners, environmental scientists, etc.) while the latter lies in the domain of computer science and is the concern of this section.

There exist a few general computational frameworks into which many current and next-generation systems, including spatial information systems, fall. Many current object-based GIS are founded, to a greater or lesser extent, upon the relational model. It has been argued that this has limited the potential of such systems and some of these limitations are discussed in the next subsection. One of the current research challenges for computer science is to apply and evaluate newer paradigms.

The underlying computational model can be divided into structural and process components, although again this division is not disjoint. Much of the work in which the authors have been involved has tried to provide a model which gives a unified treatment of spatial and temporal data. Temporal references are in general multi-dimensional, since many applications require independent consideration of at least database (transaction) time and valid (event) time. There is also the related issue of version control and special problems arising from the long transactions which characterize applications in this field. The strategy which we have adopted begins with a model of purely spatial entities, inhabiting two-dimensional Euclidean space (this can be generalized to more dimensions and non-Euclidean spaces). A model of such an object space is given by Worboys (1992). This model has been extended to allow references to attributes in one or more temporal dimensions (so-called bitemporal elements) to associate with atomic spatial references in such a way that the global spatiotemporal structure is internally consistent (Worboys, 1993).

When working on such models, the richness and complexity of spatiotemporal information quickly becomes apparent. It is not surprising that the traditional relational model, which was originally constructed for a limited range of systems handling corporate information, is inadequate in its pure form for GIS. What is not so clear is the best way forward. In the last decade or so, two general approaches have shown promise for systems holding information with rich and often deeply hierarchical semantic structures. These are the deductive and object-oriented approaches. Both are now nearing the point where the technology is ready for the general market-place. Before considering these in more detail, we briefly discuss the role of the relational model in spatial information management.

6.2.1 Approaches using the relational model

Initially, relational databases (RDBs) seemed good candidates for storing spatial data: compared to ad hoc systems, they offer many advantages. Applications can be developed on top of an RDB, leaving it to take care of important issues such as maintaining database integrity, supporting concurrent access (thus maximizing the benefit of valuable data) and insulating developers from low-level issues of data representation and storage. The relational model is founded on well-understood principles and this is a great help in such issues as the construction of query optimizers. Also, proprietary RDBs are widely available.

Although they have proved very successful in many fields, their application to GIS has been less so, and researchers have documented and debated a number of problems (for example Healey, 1991). Their major failings are ones of expressive power and efficiency. Spatial data tend to have complex and hierarchical structures that is hard to capture in tabular form. Thus, application developers have to work with data structures that do not relate closely to the natural structure of their problems, making development unnecessarily complex. Embedding the query language in another that is more expressive can compensate for some deficiencies, but this results in hard to maintain, ad hoc systems.

These problems lead to the production of so-called hybrid GIS such as ARC/INFO, which store spatial and aspatial data separately, only using the relational model for data that they can easily represent. This splitting of the database compromises database integrity and performance, as it is hard for the optimizer of one module to take account of the behaviour of the other.

While such systems are usable, they are clearly undesirable. Fortunately the computer science community has recognized the need for more advanced, more expressive databases with better analytical tools and has produced a number of alternative technologies that are applicable to GIS, including object orientation, extensible databases, expert systems and deductive databases.

6.2.2 Approaches based on object orientation

Object orientation has been successfully applied to application domains requiring support of complex objects (e.g. CAD/CAM, CASE tools, office automation applications). Thus, several GIS researchers have proposed the use of object orientation in spatial data modelling (Manola and Orenstein, 1986; Egenhofer and Frank, 1989; Worboys *et al.*, 1990) and have applied object-oriented approaches to GIS (Worboys, 1993; Choi and Luk, 1992; Worboys and Bofakos, 1993). The usefulness of object orientation for GIS is two-fold. Firstly, a spatial database needs to handle complex spatial objects and, therefore, requires a semantically rich data model. Secondly, a GIS is a large software system and, hence, good software engineering practices (e.g. modularity, information hiding, minimal dependency among software layers) are of great importance. Object-oriented (O-O) concepts and techniques have been investigated in the following fields.

- **O-O languages** Programming-in-the-large requires optimal modularity and a clear distinction between the interface and the implementation of software layers.
- **O-O databases** Advanced database applications require semantically rich database models.
- **O-O analysis and design methodologies** The principles of the O-O paradigm can facilitate the natural modelling of real world complexity.

The concept of the object is central to object orientation. An object has an *identity*, which is independent of any of its attribute values, a private *state* (a set of attribute values) and a public *interface* (a set of *operations*, the implementation of which is private). The only means of interaction between objects is through *message passing*. Objects having the same behaviour and structure are organized into *classes*. The O-O paradigm includes the following concepts.

- **Encapsulation or information hiding** The state and methods (i.e. the implementation of the operations) of objects are encapsulated. At a higher level, design subsystems or implementation modules are encapsulations having a public interface to other modules. Thus, layers of software have a great degree of independence and changes to the implementation of one layer do not affect the semantics of the layer and its interface to the rest of the system.
- **Inheritance** Object classes are organized into a *kind-of* hierarchy, where the descendant of a class (the *subclass*) in the hierarchy inherits all the characteristics of its parent classes (its *superclasses*) and has some additional properties. Furthermore, a subclass might redefine some of the properties of its superclasses for either enhanced performance or refined functionality. In *single inheritance* a class has one immediate ancestor (the hierarchy is a tree), while *multiple inheritance* permits several ancestors (the hierarchy is a lattice). *Semantic inheritance* is a design construct that implies a conceptual dependency between the object classes, while *syntactic inheritance* is a programming technique which is used for code re-usability.

- **Class relationships** Additional to inheritance, the object model provides a rich set of other class relationships. Most notably, the composition (*part-of*) relationship provides the mechanism for modelling aggregate objects as collections of their parts. Further, objects belonging to the same class might be grouped together in order to form a higher-level object. For this purpose several *container classes* might be used (e.g. sets, bags, arrays etc.). A detailed discussion on class relationships is provided by Booch (1991).
- **Polymorphism** Polymorphic operations apply to several object classes and, in conjunction with inheritance, facilitate smooth and generic programming. The operation is bound to its actual implementation code on run-time (this is known as late binding) depending on its parameters. Although this results in an efficiency overhead, it provides programming flexibility, since an operation implemented within several subclasses can be added to the interface of their superclass.

The above characteristics of the object model facilitate natural modelling of real world complexity, design and implementation modularity, design and code re-usability, facilitation of system maintenance and system evolution. However, the object-oriented paradigm does not lack problems (Atkinson *et al.*, 1989; Varharen, 1991). As yet, there is no agreed underlying formal foundation and, consequently, there is no standard object-oriented model. Thus, it is not clear which are the features that an object-oriented system should have and there is no general agreement on definitions of object-oriented concepts. Nevertheless, efforts for a consensus are being made (X3/SPARC/DBSSG/OODBTG. Final technical report. Technical report, National Institute of Science and Technology, Gaithersburg, MD 20899, USA, September 1991).

Construction of an object-oriented TGIS

Geographic information is hardly ever purely spatial information. It almost always consists of a mixture of spatial, temporal and attribute (textual and numeric) components. This subsection provides some details on the construction of a system for handling all these components, which we shall term a temporal GIS (TGIS) to emphasize the incorporation of the temporal dimensions.

The underlying data model for one such system constructed by the authors unifies the spatial and temporal components into a self-consistent framework. The principles behind this are described in (Worboys, 1993). For the purposes of this study, the following properties of the ST-model are noted.

- The model handles two independent temporal dimensions, namely *database time*, which measures transaction events with the system, and *event time*, along whose axis are measured events occurring in the observation domain. In both dimensions, time is assumed discrete and linearly ordered.
- The model handles two orthogonal spatial dimensions, which might correspond to measurements along orthogonal axes of the Euclidean plane.
- The model handles the four spatiotemporal dimensions in a unified manner, in which none of the dimensions takes precedence over the others.

The object modelling process requires identification of spatially referenced object classes in the real world related to the study domain. In order to maintain generality within the geographical data model, it is necessary to apply some generalization and abstraction concepts to these real world object classes. For example, the class *lake coverage* may be generalized to class *area*. These generalized object classes must then be expressed in computationally meaningful terms. For example, an object in the generalized spatially referenced object class *area* may be approximated by an object in class *polygon*. Figure 6.1 shows a portion of the spatial object class hierarchy in Euclidean 2-space (for details, see Jawayardena and Worboys, 1993; Worboys, 1992).

We now proceed to the incorporation of this purely spatial model in a TGIS. Our aim is to implement a temporal GIS within the framework of an existing non-temporal GIS. Snodgrass and Ahn (1986) suggest two alternative methods of development for a temporal database.

1. The temporal DBMS might be built on top of the existing conventional DBMS. In this case a new layer of code is added between the user-interface and the DBMS.
2. Major parts of the DBMS are changed (e.g. query optimizer, file indexing methods).

The first approach provides quick and easy development, while the second approach provides efficiency but requires considerable effort. This chapter describes the first approach, since it offers the opportunity for a rapid evaluation of the model. At this level of abstraction, the base GIS is treated as a black box which provides a well-understood interface (i.e. set of operations with their functionality).

The prototype system is developed in stages. The first stage constructs a system which can capture event and database time information stored in nonspatially and spatially indexed files using and extending the data structures and file organizations provided by the underlying GIS. In addition, a query interface is developed which extends the default query interface provided by the base GIS and incorporates time-based predicates. The second stage enhances the query engine and display functions with particular emphasis on an expressive query language. The construction of spatio-temporal indexing methods is necessary, due to the large data volumes and the problem of performance degradation in an append-only database.

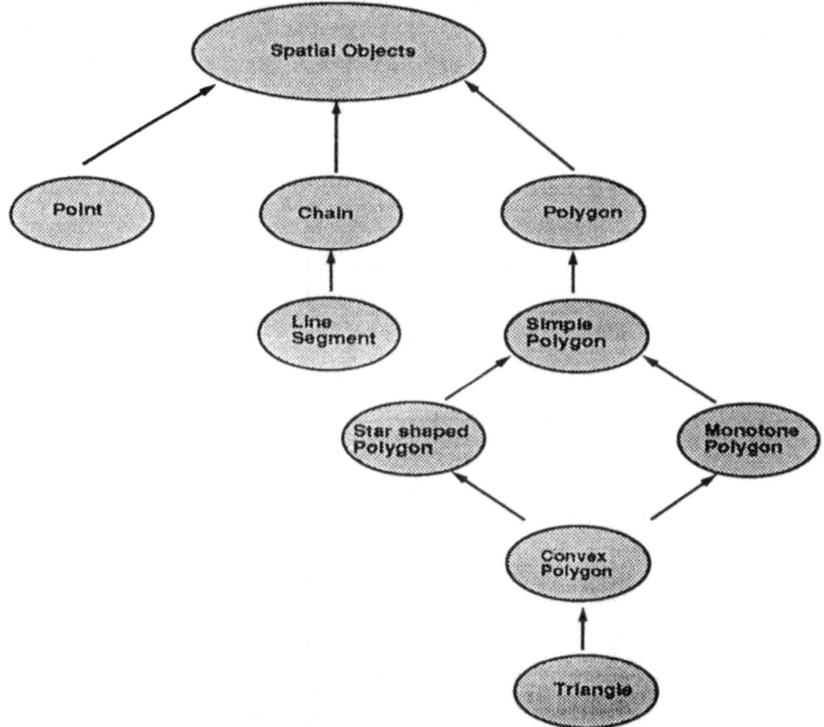

Figure 6.1 Spatial object class hierarchy

The Smallworld GIS was chosen for this project. This GIS provides an object-oriented environment through the object-oriented language Magik. Magik allows a development environment for customization. Data is stored in a version-managed relational database. The graphics interface is controlled via a set of graphics objects. This results in a set of default windows which can be altered as required by an applications programmer. We illustrate how the basic data model of Smallworld may be extended to handle database time. The structure of Smallworld GIS is shown in Figure 6.2. The area enclosed within the dotted line represents the Magik image, which is copied into main memory when the system is initialized. The Magik image consists of a set of object classes and associated methods. It includes the core Magik classes (such as the primitive and container classes). The GIS class libraries contain objects for manipulating the spatial data-store and the graphical interface. On top of this, another software layer is built, as required for the particular application. The data-store contains system tables which hold the data dictionary and the topological objects. These tables are transparent to the user and constitute the implementation of the Smallworld spatial data model. In addition, the data-store includes application-specific tables which define the database schema of the particular application.

The system architecture for the proposed TGIS is shown in Figure 6.3. Additional system tables are incorporated to enable the storage of temporal data. They provide the means for storing topological and spatial object identifiers alongside their associated bitemporal elements. It should be noted that these tables are transparent to the application programmer and the end user. Also, a new software layer is built on top of the basic Magik image. This layer defines the temporal objects and functionality required to define a spatio-temporal system. At this stage, the system provides the application programmer with an enhanced set of tools to enable the development of a temporal GIS application.

A temporal GIS offers several possible options for querying its data. The database may be queried via a query window providing a similar mechanism to that of the database interface, Query By Example. Three types of predicates might be specified: predicates on non-spatial data; predicates on spatial data (locational, topological) and predicates on the time domain.

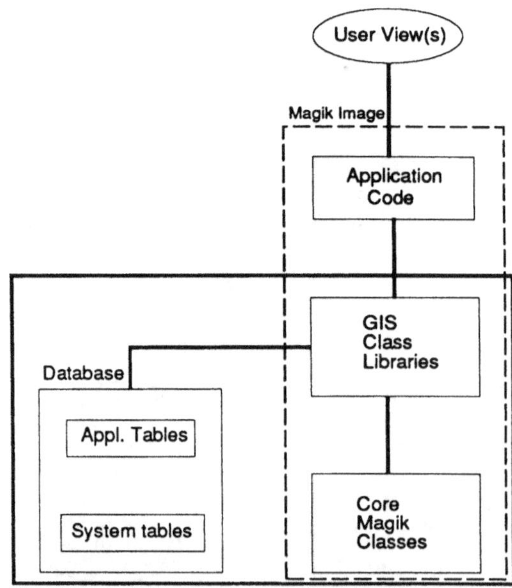

Figure 6.2 Smallworld GIS structure.

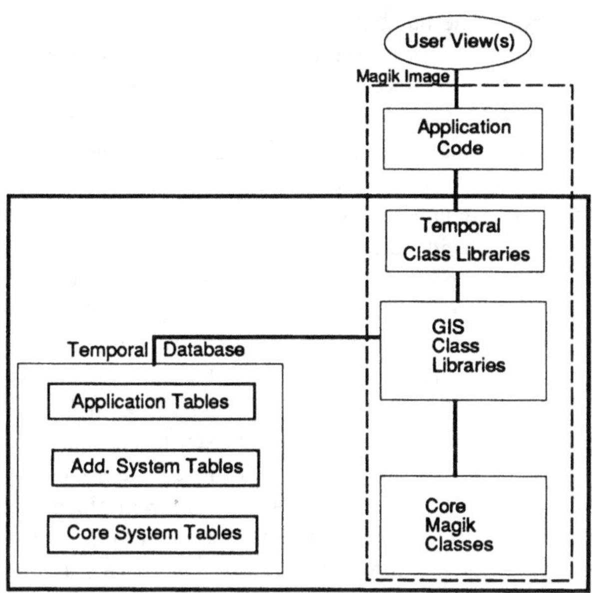

Figure 6.3 Enhanced Smallworld GIS structure.

The base GIS can cope with queries involving spatial and textual predicates. Within the Smallworld system (version 1.9) these are handled by separate editors, namely the query and object editors. It is necessary to extend both of these in order to fully incorporate time-based predicates. The construction of a complete spatiotemporal query language is still the subject of current research.

6.2.3 Logic-based approaches

Deductive databases play a key role in current logic-based approaches to information systems. Deductive databases can be thought of as generalizing the concepts of relational databases (Thayse, 1993). They comprise an *extension* (similar to a classical relational database) and an *intension*. The intension consists of virtual relations that are defined in terms of other relations using logic. Thus deductive databases allow the manipulation of data that is implied by the explicitly stored facts, allowing the meaning of the data to be considered in ways that conventional databases cannot. This ability is potentially relevant to GIS, in which much information is stored in implicit form (Smith and Yiang, 1991).

The use of logical deduction over a database allows the representation of data with complex structure and many inter-relationships. It also allows the construction of knowledge, perhaps of some domain expert. Deductive databases can therefore be seen to have goals in common with such areas as semantic modelling, object-oriented databases and expert systems, while being distinct from all of these.

A conventional database consists of a body of data in some restricted format, defined and accessed with some database language. Most GIS store their information in this way, either in tabular form in a relational database or some more specialized database structure. In contrast, a deductive database (DDB) aims to use logic as the basis for database definition, query and manipulation. It handles a body of expressions in a logic-based language, such as one related to PROLOG. This language allows the expression of facts, description of their interrelationships, and management of their storage. It both defines the structure of the data and acts as query language and control mechanism. Relationships can be hierarchical and, unlike relational databases, DDBs do not suffer from an inability to compute transitive closure. The database operates by applying deduction to the logical expressions to evaluate their truth or to find values that satisfy them. This is not as abstract as it might seem, as relational databases are themselves closely related to logic. The significance of DDB technology is in the way it does not distinguish between logic that simply states a fact, and logic that expresses a relationship or rule. Thus it greatly reduces the distinction between the data and the programs used to manipulate that data. Deductive databases can enhance simple data with a few rules, perform advanced deductive operations using many rules, or anything in between.

Logic and deduction allow representation of the behaviour of a system, and information about how the database may process it. This could provide sufficient expressive power in the database to allow most or all of an application to be developed within one system that could then meet a wide variety of users' needs. This increase in flexibility is due to the use of deduction to generalize the concept of data retrieval. Rules can define new information in terms of old, and both can be processed in the same way. Consider the simple example given by Bocca (1991). We could define a database of flights for an airline in terms of departure time and date, flight time and so on. With the use of deductive rules we could specify information for a given route, and state that there is, say, one such flight twice each working day, thus making it unnecessary to store a record for each. This information could be held in a deductive relation and queried as if it was a fully populated table in a conventional database. To add a spatial dimension to the example, suppose flight path information was stored, along with data on land-use around an airport. Topology can be represented in terms of rules, and using relationships such as 'adjacent to' or even 'downwind of' the database could answer queries about risk of noise or air pollution. This ability to represent complex spatial models, and to manipulate high-level properties of data using deduction, suggests that a suitable deductive database could be a useful tool for GIS construction and data modelling.

Deductive databases are also applicable to problems of data analysis. Many geographic problems are complex and not well suited to a mechanistic, algorithmic approach. They can reveal fundamental difficulties in some current approaches. For example, the difficulties expert systems can have with very large data volumes, the inflexibility and inappropriateness of languages like FORTRAN, and the lack of expressiveness of many databases. Thus deductive databases are also applicable to providing advanced analytical and modelling capabilities.

Deductive databases are not without their own problems. They constitute an active research area and there are several different approaches to their construction. Perhaps the simplest is to take a deductive knowledge-based system and use a database to provide large-scale permanent storage. This is known as a loosely coupled deductive database. Its great advantage is that it is comparatively easy to construct, but achieves little more than persistence of the initial rule base or of final results. Integration with the database is not close enough to enable the data in use to exceed the capacity of the computer's main memory. A better approach is to make either the database or the rule-based system a shell about the other to facilitate communication between them. This still has many flaws: for example, poor performance, and the ability of deductive rules to express things beyond the ability of the database to process in a reasonable time. What is needed is for the processing of data and its storage to be carried out as part of a single operation, but from a computing point of view building a logical language with persistent store much larger than main memory is a non-trivial task.

Database integration is not the only problem. Designers of a DDB must choose which of the wide variety of logics to implement. Different forms of logic vary in expressive power and in the ease with which they can be implemented. For spatial applications representation of uncertain information is useful, and a number of logics address this issue directly. As discussed earlier, geographical information has spatial and temporal components, and so appropriately extended logics will be applied. For GIS, we really want not just deductive, rule-based tabular data but rule-based complex objects: a system that takes the best of object orientation, databases and deduction and adds support for powerful graphical interactive user interfaces. The system of logic used must be sufficiently expressive, yet not slow the database down with its computational demands, and it should allow a wide variety of possible data models. Data should be stored in a format that is both flexible and compact. Deductive database technology holds out hopes of being able to address and solve these complex problems. Object orientation and logic are not mutually exclusive, and there have been some attempts (Williams *et al.*, 1993) to unify them in the context of GIS.

6.3 Algorithmic aspects of GIS

In the previous sections we discussed high level conceptual frameworks in which GIS concepts could be expressed. This section considers the support provided at a lower level by spatial algorithms. A typical query encountered in geographical database applications involves finding arrangements of (one or more) spatially-referenced objects under a certain set of constraints. In practice, the algorithms in computational geometry form the foundation for processing spatial data. The efficiency of the GIS query interface will depend on the selection of suitable algorithms. Moreover, the set of constraints can be complicated and may also involve fuzzy operators. Thus, the problem solving criteria are not straightforward.

Computational geometry has systematically been developed as a subject within the last two decades. However, algorithmic studies of geometric problems have appeared in the literature long before this. Such studies were not much concerned with the measures of efficiency and computational complexity. Measurements of efficiency and computational complexity of algorithms for geometrical problems made their appearance in the literature in the late 1970s. As a result, many improved algorithms were developed and a new research area was created. Thus, the newborn discipline of computational geometry added a new dimension to the traditional approach to algorithmic studies of geometrical problems. After rapid development over the last 15 years, computational geometry is now widely used to solve problems in areas such as GIS, computer graphics and robotics.

A large number of algorithms for solving geometric problems may be found in the computational geometry literature. In particular, the textbook by Preparata and Shamos (1985) provides a very good foundation for this area of study. The work done by Lee and Preparata (1984) classifies the geometric problems into five major problem domains: convex hulls, intersections, searching, proximity and optimization – according to the nature of the geometric objects and operations involved. Moreover, it provides discussions on several geometric problems under each domain.

In the field of GIS, the use of geometric algorithms is wide and varied. This study focuses on a specific issue: the role of the triangulation in spatial data processing algorithms. The authors have focused upon triangulation because of the simplicity that this provides for the underlying conceptual models. In a spatial database application we often deal with polygonal objects. We have two possible approaches to algorithms involving polygonal objects (Figure 6.4). We may treat polygons as non-triangulated objects or as triangulated objects (collections of triangles). Each has benefits and disadvantages. The following are a few examples of the type of geometrical problems involving polygonal objects, for which each of the approaches may be used.

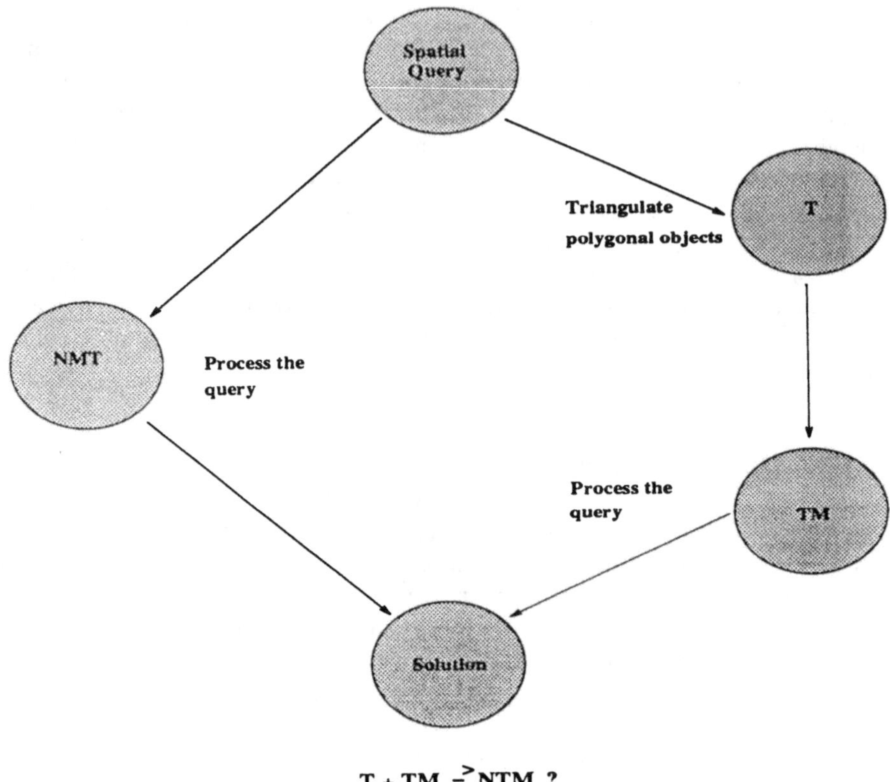

Figure 6.4 Two approaches to dealing with polygonal objects.

- **Polygon inclusion** Given a simple polygon P and a query point q in the plane, determine whether the point q is in P,
- **Set operations** Given a set of simple polygons in the plane, find their union, intersection, etc.
- **Point location** Given a set of polygons and a query point in the plane, find the polygons which have a non-empty intersection with the query point.

For each type of problem, we may pre-process the data by triangulating the polygons. This may be computationally expensive, but such pre-processing might allow the application of more efficient algorithms on the triangulated data. In order that the advantages and disadvantages of the two approaches can be evaluated, we first need efficient triangulation algorithms. Such algorithms are discussed next.

6.3.1 Triangulation algorithms

A *simple polygon* in the plane may be defined as a set of n vertices and n edges between vertices, such that no pair of non-consecutive edges have a point in common. A *triangulation* of a simple polygon is a partition of its interior into a finite set of non-intersecting triangles. It is clear that any complex polygonal structure can be partitioned into a finite set of triangles. Such a triangulation can be obtained with or without the introduction of additional points, which are usually known as Steiner points. A triangulation obtained without introducing Steiner points is termed a *diagonal triangulation*. A diagonal triangulation of polygon is obtained by introducing a set of non intersecting diagonals.

In GIS, triangulation may be used as a preliminary step for query processing. The resulting triangles may need to be processed further in order to answer spatial queries. Once a polygon has been triangulated, it can be stored in the system as a set of triangles, which can be used for query processing. Thus, the objective of the triangulation process is three-fold.

- Optimal regularity of the constituent triangles. As suggested by Preparata and Shamos (1985), in many applications it is desirable to avoid a preponderance of long, thin triangles in a triangulation. The storage of long, thin triangles may not be economical. (For example, their coordinate values may not be stored in the same data page). The following measurements of the resulting triangles may be used as criteria for regularity.
  - the average minimum angle If the average minimum angle of the triangles of a triangulation is close to 60°, then the triangles are more or less equilateral.
  - the ratio (average perimeter to average maximum edge length) If this is close to three, then the regularity of the resulting triangles is high.
  - the average variance in edge length A low average variance in edge length of the triangles implies high regularity.
- The number of resulting triangles should be minimized. The number of resulting triangles depends on the number of vertices of the original polygon and on the number of Steiner points.
- The time complexity of the triangulation algorithm should be optimized.

The triangulation algorithms developed before 1978 had the common feature of having the time complexity of at least $O(n^2 \log n)$; see, for example, the greedy triangulation algorithm of Preparata and Shamos (1985). Garey *et al.* (1978) constructed an $O(n \log n)$ algorithm based on *monotone polygons*. In the 1980s, triangulation was an active research topic within the computational geometry research community. During this period, several new triangulation algorithms have been developed, with time complexity in the range of $O(n \log n)$. Tarjan and Van Wyk (1988) constructed an $O(n \log \log n)$ algorithm. Kirkpatrick *et al.* (1990), independently constructed an algorithm with the same complexity. Followed by this, Chazelle (1990) developed a linear time algorithm for triangulating a simple polygon. A detailed discussion of the major triangulation algorithms mentioned above, including their relevance to GIS, may be found in (Jawayardena and Worboys, 1993).

When we compare these algorithms, it becomes clear that there is a trade off between the efficiency of the triangulation and the regularity of the resulting triangles. For example, the greedy algorithm is highly inefficient in terms of CPU time, but produces comparatively regular triangles. In contrast, the algorithm of Garey *et al.* (1978) is efficient in terms of CPU time, but does not provide good regularity.

6.3.2 Arguments for and against triangulation

There are a number of issues to be considered when evaluating the benefits of the role of triangulation for spatial data processing.

- **Pre-processing** Triangulation requires pre-processing, while no preprocessing is required when polygons are treated as atomic spatial objects.
- **Storage overhead** Triangulated polygons have the overhead of increased storage (of both data and index files).
- **Representational aspects** There are many indexing structures (hierarchical and non-hierarchical) available for general polygonal objects. A triangulated polygon, being

a polygonal map, may make use of the advantages of these indexing methods. However, these methods have not been designed for triangulated polygons and therefore do not take full advantage of the properties of a triangle. The PM_2 quadtree structure (Samet, 1990) is an appropriate hierarchical indexing structure for triangulated polygons. For point and range queries, a suitable access structure is a combination of an R-tree (holding the minimum bounding rectangles of the triangulated polygons) and a PM_2 quadtree for the triangulation of each polygon.

- **Algorithms efficiency** Algorithms defined on triangles are often simpler than their counterparts defined for polygons. Moreover, several topological operations are more efficient if they are performed on triangulated polygons.

For an example of the last point, consider the point-in-polygon operation. For non-triangulated polygons the point-in-polygon algorithm presented in (Preparata and Shamos, 1985) is standard. For triangulated polygons, the use of PM_2 quadtrees may simplify the complexity of this algorithm. Figure 6.5 explains this graphically. To check whether point P is in the polygon traverse through the tree and identify the quadrant containing P: perform a point in triangle operation only on triangles which intersect with that quadrant (i.e. triangle ABC in Figure 6.5).

This is an improvement on the non-triangulated algorithm, even though the height of the index for the triangulated polygon is higher than that for the corresponding non-triangulated polygon. However, not all geometric operations are simplified by the triangulation of the polygonal structures.

6.4 Conclusion

In this chapter, we have discussed some of the ways in which computer science is able to contribute to the design and implementation of spatial information management systems. We have focused upon modelling, suggesting that newer, richer computational models will have significant effects on next-generation systems. Of course, any such system can only be as good as its underlying algorithmic support, and part of the chapter was devoted to a discussion of some of the issues surrounding this support. The chapter was not intended as an exhaustive coverage of support from computer science for spatial information handling. Our perspective is necessarily subjective. We conclude by itemizing some other

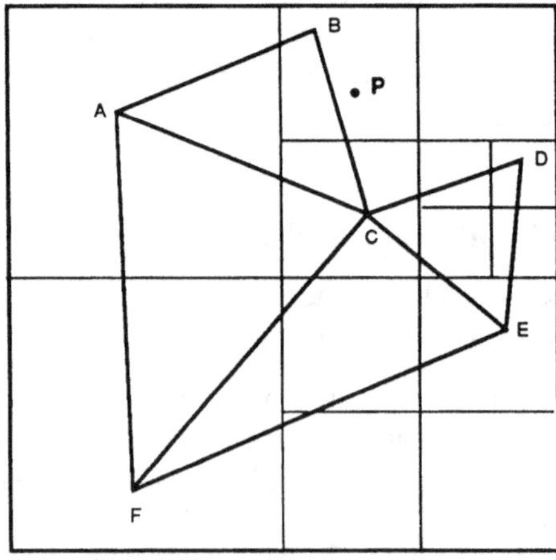

Figure 6.5 Point-in-triangulated polygon.

areas of computer science which are already having an impact on the field, and are likely to continue to do so.

- **New architectures** For example, parallelism is already beginning to be implemented and new parallel geometric algorithms constructed. Neuro-computing offers further possibilities for spatial analysis.
- **Human computer interface (HCI)** System complexity is a major drawback with current GIS, which often require many hours of self-study and attendance at training courses to master. HCI offers approaches to future system design which will make them more transparent and intuitive to use, thus accessible to wider classes of users. Linked to this is the problem of the complexity of the datasets used in many applications. Already, attention is being given to the support of users, not only at the level of the data but also with the meta-data.
- **Software engineering** A GIS is a large software system and its engineering must have elements in common with the development of any such system. However, it also has its own unique flavour. At present, there is no agreed, published methodology for GIS construction. A mix of software engineers and GIS researchers could arrive at useful results in this area.
- **Standard database issues** For example, distribution, interoperability, concurrency and transaction management.

Such a list is necessarily incomplete. However, we hope that this chapter provides further evidence of the fruitful past, present and we hope future contributions that computer science can make to this subject.

References

Atkinson, M. *et al.*, 1989, The object-oriented database system manifesto, in Kim, W. and Nishio, S. (Eds), *First international Conference on Deductive and Object-Oriented Databases*, 40–57.

Bocca, J.B., 1991, Megalog, a platform for developing knowledge base management systems, in *Proc. International Symposium on Database Systems for Advanced Applications*, 374–80.

Booch, G., 1991, *Object-Oriented Design with Applications*, The Benjamin/Cummings Publishing Company.

Chazelle, B., 1990, Triangulating a simple polygon in linear time, in *Proc. of the 31st Annual Symp. on Foundations of Computer Science*, 220–30.

Choi, A. and Luk, W.S., 1992, Using an object-oriented database system to construct a spatial database kernel for GIS applications. *Computer Systems Science and Engineering*, **7**(2), 100–21, March.

Egenhofer, M.J. and Frank, A.U., 1989, Object-oriented modelling in GIS: inheritance and propagation, in *Proceedings of the 9th Symposium on Computer-Assisted Cartography*, 588–98.

Garey, M.R., *et al.*, 1978, Triangulating a simple polygon. *Information Processing Letters*, **7**, 175–9.

Healey, R., 1991, Database management systems, in Maguire, D.K., Goodchild, M.F. and Rhind, O.W. (Eds), *Geographical Information Systems*, volume 1, 251–67. Longman.

Jawayardena, D.P.W. and Worboys, M.F., 1993, On the role of triangulation in spatial databases (part I). Technical report, Department of Computer Science, Keele University, Keele, Staffordshire, UK.

Kirkpatrick, D.G. *et al.*, 1990, Polygon triangulation in O(n log log n) time with simple data structures, in *Proc. of the 6th Annual ACM Symp. Computational Geometry*, 34–43.

Lee, D.T. and Preparata, F.P., 1984, Computational geometry – a survey. *IEEE Transactions on Computers*, 1072–101.

Manola, F. and Orenstein, J.A., 1986, Toward a general spatial data model for an object-oriented DBMS, in *Proceedings of the 12th International Conference on Very Large Data Bases*, 328–35.

Preparata, F.P. and Shamos, M.I., 1985, *Computational Geometry – An Introduction*. Springer-Verlag.

Samet, H., 1990, *The Design and Analysis of Spatial Data Structures*. Addision-Wesley Publishing Company.

Smith, T.R. and Yiang, J.E., 1991, Knowledge-based approaches in GIS, in *GIS Geographic Information Systems (Vol 1)*. Longman Technical.

Snodgrass, R. and Ahn, I., 1986, Temporal databases. *IEEE Computer*, **19**(9), 35–42.

Tarjan, R.E. and Van Wyk, C.J., 1988, An O(*n* log log *n*)-time algorithm for triangulating a simple polygon. *SIAM Journal of Computing*, **17**, 143–78.

Verharen, E., 1991, Object-oriented system development; an overview, in *Proceedings of the Workshop IS-CORE 1991*, 202–34.

Williams, M.H., Abdelmoty, A.I. and Paton, N.W., 1993, Deduction and Deductive Databases for Geographic Data Handling, in Abel, D. and Ooi, B.C. (Eds), *Advances in Spatial Databases, Lecture Notes in Computer Science 692, Proc. 3rd Int. Symp., SSD*, 443–64. Springer-Verlag, June.

Worboys, M.F., 1992, A generic model for planar geographical objects. *International Journal of Geographic Information Systems*, **6**(5), 353–72.

Worboys, M.F., 1993, A data model for information with spatial and bitemporal components. Technical report TR93-06, Department of Computer Science, Keele, University, Keele, Staffordshire, UK.

Worboys, M.F. and Bofakos, P., 1993, A canonical model for a class of areal spatial objects, in Abel, D. and Ooi, B.C. (Eds), *Advances in Spatial Databases, Lecture Notes in Computer Science 692, Proc. 3rd Int. Symp., SSD*, 36–52. Springer-Verlag.

Worboys, M.F., Hearnshaw, H.M. and Maguire, D.J., 1990, Object-oriented modeling for spatial databases. *International Journal of Geographical Information Systems*, **4**, 369–83.

7

Incorporating the temporal dimension in a GIS

Zarine Kemp and Arek Kowalczyk

In the field of spatial data handling, much research has focused on providing temporal support for GIS. It is clear that GIS could benefit greatly from the ability to store and maintain changes in data over time and some of the practical implications of this are outlined in this paper. This chapter first looks at different problems and solutions for incorporating temporal support in a practical manner, by describing the implementation of a land terrier system for a local authority using the relational model. It then considers how these temporal requirements may be met using an object-oriented approach, enabling the consideration of alternative strategies.

7.1 Introduction

The evolution of processes and phenomena through time has always been recognized as a fundamental fact underlying information systems. However, traditional database models, such as the relational model, which are widely known and used today are atemporal. The state of the database at any one time represents only a snapshot of the real world and not the continuum that it really is. As databases are being used for a wider range of applications than before, the requirement to record change in database entities is becoming crucial. Now that the cost of computer resources is rapidly decreasing, it has become possible to implement some of the theoretical work that has been done to address the temporal problem. These include creating new database structures and semantic data models (Lum *et al.*, 1984; Ariav, 1986; Schiel, 1982), extended relational algebras and logic that can cope with the extra dimension (Lorentzos and Johnson, 1987; Gabbay and McBrien, 1991; Allen, 1983) and augmented query languages (Snodgrass, 1987; Sadeghi *et al.*, 1988) that allow easy extraction of the temporal data.

These research attempts to bring time to the forefront of information systems have focused on the following related facets of the problem:

- conceptual data modelling with a focus on the representation of time;
- database models and structures affecting the management and retrieval of temporal information; and
- temporal deductive capabilities, involving reasoning about time.

When the temporal dimension is applied to geographical information systems (GIS), the problem is compounded by several orders of magnitude. However, the requirement to incorporate the temporal dimension in GIS is undisputed and well-documented (Lagran, 1989). Temporal GIS encompass a very wide range of problems such as facilities management, capturing changes in their states over time, environmental monitoring to capture the evolving patterns of natural phenomena and decision support systems which use data about past behaviour to make predictions about future trends.

This chapter concentrates on the second aspect mentioned above, namely the data models and structures appropriate for dealing with spatio-temporal information. A 'hands-on' approach is used in an attempt to generalize about wider issues by attempting to solve a particular problem. The chapter addresses widely-applicable principles by discussing how a particular problem can be solved using different data models. The chapter is structured as follows: the rest of this section gives a brief description of the vector-based problem used to illustrate the arguments, section 7.2 surveys the conceptual issues involved, section 7.3 describes the way it was implemented using the geo-relational model, section 7.4 describes how an extended relational data model could be used, section 7.5 discusses an object-oriented approach and section 7.6 concludes the chapter.

7.1.1 An example application

The application that we will be considering is a land terrier system maintained by a local authority; a similar problem has been described in (Hunter and Williamson, 1990). Currently, it contains records of all the lands, properties and other relevant information about the area that it controls using an entirely manual system which consists of three main elements: maps, which are subdivided into rural and urban areas; a manual register, consisting of data sheets and map photocopies for each individual deed packet; and deed packets themselves, each of which refers to an area of land, and consists of papers and legal documents concerning it and properties on it.

As well as having the capability to access existing static information from the current deed packets (by simply pointing to the required land parcel on a digitized map), there is also a necessary requirement for temporal support. It is required, for example, to look back at past records and see how certain land parcels have evolved over time to assist with future planning. The application is also required to show ownership histories, previous leases and licences as well as any easements. Therefore the application will be required to cope with queries on the spatial as well as aspatial properties such as: who owns this piece of land?; is it for sale?; and how many other pieces of land does person X own?

The application will also be required to cope with specifically temporal queries such as: when was this land last sold?; how much did it cost in 1977?; was this land ever used for industrial purposes involving toxic chemicals?

7.2 Temporal GIS

It is worth emphasizing at the outset that the entities that GIS manage consist of a combination of spatial (geometric and topological) properties and aspatial ones (the latter being frequently referred to as attributes). For practical, implementation reasons the two types of properties are often handled differently, but conceptually they should be considered as properties attributable to the same entity. Although differences can be noted, GIS defy attempts to strictly categorize them according to their temporal requirements. The observable differences in temporal GIS refer to aspects such as the extent to which they rely on spatial data (for example, in forest management systems the spatial attributes are vital whereas they may be less critical in a cadastral system which merely records the history of ownership of land plots). Those that are spatially oriented can be further classified into systems which deal with constant geographic space (such as land terrier) and systems where the spatial attributes change over time (such as pollution control systems). Furthermore, spatially oriented systems may use absolute geographic locations (e.g. a map of a particular area of earth) where it is important to the user to know where features are located, or relative spatial data (e.g. in an air traffic-control system), where the user is not concerned with the exact location of entities but with their relative positions. Temporal applications may also be classified according to the occurrence rate of events (for example, a system such as a coastguard monitor system where vessels are constantly changing position would have a high occurrence rate, whereas in a land use monitoring system the occurrence rate is likely to be low). Systems may also require the animation of previous states (as in a weather monitoring system), or the ability to call up any number of historical states at once, in order to compare them (as in wildlife monitoring systems).

Any one application will normally fit into one or more of the categories mentioned above but a completely generic temporal GIS should be capable of handling any type of application. There are several concepts pertinent to the design of temporal GIS which are briefly discussed below.

7.2.1 Concepts of time

There are two types of time that need to be considered. Effenberg (1992) summarizes some of the terminology used and even adds a third type (which will not be considered here). We will consider world time and database time. World time refers to when events actually occur in the real world, and database time refers to when these events are recorded in the database. For example, in the example application a particular piece of land may have been sold in January 1981, that would be the world time, but the information may not actually be entered in the database until June 1981, which would be the database time.

It is useful to consider how these two types of time will relate to each other in an application. In most cases world time will be less than database time, as there will undoubtedly be a period between an event happening and that event being recorded. If the two times were equal, that would suggest that the real world event is also a database transaction itself, e.g. 'delete this piece of land from the terrier'; the world time at which that event happens, is also the database time. The final combination is where world time is greater than database time. Ths means that the system contains information about future events. These may be facts, e.g. 'this property's postcode will change to. . .' or they may be predictions based on past trends, e.g. 'this piece of land will become a housing estate'. Lum *et al.* (1984) discuss the handling of future time.

7.2.2 Granularity of time

Due to the nature of digitized information, continuous entities and phenomena have to be dealt with in a discrete fashion. In information systems, it is not possible to store time as a continuous entity, it must be represented as a series of discrete events. This does of course mean that some information may be lost and an inaccurate picture of the real world may be stored. In systems that sample the state of the entities at regular intervals, the smaller the time period between events (variously referred to as ticks (Gabbay and McBrien, 1991; Theodoulidis *et al.*, 1991) and time points (Lundberg, 1983) the more accurate the representation will be. For example, if two events occurred at different times on the same day in a system that used days as its time granule, then, to the database user, it would appear as if the events occurred simultaneously. On the other hand capturing data at time granules that are too small for the requirements of the application would result in wasted memory.

Information systems may, however, choose to record temporal information only when entities change their state. In that case, the time intervals will be variable and determined by events. In either case, the semantics of temporal resolution are determined by the requirements of individual applications. Due to the nature of the events in the example application, such as when a piece of land is sold, or when its usage changes, it is sufficient to use a day as the smallest time period. When a property is sold by the council, it is not necessary or practical, to know the exact time when the final contract was signed, only the day the sale occurred.

7.2.3 Time-stamping

Another consideration to take into account is the way in which time is actually attributed to events. There are two broad trains of thought on this issue. Should a time-stamp merely record the time an event occurred or should it reflect the lifetime of an event?

The first method will naturally use less storage, and will carry no redundant data or null values (sometimes considered bad practice in database circles). However, economy of memory requirements may have to be balanced against additional processing time required to respond to a query. When an event occurs in the real world, it is reflected in the database by a change in the value of one or more attributes of an entity; in other words the information captured is the changed state and the time that the change occurred. This means that if a temporal query requires information about the time span of that particular state, the database has to sift through several records until it comes to one which enables it to determine when that state was terminated. In the example application, a record (depicting a parcel of land) has several attributes including its current usage. To answer a query such as 'how long was parcel A used for manufacturing?', the database would have to be searched for records pertaining to parcel A (in chronological order) and the attribute referring to 'land use' checked for change, to derive the required information.

If frequent temporal range queries are anticipated, a better method would be to store a set of two time-stamps for each state; *since* and *until* values, to depict the time period over which a particular state is valid. This mechanism makes it convenient to answer queries such as 'how long has this property been owned by. . .' or to perform functions like, 'highlight on a map all the land that was used for agriculture bettween 1945 and 1980'.

7.2.4 Response time

A factor worth mentioning before we proceed to discussion of implementation strategies, is that of response time. GIS invariably deal with vast volumes of spatial and aspatial data so it is in the interests of efficiency and economy to use as little extra storage space as possible in implementing temporal support. On the other hand, acceptable response time to certain types of temporal queries may require controlled data redundancy, thus increasing the volume of data held in the database. Thus, decisions regarding the design of temporal support often involve a trade-off between the additional volume of data and the speed of information retrieval.

7.3 *The geo-relational model*

There have been many attempts in the past to design suitable mechanisms to cope with the extra time dimension. Most of them have involved adding temporal capability to the relational model as that is the most ubiquitous database model. The relational model is also probably the most popular one underpinning several widely-used GIS. It is therefore appropriate that the discussion of alternative temporal capabilities should, in the first instance, consider the geo-relational model. (Another, pragmatic consideration was that the local authority where the problem originated were committed to using ARC/INFO.) A complete survey of all the research attempts in this area is beyond the scope of this paper, but some of the more popular ones will be considered, looking at their advantages and drawbacks from the point of view of the example GIS application.

7.3.1 Archiving

One of the crudest but simplest methods of maintaining a temporal database is to back-up all the data stored in the database at regular intervals, maybe weekly, or even daily if resources permit. However, this is unsuitable because: first, any events that occur between back-ups would not be recorded, and it is essential for the land terrier to keep track of all changes (such as sales of land, which would cause problems if not recorded); second, retrieval of the archived information would be slow and clumsy. If the application needed to refer to an attribute value from the past, it would be necessary to reload each back-up separately and examine it (even if the archiving medium was physically labelled with a date, this process would be unacceptably slow).

Another important reason why the archiving method would be unsuitable, is the fact that much of the data would be repeatedly archived even though the attribute's values had not changed, thereby wasting scarce resources. The following mechanisms remedy this; they all store changes only when they occur.

7.3.2 Time slicing

This method is described in many works on temporal support, most notably (Clifford and Warren, 1983), and will work where the database information is stored in flat files or tables, as in the relational model. When an event occurs, the database must alter at least one of the attributes in at least one of the records from a particular table. Time-slicing simply stores the table (relation) prior to the event and gives it a time-stamp. Then a duplicate, but updated copy is created and held as the current state. Figure 7.1 illustrates the operation and shows the time cube that results.

Although there is an improvement in efficiency with this method when compared with archiving, there is still a lot of data redundancy, due to the fact that entire relations are being duplicated when perhaps only one attribute value has been altered.

When all possibilities are considered, this method would not be a bad choice for the example application, as events relating to council land parcels are relatively infrequent.

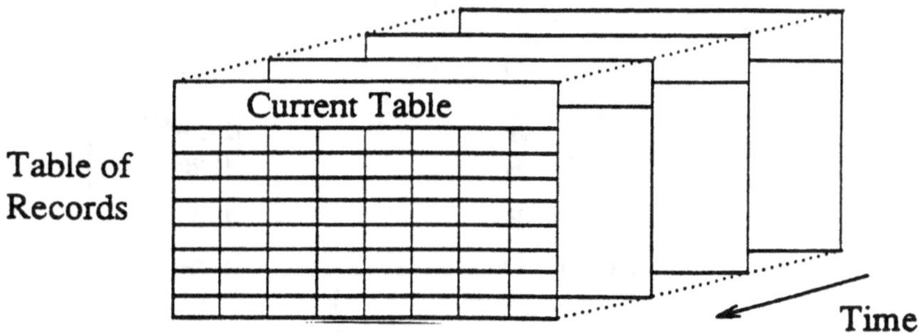

Figure 7.1 Time-slicing.

However, there is still a problem when the application searches through the database with regard to a period query (the user wants to know how long a particular database state lasted). This arises due to the fact that this method stores only one time-stamp value, the *since* value (see section above on time-stamping) and does not keep track of the *until* value. The second time value is derived by searching through past tables, in chronological order. This situation is avoided in the method described in (Clifford and Warren, 1983) by the addition of an extra time-stamp attribute to each record, but this still requires the application to examine all possible tables to find the right attribute change.

7.3.3 Time-stamping at the record level

This option overcomes the problem discussed in section 7.3.2 by keeping track of both *since* and *until* time-stamps for each object. This is done by adding two extra attributes to each record, and is known as time-stamping at the record (or tuple level) because time attributes are applied to the record (rather than to the entire relation). Figure 7.2 below illustrates this method.

One of the drawbacks of this method is that some time-attributes will be null as their values will not be known at the time of entry, i.e. when an event occurs producing a new table state, it may not be possible to know when it will end. A related problem is that the very first states in the database, may not have *since* values. For example, in the application, there may not be records that go back far enough to indicate when a particular parcel of land commenced being used for a particular purpose. It may be possible to use a 'null' or 'dummy' value but this may not be suitable or acceptable in all cases.

The fact that temporal values are being attached at the tuple level makes it unnecessary to duplicate entire relations; all that is required is to add tuples to the existing relation when events occur. This is done by altering the null value of the *until* attribute of the current record to the correct time at which the event occurred (in this case it will be *world time*), and creating a new record with the updated attributes reflecting the event, with *since* time-stamped at the same event time as above.

These new records may be added to the table in three different ways. The simplest but least efficient method would be to append the new records at the end of the table. This would give a neat chronological view of the history, but it means that the application would need to make a sequential search through the table to respond to a query.

Deed Packets					
id.	Area	Use	etc ...	since	until
1	1000	A		-	-
2	3000	B		-	1970
2	3500	D		1971	-
3	1230	C		1750	1900
4	500	B		1880	1900
4	700	B		1901	-
5	4000	D		1990	-
⋮	⋮	⋮	⋮	⋮	⋮

Records (arrows pointing to rows)

Figure 7.2 Time-stamping at the record level.

A better method, described in (Snodgrass and Ahn, 1985) is to keep all related objects (i.e. the records) together, in chronological order. This clustering makes it easier for the application to find relevant subsequent records when trying to calculate the period of a particular state.

Another variation is to sequence the records in the table the same way for each time slice (as described in Ariav, 1986). For example, if the table contained ten different objects (deed packets in this case), it would be divided into segments of ten records in identical order. The problem with this method is that if a particular object had not changed at an event time, then it would need to be represented by either duplicating it, or filling the unchanged records with blanks or null values. The biggest drawback with all these methods is that relations can get very long quickly, thus resulting in a decrease in response time.

7.3.4 Chaining

Another tuple level temporal database design, suggested in (Lum *et al.*, 1984), goes a long way towards solving some of the problems discussed above. It is called chaining and works on the principle that only changed values are stored. Temporal entities in the database are represented by two relations instead of one. The first holds only the current state, and is therefore updated at every event. The second relation holds all the history records, in the form of a linked chain, (analogous to the programming construct of a dynamic data structure called a linked list). Each past entry, in the history table has an extra field which points to the next entry in chronological order. This method is particularly suitable for GIS applications where the most frequent requirement is for access to current data (this is true of the land terrier application, as most of the queries are of the form 'who owns this land?').

Each time an event occurs, the relevant record in the first table is appended to the history table with the correct reference to the next entry in order. Then a new current entry is created and put in place of the old one in the current state table. This method creates easily traversable access paths between related records, thus improving efficiency. For the same reason, deletion of records is very simple too; only a simple change in the next reference field is needed to point to a different record. The chaining method makes it convenient to access single attributes but not if entire records are required.

A careful analysis of the land terrier application revealed that not all attributes are time-variable, hence there is no need to duplicate all the data for every event. The chaining method can therefore be taken further, by only storing in the history table those attributes that change with time. There are very few time variable attributes in the example application, so the method illustrated in Figure 7.3 was considered to be the most appropriate.

The data model separates the time-variable and non time-variable attributes, uses as little memory as possible, allows fast access to the historical data (making the application efficient) and is easily updated with minimum effort.

7.3.5 The temporal dimension applied to spatial attributes

So far, the discussion has dealt with capturing aspatial attributes of an entity as they evolve through time. However, it may also be a requirement to capture change in the spatial properties of entities. For example, in the land terrier application, land parcel boundaries may change due to fragmentation or concatenation or both. Fragmentation can occur when a packet is physically divided up, for example, for building purposes on a housing estate, thus creating fragments that have to be treated as separate objects. Concatenation is the opposite and produces a composite land parcel made up of two or more adjoining parcels.

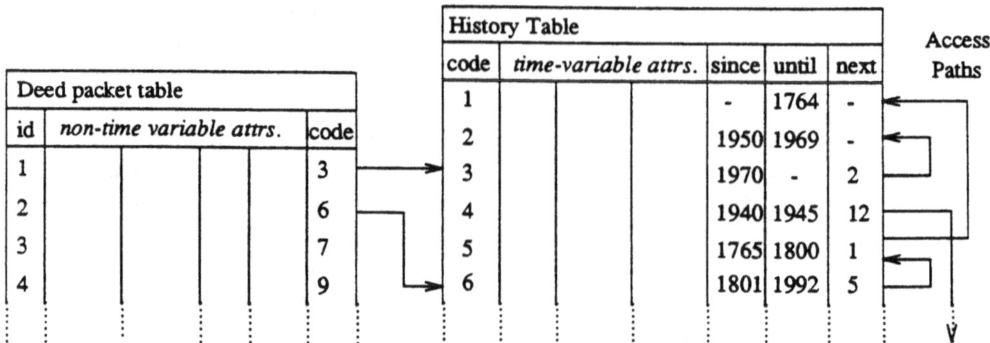

Figure 7.3 Chaining; each record in the history table points to its predecessor.

It is less easy to maintain a record of the changes in spatial attributes as most geo-relational GIS use non-standard internal formats to store the spatial attributes and link them to the aspatial attributes. Attempts have been made to deal with this problem (Stickler *et al.*, 1992); here, we shall consider the issues involved and suggest a possible solution.

At any time a particular land parcel may be associated with one or more polygons which define its boundary. The spatial data, which define the geometry and the topology is held in tables (in ARC/INFO format in this case). As there is an m:n (many-to-many) correspondence between polygons and land parcels, an intermediate link table is required so that it is possible to determine which polygons make up the composite land parcel represented in the deed packet (and vice versa). Moreover, as the temporal dimension is being considered, there is also a requirement to capture the changes that occur in the spatial data. The solution eventually arrived at is to construct a 'shadow' history table associated with the intermediate link table, using a mechanism similar to that used for the aspatial data. In other words, at any time T_n, the link table will contain information about the current state, i.e. the mapping between polygons and land parcels; each record will also contain information linking it to the previous state held in the history table. The history table will contain information about land parcel/polygon mappings that existed at previous time periods, T_{n-1}, T_{n-2} etc. including time stamps indicating when that state came into existence. Thus the history table can be traversed (backwards through time), to extract information about previous states of the land terrier. The polygon table will contain spatial information about the boundaries of all land parcels that arise through the fragmentation and concatenation process. Figure 7.4 below illustrates the evolution of a typical land parcel through time.

At time T_1, land parcel A consists of polygon 1; at time T_2, part of the land parcel has been sold off, giving rise to the situation where the boundary of A has changed shape, and there are now two polygons, 2 and 3, in the database. When the change occurs, the previous state is written to the history table with the appropriate time stamp. At time T_3, land parcel A has been augmented by an additional piece of land, polygon 4; parcel A now consists of two polygons, 2 and 4. At time T_4, parcel A fragments into polygons 4, 5, 6 and 7; at time T_5, A consists of polygons 4 and 8 and a new land parcel B, consisting of polygons 6 and 7 comes into existence, and at time T_6, polygon 6 is transferrred from parcel B to parcel A. To summarize, whenever an event occurs that results in a change in the spatial data, three procedures are carried out; the newly generated polygons are added to the polygon table, the land parcel records from the link table are transferred to the history table with relevant time stamps and the revised land parcel/polygon mappings are added to the link table.

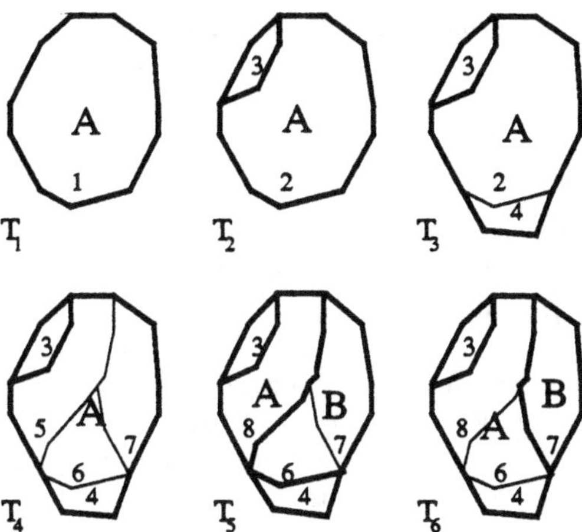

Figure 7.4 An example of spatial data changes through time.

A few comments on the suggested mechanism might be helpful at this stage. The method succeeds in enabling previous states of the land terrier to be reconstructed, but it is cumbersome. It is acceptable as far as this paricular application is concerned, because events generating spatial attribute change are infrequent and the requirement to access previous states does not arise frequently. Therefore, the history table and the tables holding the spatial data are unlikely to become unacceptably large, and system response time would not be adversely affected. It should be noted in passing, that another option was considered, which required storing all the polygon data as well in a history table, to enable fast access to previous states of the terrier. On balance, however, it was rejected because of the vast volume of data that would be generated.

7.4 An object-oriented approach

The different database designs that have been considered so far have been based on the traditional relational model approach, which enforces certain limitations on the modeller. Due to lack of expressive power and limited semantics (simple data types and relationships only), many cases arise in GIS, where real world entities and constructs do not easily map into the relational model. In a field where the problem domains are rapidly expanding, there is a need to cope with different representations of spatio-temporal entities and the structures and constructs needed to support them. Much of the research effort in recent years has been directed towards this problem of modelling the real world as closely as possible without forcing unnatural relationships between and within entities. An approach that has aroused much interest is the object-oriented one (Worboys *et al.*, 1990) where entities may be represented by complex objects which can encapsulate the structural and behavioural properties of the entities they represent.

In order to realise the object-oriented paradigm, databases have to be implemented with a rich set of tools that can handle the semantic constructs that are needed, such as inheritance, structured data types, support for shared objects and extensions such as version control and temporal capabilities. Some of these object-oriented databases are extensions of the relational model, one of the factors being cited in their favour being that they provide a natural evolutionary path from relational databases to object-oriented ones. Guptill (1990) describes the design of a GIS database application that is based on two of these, POSTGRES and EXODUS.

7.4.1 An extended relational model

POSTGRES (Rowe and Stonebraker, 1987; Stonebraker and Rowe, 1986; Stonebraker, 1987) is an extended relational model which subsumes the traditional relational model and includes certain object-oriented features. These include: extra built-in types and the ability to define abstract data types, support for inheritance, support for shared objects and the capability to support rules and functions to manipulate data more easily. However, the main reasons for exploring the capabilities of POSTGRES as a suitable platform for a spatio-temporal GIS, were its built-in capabilities for manipulating time (both real-world and database time), its support for geometric data types and associated spatial operators, as well as its ability to manage 'large object' types which can include bitmaps for raster based data. The geometric and temporal types are summarized below. The geometric types are:

- **point**: consists of a coordinate pair, describing a location in two dimensional space; it can be used to identify topological features like the nodes of a polygon, or higher level geographic features like lamp-posts, railway stations and towns
- **lseg**: represents a single straight line segment, consisting of a start and an end point; used mainly as a part of a larger, more complicated of line feature
- **path/polygon**: consists of an array of lsegs
- **box**: describes a rectangle.

There are three main temporal types; they can either be used as part of a query, in which case POSTGRES assumes it to be database time, or a type can be defined inside a relation which corresponds to real-world time.

- **absolute time**: consists of a month, day, hour, minute, second, year and timezone; it is used to time-stamp data with absolute times (a shortened form 'now' represents current time)
- **relative time**: allows time to be specified in relative terms, (for example, '2 days ago'); it consists of a quantity, a unit and a direction
- **time range**: allows a time period to be specified consisting of two absolute time types, representing the start and end (inclusive) of the time period; at the outset this type seemed the most appropriate for a temporal GIS.

The spatial operators enable testing of polygons for overlap, containment, position relative to another polygon and equality. The *box* related operators are similar. The *point* related operators enable testing of relative positions of two points, retrieval of distance between two points and whether a point lies on a *path*.

The temporal operators enable both absolute and relative time types to be tested for equality, manipulated arithmetically and tested to check whether a given time value (or range) falls within another time range.

7.4.2 Using POSTGRES to model a spatio-temporal GIS

Initial attempts at designing a spatio-temporal GIS made extensive use of the built-in types. However, it soon became apparent that using the standard geometric types was not memory efficient as a lot of duplicate data was required to be stored. Moreover, low level manipulation of the data was difficult; for example, it was not easy to select a particular segment of a polygon boundary and display it in a different colour (perhaps to indicate a different type of boundary). Therefore, in order to provide the requisite topological capabilities and flexibility to manipulate spatial objects, the final model used only the built-in point type and user-defined line and area types.

Temporal aspects

The POSTGRES extensible DBMS maintains historical data, by saving data deleted or modified in a separate relation, thus enabling queries to be executed on past states of the database. This capability involves the database maintaining a record of all the tuples that have been updated along with appropriate time stamps (to the granularity of a second, which is unalterable), which can subsequently be queried using the standard query language. This facility seems to be what is required for a temporal GIS, but certain factors have to be considered. This built-in historical facility uses database time only rather than real-world time (i.e. the database transactions themselves are time-stamped rather than the changes in the data). Moreover, any history of events which occurred before the database is first set-up cannot be recorded in this way as POSTGRES does not know the meaning of time before initialization and will return the current state by default.

Another temporal capability that was considered is the explicit versioning capability provided in POSTGRES. This enables existing relations to be versioned at specific, user-determined points in time. The version relation thus created may be accessed (and updated) as required and may optionally have updates from the base relation propagated to it. A related facility enables versions to be created from 'snapshots' of a relation, in which case it is merely a historical record of an entity at a particular point in time; it may be accessed (and updated independently) but is not affected by updates to the base relation. The relation level granularity of the temporal aspect in this case renders it unsuitable for use in the land terrier system as the requirement in that application is for the temporal dimension to apply to a few attributes only and not the entity as a whole. Experience has indicated that relational level versioning is not suitable for most GIS applications.

After consideration of all relevant factors it was concluded that the concept of world time needs to be built into the model explicitly, by adding attributes (of type absolute or interval) directly into the tuples involved. Then, the built-in POSTGRES time functions may be used, to manipulate and access relevant temporal data. It was decided that a generic temporal capability required time attributes to be applied to the spatial data at the lowest possible level and within the control of the application rather than at the system level. For these reasons, the final model arrived at was the one depicted diagrammatically in the following section.

The POSTGRES model

This section discusses some of the pertinent aspects of the model illustrated overleaf. The vector-based spatial data is contained in separate POINTS, LINES and AREAS relations. Each point has an identifier, an *x-y* location and an attribute describing what object the point is representing. There are no time attributes as each point can either represent a point-feature (e.g. a town) and will therefore have time attributed separately (e.g. via a history table) or it will be one of two nodes making up a line segment, in which case it will be time attributed as part of the line. The LINES relation consists of unique identifiers, start and end nodes (which are referenced as point identifiers), left and right area identifiers

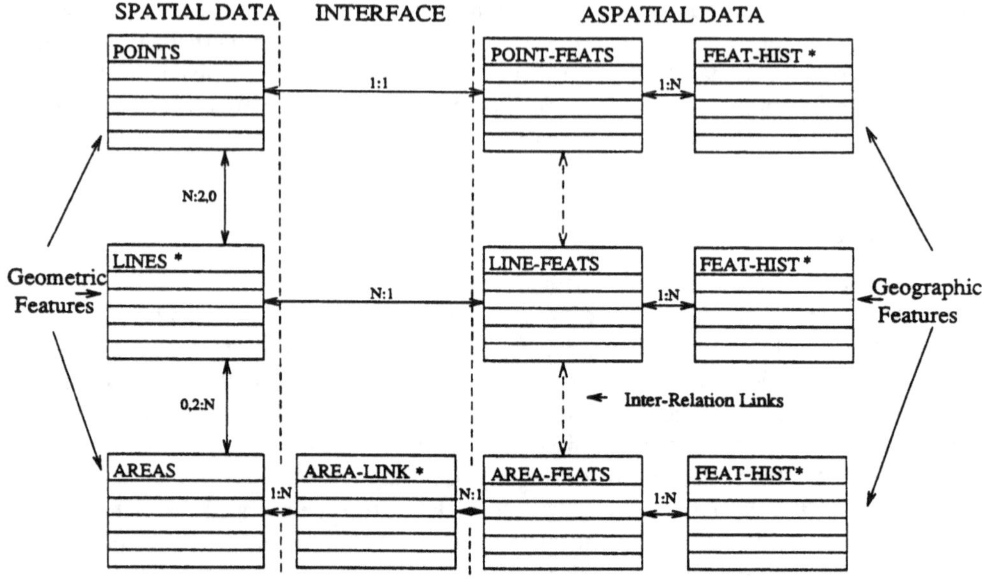

Figure 7.5 Design for a spatio-temporal GIS in POSTGRES.

(for topological information), a time range attribute representing the life-time of the line and finally a feature field, which references a line feature (if the line is not just a polygon boundary otherwise it is set to zero). As areas are implicitly stored with the line tuples, the AREAS relation merely keeps track of what type (e.g. land parcel, pollution area) relates to each area identifier.

A set of relations is then required to enable the aspatial information to be associated with each spatial feature. In the case of points, what is required is a relation with records containing: the feature's identifier, the identifier of the point which represents it in space, followed by required aspatial attributes. Depending on the method used to capture temporal changes in the non-spatial data, extra history tables containing past states may also exist in the database. Note that the point identifier is a time-variable value, and would therefore need to reside in a history table.

In the case of line features, there is no need to keep separate identifiers for the feature and the physical line segment used as part of it, as each line feature (e.g. a road) will almost always be made up of more than one segment, which is why each segment references the feature it is part of. Thus for each line-feature relation, the feature identifier is included as well as all its aspatial attributes.

Areas are treated slightly differently because each area feature may be made up of more than one disjoint polygon, and these can change over time. This situation does not arise with points (because they are dimensionless and simply have two states, existence or non-existence), nor with lines, because the feature being represented is stored in the line tuple itself, and time-stamped directly, (therefore any change in the link between geometric feature and geographic feature is recorded explicitly). Therefore, the areas relation is associated with a link relation which includes appropriate time-stamps.

Note that each feature relation in the diagram represents one type of geographic feature, and although not shown, there can be more than one set of relations for each non-spatial feature (e.g. lines can be roads, rivers, boundaries etc.).

Note also that the AREA-LINK relation does not show an associated history table. This is purely to simplify the diagram. Depending on the application, if a lot of changes occurred in the relationship between geometric and geographic features, then a history table (using the chaining method) would be suitable.

The model presented here is a single layer model. This simply means that the topology of different feature types (e.g. roads, lakes, towns) is established in a unified manner. Other systems separate features into different coverages which then have to be manipulated as required. Using multiple layers thus has the disadvantage of needing processor intensive algorithms in order to be able to operate on spatial features across coverages. Using the POSTGRES model, we can explicitly hold point-area containment information (fundamental to nearly all spatial operations (Hoop and Oosterom, 1992). It actually does not require much effort to turn the single-layer model into a multi-layered one; the only addition needed is the implementation of spatial operators, using a point containment algorithm and the (logical if not physical) partitioning of different features.

7.4.3 The Zenith object-oriented model

Research (Langran, 1989) has highlighted certain factors that could be barriers to successful implementation of temporal GIS. For example, how to allocate a constant identity to a changing object, how to deal with non-linear temporal topology and the need for adaptable clustering mechanisms to cope with a diversity of access requirements. It appears that some of these problems can certainly be addressed using object management systems and therefore one particular object model was investigated.

One such system is the Zenith object management system (Kemp and Oxborrow, 1992) which provides a flexible, tailorable version control capability. Zenith provides unique object identity which enables entities to be immutably identified through various changes of state. Zenith objects are totally encapsulated and the system provides the infrastructure that enables inter-object associations to be declared flexibly with associated semantics. This relationship capability is used to provide a flexible version control mechanism.

Versioning in Zenith (Kemp and Thearle, 1992) is achieved by using a *has_version* relationship to connect the object in its current state to previous states of the object at different points in time. Each version is also linked to the previous and successive versions using *predecessor/successor* relationships. An advantage of this mechanism is that the object and all its versions can be accessed as a set of related objects, or each individual version may be accessed, depending on the requirements of the application. The version capability can be used to realize the temporal dimension by attaching a time component to the object at a suitable level in its structural hierarchy. Moreover, the model is not restricted to a linear temporal topology; an object's states can be captured in a version tree thus enabling multiple versions to exist during a particular time span.

With reference to the land terrier application, the deed packet is central to the design (though the suggested design is not the only possible one) and thus encapsulates most of the other entities as well as its own attributes. Figure 7.6 illustrates part of a possible design for the land terrier which uses the versioning capability to model the temporal dimension. For example, if it was required to know when deed packet number 4 was used as a car park, in the relational model we would need to sequentially search the history table and match up values for the particular deed packet attribute value and the *use* attribute. In the object model the historical data is already directly related to the relevant deed packet object. This is one of several alternative strategies that are being considered as part of on-going research into the applicability of the object-oriented paradigm.

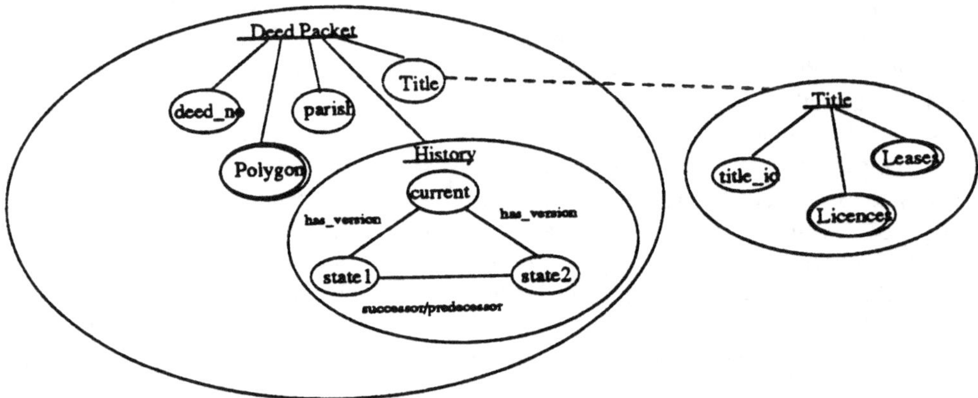

Figure 7.6 A portion of the land terrier model.

7.5 Conclusions

We conclude the chapter by mentioning some factors pertaining to the inclusion of generic spatio-temporal capabilities which are relevant to this research. An important factor to consider when designing a temporal GIS, is how to query the database. One way to achieve this is to create a language which is rich in relevant temporal operators like *when* and *during* as described in (Snodgrass, 1987; Sadeghi *et al.*, 1988). Attempts have been made to modify existing query languages, for example, Ariav (1986) describes an extension to the popular SQL query language called TOSQL and proposals are being discussed to extend SQL3 with spatial and temporal operators.

If an attempt is made to categorize queries it becomes apparent that one way is to consider what is returned to the user and in what form. For example, if the user asks a question such as 'was this state true at time T'?, although the result at a low level will be one or more objects (containing the state in question) the system will need to analyse the data in order to return the correct reply type, namely a simple Boolean value. In another case, the user may wish to see the result graphically, hence, once the relevant objects have been extracted they will have to be appropriately displayed. Another point to be borne in mind is that as a temporal GIS is a superset of a non-temporal GIS which in turn is a superset of a non-spatial database, a generic query capability should be capable of handling any combination of multidimensional queries. It is possible that the user may also be allowed to control overall 'policies' to tailor the functioning of a temporal GIS to the demands of individual environments.

It is apparent that when adding temporality to databases, in particular with respect to a GIS, many considerations have to be taken into account before an appropriate method is selected. In the application discussed in this paper, the method using chaining was considered the most appropriate, due to its simplicity, efficiency and relatively small storage requirement. In other cases it may be wiser to adopt one (or may be a combination of more than one) of the other mentioned methods, giving greater flexibility and power. The next generation of GIS are likely to embody generic features which will enable users to select the type of temporal support most appropriate for specific applications.

References

Allen, J.F., 1983, Maintaining knowledge about temporal intervals, *Communications of the ACM*, **26**(11), 832–43.

Ariav, G., 1986, A temporally oriented data model, *ACM Transactions on Database Systems*, **11**(4), 499–527.

Clifford, J. and Warren, D.S., 1983, Formal semantics for time in databases, *ACM Transactions on Database Systems*, **8**(2), 214–54.

Effenberg, W.W., 1992, Time in spatial information systems, *Proceedings of the First Regional Conference on GIS research in Victoria and Tasmania*, Ballarat Victoria.

Gabbay, D. and McBrien, P., 1991, Temporal logic and historical databases, *Proceedings of the Seventeenth International Conference on Very Large Databases*, Barcelona.

Guptill, S.C., 1990, Multiple representations of geographic entities through space and time, *Proceedings of the 4th International Symposium on Spatial Data Handling*, Zurich, Switzerland, 859–68.

Hoop, S. de and Oosterom, P. van, 1992, Storage and manipulation of topology in POSTGRES, *Proceedings of EGIS '92, Third European Conference on Geographical Information Systems*, Munich, Germany, 1324–36.

Hunter, G.J. and Williamson, I.P., 1990, The development of a historical cadastral database, *International Journal of Geographical Information Systems*, **4**(2), 169–79.

Kemp, Z. and Oxborrow, E., 1992, An object model for distributed multimedia geographic data, *Proceedings of EGIS '92, Third European Conference on Geographical Information Systems*, Munich, Germany, 1294–303.

Kemp, Z. and Thearle, R., 1992, Modelling relationships in spatial databases, *Proceedings of the 5th International Symposium on Spatial Data Handling*, 313 Charleston, USA, 313–22.

Langran, G., 1989, A review of temporal database research and its uses in GIS applications, *International Journal of Geographical Information Systems*, **3**(3), 215–32.

Lorentzos, N.A. and Johnson, R.G., 1987, TRA: a model for a temporal relational algebra, *Proceedings of the Temporal Aspects in Information Systems Conference*, France.

Lum, V., Dadam, P., Erbe, R., Guenauer, J., Pistor, P., Walch, G., Werner, H. and Woodfill, J., 1984, *Designing DBMN support for the Temporal Dimension*, ACM SIGMOD Record, **14**(2), 115–30.

Lundberg, B., 1983, An analysis of the concept of event, in Schneider, H.J. and Teubner, B.G. (Eds), *Computer Symposium 1983 on Applications Development*, Stuttgart, 135–50.

Rowe, L.A. and Stonebraker, M.R., 1987, The POSTGRES data model, *Proceedings of the 13th VLDB Conference*, Brighton, 83–96.

Sadeghi, R., Samson, W.B. and Deen, S.M., 1988, A historical query language, in Gray, W.A. (Ed.), *Proceedings of the Sixth British National Conference on Databases (BMCOD 6)*, Cardiff, Cambridge University Press, 69–86.

Schiel, U., 1982, A semantic model and its mapping to an internal relational model, University of Stuttgart, Germany.

Snodgrass, R., 1987, The temporal query language TQuel, *ACM Transactions on Database Systems*, **12**(2), 247–98.

Snodgrass, R. and Ahn, I., 1985, A taxonomy of time in databases, *Proceedings of the SIGMOD '85 Conference in Austin*, Texas, 236–45.

Stickler, G.J., Shearn, V.J. and Browning, G.P., 1992, Handling overlapping parcels in ARC/INFO, ESRI (UK) Technical Paper 4.

Stonebraker, M.R. and Rowe, L.A., 1986, *The design of POSTGRES*, Department of Electrical Engineering and Computer Sciences, University of California, Berkeley, CA 94720.

Stonebraker, M., 1987, *The design of POSTGRES storage system*, Department of Electrical Engineering and Computer Sciences, University of California, Berkeley, CA 94720.

Theodoulidis, C., Loucopoulos, P. and Wangler, B., 1991, A conceptual modelling formalism for temporal database applications, *Information Systems*, **16**(4), 401–16.

Worboys, M.F., Hearnshaw, H.M. and Maguire, D.J., 1990, Object-oriented data modelling for spatial databases, *International Journal GIS*, **4**(4).

8

Towards temporality in GIS

Monica Wachowicz and Richard G. Healey

Spatio-temporal capabilities in GIS are increasing in importance for many applications. A spatio-temporal data model is proposed to support spatio-temporal changes occurring in real-world phenomena. Particular attention is given to modelling the semantics of time in an update process as well as defining a version management strategy.

8.1 Introduction

Expectations about exploring temporal capabilities have become more significant in GIS applications. Some of the dynamic functionality to be offered by forthcoming temporal GIS includes the use of animated sequences in order to represent changes occurring in real-world phenomena over a period of time (Davis and Williams, 1989, Itami, 1988). Implementing hypermedia coordination (Armenakis, 1993) as well as multimedia functions (Fonseca *et al.*, 1992) will enable temporal GIS to handle the dynamics of real-world changes in a more realistic manner with animated maps, images and special effects.

Moreover, a wider range of manipulation tools will be offered by having integrity constraints to maintain update processes that are coherent with the former stored states of real-world phenomena (Langran, 1992). By producing a lineage of data to track the historical information associated with real-world phenomena, temporal GIS will provide analytical tools for the recognition of patterns of change through time as well as the prediction of future changes, by implementing dynamic simulations (Langran, 1989; Price, 1989; Vrana, 1989).

In spite of this accelerating trend, most of the reviews of temporal research (Snodgrass *et al.*, 1993, Soo, 1991, Snodgrass, 1990) reveal the inherent challenges involved in incorporating time in GIS. These challenges are mainly concerned with essential issues such as modelling the semantics of time, temporal database design, performance of access methods for temporal data, temporal query optimization and version management approaches.

Two of these issues, namely the semantics of temporal models and version management approaches, have been chosen for further investigation about establishing a spatio-temporal data model within a GIS. This spatio-temporal data model has primarily been developed for treating specific spatial changes that can occur with real-world phenomena over their lifespan within a GIS. The model can be examined from the following viewpoints:

- From the application point of view, the model is a general framework that can be used, for example, to support updates of boundary definitions conforming to the statutory requirements of a national mapping agency or a public sector. Four main update procedures have been selected on the basis of their evolving spatio-temporal changes.

- From the methodological point of view, the dimensional representation of time is proposed as the semantic construct underlying the spatio-temporal data model. Besides, the levels of object versions and version configurations are proposed for version management support.
- From the implementation point of view, an object-oriented database paradigm is suggested as a natural way to implement the semantic constructs and version hierarchies which have been identified in the spatio-temporal data model.

In the following sections of this chapter, each of these perspectives will be briefly presented in turn. In section 8.2 the framework of the overall update process is introduced. Section 8.3 outlines the main aspects of the spatio-temporal data model applied to four update procedures which have been previously identified. The implementation issues are covered in section 8.4. Finally, a brief discussion dealing with some of the issues raised completes the chapter.

8.2 The application domain for the spatio-temporal data model

> 'The passage of time is important only because changes are possible with time.'
> (Shoham and Goyal, 1988, p.420)

The concept of time implies that changes occur throughout the present, past and future of the life span of a real-world phenomenon. A temporal GIS will aim to understand these changes and their effects over time rather than simply reproducing them by displaying a sequence of snapshots.

The update process is one of the domains involved in temporal GIS applications, others include inventory descriptions, quality control and evaluation of uncertain data. Goodchild (1982) cites that the effort required for updating data in GIS may be equivalent to 80 per cent of the original capture effort, thus revealing an application area of considerable operational importance.

Basically, an update process is characterized by a set of procedures which can generate one or more specific spatio-temporal changes according to the spatial data representation being used. Among the spatial data representations available within GIS, the vector-based representation is the most complete because its characteristics of geometry, topology and thematic properties of a real-world entity can all be employed to describe spatio-temporal changes. For a vector representation, Armstrong (1991) defines eight possible combinations of spatio-temporal changes which can occur in an updating process. In contrast, a grid cell representation allows the description of spatio-temporal changes by employing only the thematic properties.

Our spatio-temporal data model will deal with the spatio-temporal changes in a vector based representation by focusing on updating procedures without losing valuable non-current spatio-temporal data. Therefore, four update procedures which can occur in a continuous section of a boundary have been selected to describe spatio-temporal changes. Boundaries can define the limits of the administrative areas, forested landscapes as well as surface water hydrology. Hence, a spatio-temporal data model is a valuable means of evaluating and testing possible boundary changes.

The following four main update procedures and their respective spatio-temporal changes have been chosen for more detailed investigation.

1. Spatio-temporal changes: none
   Update procedure involved: creation of a new boundary
2. Spatio-temporal changes: geometry, topology, thematic
   Update procedure involved: creation of a new object from a previously existing one
3. Spatio-temporal change: thematic
   Update procedure involved: description updating of existing boundaries
4. Spatio-temporal change: geometry
   Update procedure involved: relocation of a boundary

8.3 Modelling the semantics of time

> '...the philosophy of object-orientation...offers both a holistic as well as modelling-oriented view of the real-world...[these] are necessary for dealing with an elusive subject like modelling temporal change...'
>
> (Ramachandran, 1992, p.57)

The semantics of time have been incorporated into data models using several different approaches. Three of these can be distinguished according to the assumption of time as a parameter, property or dimension (Effenberg, 1992).

In the parameter representation, time is used as a control argument while some possible effects of other variables are investigated. Its use in temporal GIS is limited.

In contrast, the property representation has been widely employed and implemented in several applications. Clifford and Ariav (1986) describe various attempts at modelling time as a property. Most of the research extends the relational model by creating new versions of tables, tuples or attributes to reflect changes occurring over time. Langran (1989) also reviews temporal research on the basis of dimensional dominance and concludes that attribute versioning is a hybrid organization which offers the most adequate approach for GIS applications requiring spatial dominance.

Recently, McKenzie and Snodgrass (1991) provided an evaluation of 12 time-ordered relational algebras on the basis of 26 criteria relating to the design decisions adopted and conceptual models of time employed. Although time is generally perceived as being continuous, the preference for a discrete time model stands out from the evaluation. Time is mostly incorporated into relational databases as discrete subsets of the real numbers ordered linearly. Therefore, changes are supposed to take place a finite number of times so that each update operation produces a sequence of historical states indexed by time.

Current database systems already support user-defined time which is incorporated as an attribute domain such as an integer field or character string. For example, SQL2 standard has explicit support for user-defined time in its datetime and interval types. However, different forms of time such as transaction and valid time (Snodgrass and Ahn, 1985), also referred to as physical and logical time (Lum *et al.*, 1984) or database and world time (Langran, 1992) have not yet been implemented effectively. Valid time is characterized as the time that an event occurs in reality, while transaction time is the time when the data concerning the event is stored in the database.

Finally, the dimensional representation of time has introduced a novel dynamic construct. The most employed concept consists of orthogonal space and time dimensions intersecting at a known origin using a multidirectional coordinate system. Ariav (1983) develops the cube view in which time is represented as a three-dimensional construct orthogonally associated with objects and attributes. Worboys (1992) also proposes the one-dimensional and uni-directional representation of time as being orthogonal to the two spatial dimensions.

After analysing some metaphors of the semantics of time, Hazelton (1992) proposes a multi-temporal 4D GIS in which time is represented as a fractal dimension.

8.3.1 The spatio-temporal data model

Time and space appear to interact in a more natural way through the dimensional representation. However, the time dimension could not assume both valid time and transaction time representation uniformly and still retain a consistent semantic for transaction time. The transaction time is mainly relevant to system domain developments for concurrency control and recovery to re-establish a consistent database state during a multi-user access or after a system failure. Consequently, the time dimension used in our spatio-temporal data model will represent the valid time used to capture the history of a real-world phenomenon over time. User-defined time is a date denominated attribute data type which may either be atomic or set valued in our spatio-temporal data model.

An important aspect of this dimensional representation is the possibility of exploring the similarities between time and space concepts in dealing with GIS applications. The outstanding similarity between both is that time is conceptualized along a linear dimension to provide a sense of the past, present and future, while space is conceptualized not only along one, but three linear dimensions. However, in our spatio-temporal data-model, space and time dimensions are not modelled orthogonally. This point allows two basic concepts to be established, namely OBJECT and EVENT (Figure 8.1).

Objects are abstractions of real-world phenomena which may exist in many different versions and have complex interrelationships. Objects are referred to an object identifier which is a unique number internally generated within the spatio-temporal data model. Events are abstractions of any action which can produce any change of the state of an object at a time T. Events provide a mechanism for controlling the propagation of the update procedures over the objects. The concept of event identifier is defined in the same way as for objects.

In our spatio-temporal data model, one event can be associated with several different objects at a particular time T. That means, the same update procedure can be applied to more than one object. The interaction between events and objects occurs when an update procedure is carried out over an object at any time T (Figure 8.2).

Furthermore, events and objects support integrity enhancements in the spatio-temporal model. Both comprise spatial and temporal data which must coexist in the same single database environment. Our spatio-temporal data model allows their coexistence without requiring any additional support. This similarity is illustrated in Figure 8.3 by referring to some possible spatio-temporal data types and operators for objects and events respectively.

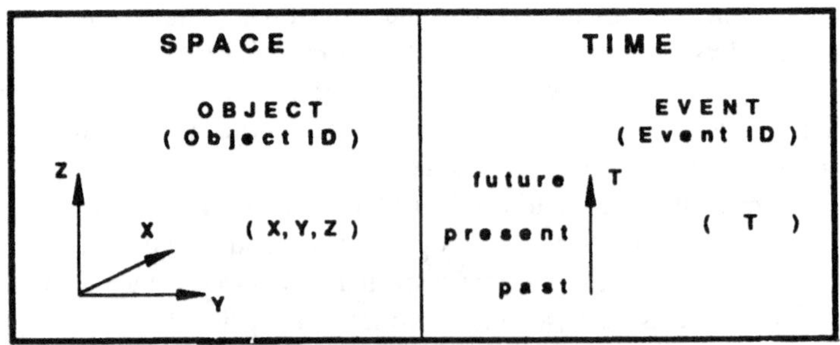

Figure 8.1 Space and time dimensions.

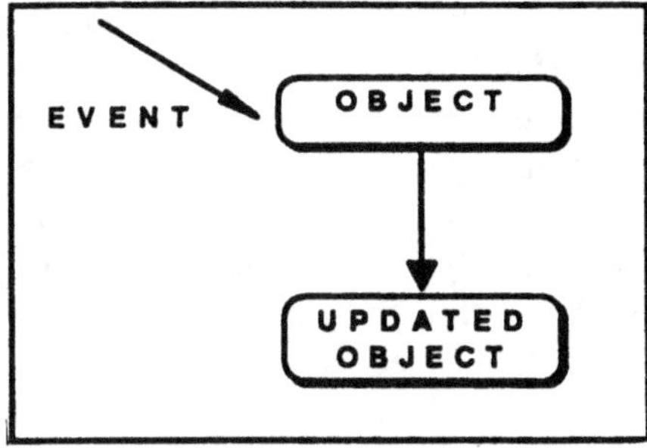

Figure 8.2 Event and object interaction at a time T.

SPATIAL DATA TYPES	TEMPORAL DATA TYPES
point node link chain polygon	chronon interval duration sequence period
SPATIAL OPERATORS	**TEMPORAL OPERATORS**
disjoint meets equals contains covers overlaps intersects	early/late meets equals during starts/finishes overlaps intersects

Figure 8.3 Proposed spatio-temporal data types and operators.

8.3.2 The version management approach

Besides incorporating the semantics of time, the concept of evolution of real-world phenomena over time is also necessary to integrate object and event elements in the proposed spatio-temporal model. This introduces the concept of version management. Because changes occur over time, whenever an object is created or modified, versions of this object should be available to provide a temporal update process in the GIS.

Ahmed and Navathe (1991) state that most problems encountered in version management are due to the absence of a concrete underlying data model in the existing approaches which have been employed.

By looking for different alternative strategies, Snodgrass (1990) argues that object-oriented data models include significant support for versioning despite the lack of research work on valid-time in object-oriented modelling.

Our spatio-temporal data model is based on object-oriented semantics in order to set up version management mechanisms. Two main versioning levels can be distinguished: object version and version configuration. At the object version level, the principal aim is to implement the versioning semantics and maintain the object versions explicitly stored in the database. On the other hand, the version configuration level is concerned with the organization of the ensemble of object versions in a database.

Basically, two approaches can be used to represent object versions. One possible approach is a single object identifier for all object versions. Time stamps or version numbers are associated with the properties of an object and then chained in an historical order. This approach can have advantages if only a few properties of an object require updating.

The most commonly used approach is the association of different identifiers with each new version of the same object. New object versions are created in relation to a base state of an object which can be located in the present or the past. This approach seems very efficient for tracking versions at the object level and thus, it has been adopted in our spatio-temporal data model. Unfortunately, comparative performance analyses of these two approaches are currently lacking.

Four basic premises underlie our spatio-temporal model at the object version level:

- every object must have an initial version;
- an hierarchical structure is imposed on the versions of an object;
- different versions of an object denote different object instances;
- among versions, a current version is always distinguished.

Figure 8.4 shows the version set representations derived from the above premises.

For each version set in the spatio-temporal data model, the object versions are managed through the concept of discrete incremental modifications (Wegner and Zdonik, 1988). This concept states that a new object version is one which inherits all the specifications from its previous version state and adds its own specifications.

At the version configuration level, two main object-oriented abstraction mechanisms, namely generalization and aggregation, are important tools for spatio-temporal modelling. The generalization abstraction mechanism allows us to assemble several classes of objects (subclasses) which have some properties and behaviour in common into a more general superclass. On the other hand, the aggregation abstraction mechanism groups objects (components or subparts) which consist of several different objects into a composite object (Egenhofer and Frank, 1989).

In terms of spatio-temporality, these abstraction mechanisms generate an ensemble of version sets which coexist and interrelate over time. The relationship between different version sets can be established by an inheritance hierarchy (generalization abstraction) in which each object version of a subclass will always inherit the state from a respective object version of a superclass and add its own state (Figure 8.5). Consequently, the existence of a subclass object version depends on the previous existence of its respective superclass

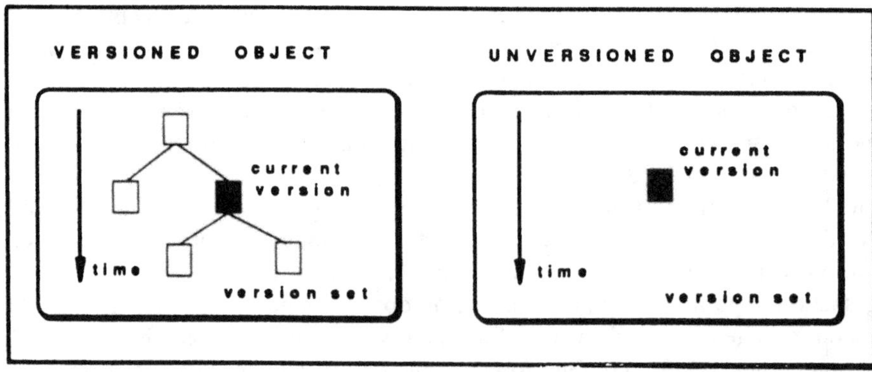

Figure 8.4 Version set representation for object versions.

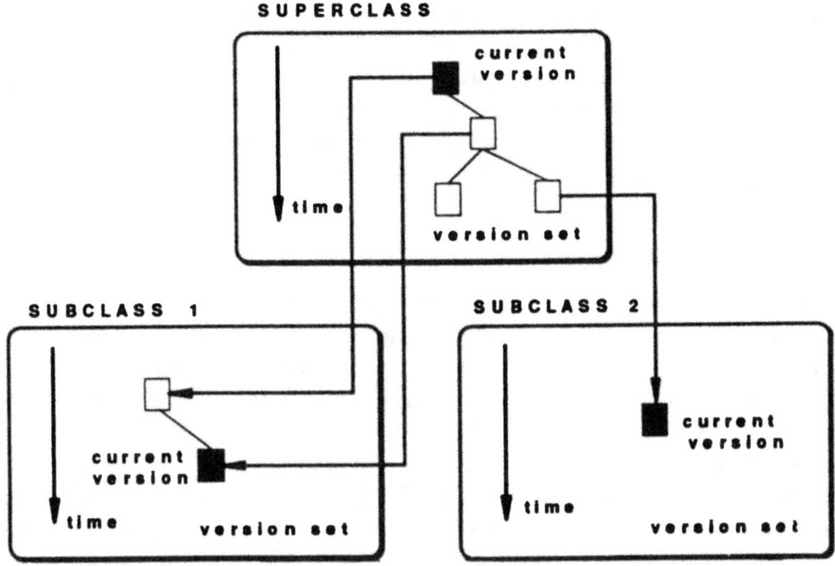

Figure 8.5 Inheritance hierarchy of object versions.

object version. That means, at any time T, a subclass object version has to be generated later than its respective superclass object version.

A propagation hierarchy (aggregation abstraction) can also be employed to establish the relationship between different version sets. In this case, the concept of dependence is applied in an opposite direction (Figure 8.6), since each composite object version owes its existence to its respective component object version(s). That means, at any time T, a composite object version can have the same or a later occurrence as its respective component object version(s).

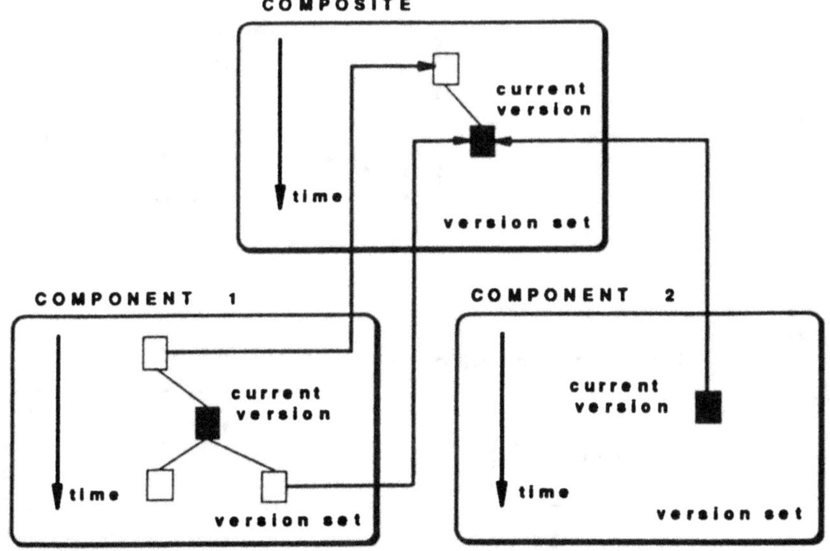

Figure 8.6 Propagation hierarchy of object versions.

There exists some debate about the implementation of version configurations in a GIS, since applications may have different configuration semantics. Instead of having a restricted version configuration, some basic mechanisms are provided to allow GIS users to implement their own version configurations such as create and delete object versions. However, this last alternative requires a substantial programming effort for developing an application version configuration that can be not helpful for many GIS users.

8.3.3 The spatio-temporal changes

The four previously mentioned update procedures and their respective spatio-temporal changes can be described using the semantic constructs underlain in our spatio-temporal data model. First of all, a version set is always generated when an update procedure creates a new boundary. In this case, the new boundary is generated within a version set with an unversioned object representation (Figure 8.7).

Once this version set exists within a version configuration, other update procedures can be carried out over the boundary at any time T. Figure 8.8 illustrates a possible state of this version set after two new boundaries have been created from the previously existing one. As a result, the version set assumes a versioned object representation.

Moreover, the version set representations can be generated from update procedures which involve thematic or locational spatio-temporal changes. Figure 8.9 illustrates an example of a version set evolution for a boundary after its description and location have been changed.

8.4 *Implementation aspects*

'The richness of an object-oriented data model is a mixed blessing. On the one hand, it makes it easier for the users to model their applications. On the

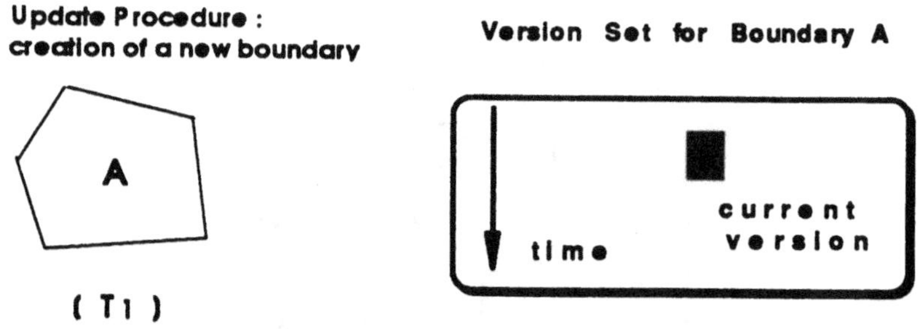

Figure 8.7 Version set representation for a new boundary.

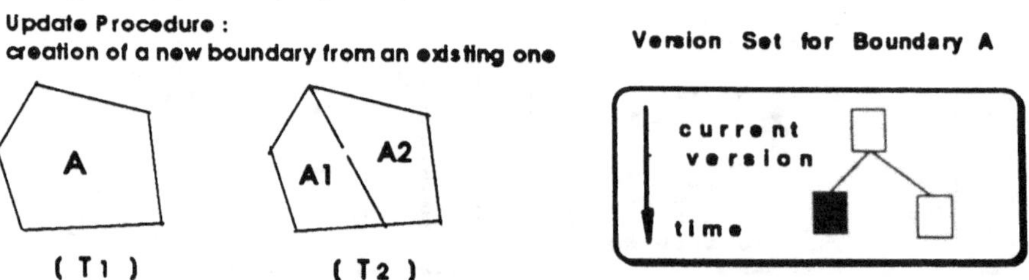

Figure 8.8 Example of a version set representation.

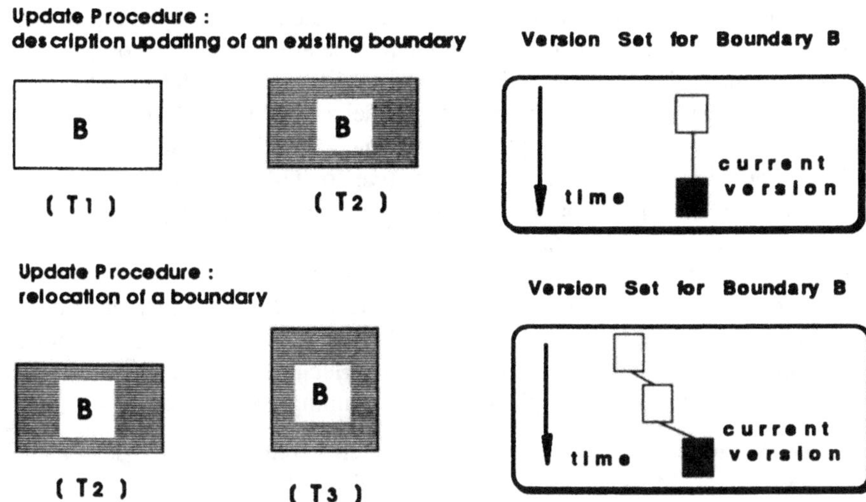

Figure 8.9 Examples of some possible object versions for update procedures.

> other hand, the complexity of the object-oriented database schema, with the class hierarchy and aggregation hierarchies, makes it difficult for both programmers and non-programmers to use the systems and compounds the difficulties of logical and physical database design.'
>
> (Kim, 1991, p.28)

Version support for transaction time has been implemented in some object-oriented databases. ORION, ONTOS and ObjectStore are some examples of object-oriented databases which provide the mechanisms to create and delete object versions. The most commonly used approach is the instance versioning in which different versions of the same object differ only in the values of some of their properties. Ahmed and Navathe (1991) identify type versioning as another possibility in which different versions of an object can have different assigned data types. However, this versioning approach proposition deviates from the traditional concepts of databases.

Three main issues must be addressed for the implementation of our spatio-temporal data model:

- at the object level: how to create an object version and how to validate which object versions are to belong to the version sets in the database;
- at the configuration level: how to implement version set interrelations using object versions.

How these issues are achieved is related to the way in which our spatio-temporal data model is implemented and integrated with other components of a GIS. Broadly, three main strategies can be identified on the basis of database systems.

1. The spatio-temporal constructs may be implemented as an extended database functionality. Embedded query languages with programming capabilities can be employed for the user-interface. POSTGRES and PROBE are examples of extended relational and functional databases. The GEO System (de Hoop and van Oosterom, 1992) is a prototype design in which spatial constructs have been implemented by developing some extensions to the POSTGRES structure.
2. The spatio-temporal application may be implemented in an object-oriented database

to handle spatial and temporal constructs in an integrated fashion by using an existing programming language. ONTOS, O2 and GemStone are examples of object-oriented databases, although their semantics differ widely. Milne *et al.* (1993) developed a first prototype system based on the conceptual data model defined in the Spatial Data Transfer Format (SDTS) which has been implemented in the ONTOS database. There have been few implementation studies using this alternative for GIS applications so far.

3. The spatio-temporal application may be constructed in a proprietary database tailored to spatio-temporal needs. An application programming environment is used to specify the query and representation capabilities of the system. The proprietary database can be used to produce relational databases or even file systems. EXODUS and GENESIS databases are two examples. Some commercial GISs are now utilizing this alternative for building the spatial constructs on top of a proprietary database. The user interacts with the database through an application programming environment.

Whatever approach is undertaken, several problems remain to be solved in the implementation phase. Most of these are related to difficulties with indexing techniques for very large databases, optimizing data access and archiving historical data.

8.5 Conclusion

This chapter has described a new spatio-temporal data model which proposes to incorporate the time dimension in an update process application within a GIS. The distinct characteristic of this model compared with former proposals is that time is conceptualized as an independent dimension providing the sense of past, present and future. This concept implies the representation of an event as an action which can produce any change of state of a real-world phenomenon. An ensemble of events will describe the updating procedures performed during the life span of a GIS. Moreover, space is conceptualized in three linear dimensions where objects are generated to represent real-world phenomena over time.

Space and time dimensions are then integrated by means of the coexistence of objects and events within the data model. The interaction between objects and events takes place when an update procedure produces spatio-temporal change in an object. Whenever a spatio-temporal change occurs over time, versions of an object are created.

Therefore, two versioning levels are set up for developing object versions and their configuration within an update process in a GIS application. Some uncertainties still remain about adopting a strategy for the object version and version configuration implementations due to a lack of performance studies on the existing alternatives.

Whatever alternatives are undertaken for implementation, our spatio-temporal data model is based on object-oriented semantics. The object-oriented enhancements adopted by the model offer useful improvements in functionality, clarity of data modelling and the potential for simplification of future application developments in temporal GIS.

References

Ahmed, R. and Navathe, S.B., 1991, Version management of composite objects in CAD databases. *Proceedings ACM SIGMOD 1991*, Denver, Colorado, USA, May, 218–27.

Ariav, G., 1983, Preserving the time dimension in information systems. PhD thesis, Philadelphia, Pa, Wharton School, University of Pennsylvania.

Armenakis, C., 1993, Map animation and hypermedia: tools for understanding changes in spatio-temporal data. *Proceedings The Canadian Conference on GIS*, Ottawa, Canada, 859–68.

Armstrong, M.P., 1991, Temporality in spatial databases. *Proceedings GIS/LIS'88*, San Francisco, USA, November 1988, **2**, 880–9.

Clifford, J. and Ariav, G., 1986, Temporal data management: models and systems, in Ariva, G. and Clifford, J. (Eds), *New directions for database systems*. Ablex Publishing, Ch. 12, 168–85.

Davis, B.E. and Williams R., 1989, The five dimensions of GIS. *Proceedings GIS/LIS'89*. Orlando, USA, November, Vol. 1, 50–8.

Egenhofer, M.J. and Frank, A.U., 1989. Object-oriented modelling in GIS: inheritance and propagation. *Proceedings AutoCarto 9*, Baltimore, USA, April, 588–98.

Effenberg, W.W., 1992, Time in spatial information systems. *First Regional Conference on GIS Research in Victoria and Tasmania*, Ballarat Victoria, September.

Fonseca A., Gouveia C., Camara A.S. and Ferreira F.C., 1992, Functions for multimedia GIS. *Proceedings EGIS'92*, Munich, Germany, March, **2**, 1095–101.

Goodchild M., 1982, Accuracy and spatial resolution: critical dimensions for geoprocessing, in Douglas, D. and Boyle, A.R. (Eds), *Computer Aided Cartography and Geographic Information Processing: Hope and Realism*, Ottawa, Canada, 87–90.

Hazelton, N.W.J., 1992, Beyond the 2-D Map: a new metaphor for multi-temporal 4-D GIS. *Proceedings GIS/LIS'92*, San Jose, California, USA, **1**, 303–13.

de Hoop, S. and van Oosterom, P., 1992, Storage and manipulation of topology in Postgres. *Proceedings EGIS'92*, Munich, Germany, March, **2**, 1324–36.

Itami R.M., 1988, Cellular automatons as a framework for dynamic simulations in geographical information systems. *Proceedings GIS/LIS'88*, San Antonio, USA, November, **2**, 590–7.

Kim, W., 1991, Object-oriented database systems: strengths and weakness. *JOOP*, July/August, 21–3.

Langran G., 1992, *Time in Geographic Information Systems*. London, Taylor & Francis Ltd.

Langran G., 1989, A review of temporal database research and its use in GIS applications. *IJGIS*, **3**, (3), 215–32.

Lum V., Dadam P., Erbe R., Guenaver J., Pistor P., Walch G., Werner H. and Woodfill J., 1984, Designing DBMS support for the temporal dimension. Proceedings SIGMOD'84 Conference,, Boston, USA. *SIGMOD Record*, **14**(2), 115–30.

McKenzie, L.E. and Snodgrass, R.T., 1991, Evaluation of relational algebras incorporating the time dimension in databases. *ACM Computing Surveys*, **23**(4), December, 501–43.

Milne, P., Milton S. and Smith J., 1993, Geographical object-oriented databases – a case study. *IJGIS*, **7**(1), 39–55.

Price S., 1989, Modelling the temporal element in LISs. *IJGIS*, **3**(3), 233–43.

Ramachandran, B., 1992, *Modelling temporal changes in the structure of real-world entities within a GIS environment using an object-oriented approach*. MSc thesis, Department of Geography, University of Edinburgh, September.

Shoham, Y. and Goyal, N., 1988, Temporal reasoning in artificial intelligence, in Shrob, H.E. (Ed.), *Exploring Artificial Intelligence: Surveys Talks from National Conferences on Artificial Intelligence*. Morgan Kaufman Publishers, Ch. 11, 419–38.

Snodgrass R.T., 1990, Temporal databases status and research directions. *SIGMOD Record*, **19**(4), 83–9.

Snodgrass, R.T and Ahn, I., 1985, A taxonomy of time in databases. *Proceedings ACM SIGMOD Int. Conf. on Management of Data*, Austin, USA, May, 236–45.

Snodgrass, R.T., Al-Taha, K. and Soo, M., 1993, Bibliography on spatiotemporal databases. *ACM/SIGMOD Record*, March.

Soo, M.D., 1991, Bibliography on temporal databases. *SIGMOD Record*, **20**(1), 14–23.

Vrana, R., 1989, Historical data as an explicit component of land information systems. *IJGIS*, **3**(1), 33–49.

Wegner, P. and Zdonik, S.B., 1988, Inheritance as an incremental modification mechanism or what like is and isn't like. *Proceedings ECOOP '88*, Oslo, Norway, August, 55–7.

Worboys, M.F., 1992, Object-oriented models of spatiotemporal information. *Proceedings GIS/LIS'92*, San Jose, California, USA, November, **2**, 825–34.

SECTION II

Two Spatial Analysis and Error

Section Two: Spatial Analysis and Error

This section is more closely related to scientific models of geographic information than the computational support discussed in Section One. Most of the work here can be categorized as falling into the field of spatial analysis. Research in spatial analysis is primarily concerned with developing approaches to manipulation and transformation of geo-information which will facilitate the construction of underlying scientific theories for geographic phenomena. The spotlight has shifted from computational to analytical models of geo-information. However there is currently a close and symbiotic link between spatial analysis and GIS, as the former can use the computational power of the latter and the latter can take in richer analytical approaches from the former. The chapters in this section show up the range of present research issues in this area. These include finding new ways to use a GIS in spatial analysis, the handling of non-Boolean values in a GIS (essential for reasoning about the many fuzzy-edged entities inhabiting geographic space), visualization and the ever-present problems with the treatment of error in GIS.

The section begins with a chapter contributed by **Stan Openshaw**, from the School of Geography at the University of Leeds in England. Openshaw is a leading researcher in this field and it is always interesting to learn of his latest ideas. In this chapter he continues his quest for a meaningful role for a quantitative geography of the late twentieth century, where the tools available include the newly developed computational models. He sees the problem now, as it was in the earlier quantitative geography revolution of the 1970s, as plenty of raw computational power but relatively shallow conceptual foundations. The need is to generate a modern theory of spatial analysis which draws on the new developments in computation. Openshaw chooses to address questions which are much less specific than in the past, of the kind 'Given a set of points in a geographic space, is there any kind of geometric or geographic regularities in their structuring?' Existing abstract theories, of which there may be a computer-based library, are fitted against the observed spatial patterns. The idea is to explore the data looking for such regularities and 'in the process develop appropriate new theoretical generalizations'. He sees hope in using the methodologies developing in artificial intelligence and computer vision to handle complex knowledge structures, supported by massively parallel hardware. There is a connection with the topic of 'data mining' in database theory, where databases are constructed which facilitate such general questions as 'I'm not sure what I'm looking for, but find something interesting in the data'. In the latter parts of the chapter, several new computational approaches such as logic-based and neuro-computational models, machine learning and computer vision are discussed.

The theme of exploratory spatial analysis is continued by **Richard Aspinall** of the Macaulay Land Use Research Institute in Scotland, and applied in the context of bioclimatic mapping of Scottish wildlife. Bioclimatic mapping attempts to model the relationship between species distribution and climate patterns. The difference in resolutions between the species distribution and the climate data sets opens up certain methodological possibilities. The method described here combines known approaches to the exploratory analysis of point data sets and an inductive pattern learning process using Bayesian statistics. It allows the systematic generation of geographic hypotheses and is an excellent example of the application of a non-trivial spatial analysis using a GIS as an essential tool.

The primary theme of the next chapter by **Jason Dykes**, from the Department of Geography at Leicester University, is visualization. The link with the previous work is that its goal is the exploration of patterns in the data. The author presents his research on the presentation of area-value data. Such data are often represented as a choropleth map, where the surface is partitioned into polygonal areas. However, there is the possibility of changing the map by varying the classification scheme for the attribute data. Dykes constructs a set of maps using various classification schemes and calculates Moran's attribute similarity index I in each case. He then draws conclusions about how visualization may be best employed and argues for the power of representations based on regular grids.

Peter Fisher, from the Department of Geography at Leicester University, very clearly makes the distinction between fuzzy and probabilistic data values. Simply, the uncertain Boolean value representing the truth or falsity of the proposition that you the reader are reading this sentence on a Tuesday is probabilistic. It must be either Tuesday or not, the boundary is clear cut, and probability theory might be invoked to calculate the *a priori* probability of the proposition being true as one in seven. On the other hand, the uncertain value associated with the proposition that Tuesday is a hot day is fuzzy since there are degrees of hotness and the boundary between hot and not hot is unclear. Fisher shows that these notions have sometimes been conflated by GIS researchers. The chapter aims to clarify the distinction in the context of the GIS viewshed function. The basic Boolean viewshed function, showing which points are and are not visible from a fixed point, is well-known to be highly algorithm dependent and unstable under small perturbations of the underlying digital elevation model (perturbations well within the expected error bounds of such DEMs). Thus there is a need for uncertainty to be built into this function. The concepts of 'probable viewshed' and 'fuzzy viewshed' are separately defined and discussed. The chapter concludes by providing the beginnings of an algebra of fuzzy and probable viewsheds. Experimental evidence related to the Malvern Hills in Worcestershire, England, is included.

The final two chapters in this section discuss aspects of the treatment of error in GIS. The first is written by **Chris Brunsdon** of the Department of Geography at Newcastle University and **Stan Openshaw**, mentioned earlier.The fundamental issue addressed in this work is the modelling and prediction of error in vector-based spatial data, where no error information is available in advance. At first sight, this appears an unsolvable problem. However, the authors consider that there are certain features of the data which give clues about the nature of the error associated with them. The thesis advanced is that the geometric complexity of a linear feature may have a bearing on its associated error. The relationship between complexity and error depends upon context. Two strategies are developed. A long-term approach would be to provide methods for interactively constructing error models at data capture time, which could be distributed with the database. Alternatively, line segment components of the database could be classified according to their complexity and error predicted. The main technical contribution of the chapter is to provide methods, some based upon neural computing, for such classifications. The final chapter in the section is contributed by **Colin Monckton** of the Department of Computer Studies at Loughborough University of Technology in England. Here, research is reported on error in digital elevation models. The author defines error in a DEM as divergence from an independent source of higher accuracy. Included in the chapter is a survey of related work, description of the research methods employed, results and conclusions.

Summarizing such a diverse range of work in this area of GIS research is not easy. It is encouraging to observe the use that spatial analysts are now making of the newer computational models. They are breaking free from a narrow procedural paradigm typified by large amounts of Fortran code. The reader will find amongst these chapters thoughts about the use of massively parallel architectures, logic-based systems, new logics, visualization methods and neurocomputing. There is still a role for procedural programming in C or Fortran, but it is not the whole computational picture. Another striking feature of these chapters is their willingness to question and challenge traditional statistical approaches when applied to this field and to seek out new methods.

9

A concepts-rich approach to spatial analysis, theory generation, and scientific discovery in GIS using massively parallel computing

Stan Openshaw

9.1 Introduction

A major criticism of geographical analysis has been its inability to provide results that are rich in a traditional conceptual sense. Quantitative geographers often claim to be concerned about developing and testing theory but this task is hard, confirmatory methods do not always mean much in a geographical setting, and there is a notable lack of major theoretical achievements that can be explicitly traced back to nearly 30 years of quantitative analysis and mathematical modelling of spatial databases. The GIS revolution does not appear to be of much assistance here because it is creating an immensely spatial data-rich situation but without providing sufficiently powerful spatial analytic and spatial process modelling tools to allow users to cope with the new opportunities. Increasingly, it seems that it is not only that we can see the forest for the trees but we cannot even see the tree we are sitting in. Quite simply we currently lack tools that are sufficiently powerful to release the empirical regularities that exist amongst the noise in our spatial databases.

Furthermore, there is a wide gap between the apparent intellectual elegance and generality of those abstract spatio-theoretical constructs that exist as a legacy from the pre-GIS years and the relatively theory-poor, empirical results obtained from the geographical analysis and the mathematical modelling of spatial data. Also those, often qualitative researchers, who are seemingly most able to develop descriptive theoretical and pseudo-theoretical conceptualizations of the processes and dynamics of spatial phenomenon, usually fail to deliver testable hypotheses and computer models capable of either validating their theories or yielding useful and testable predictions in any remotely scientifically acceptable manner. As a result much of the current process knowledge base in geography is extremely fragile due to this history of untestable hypothetical speculation about the nature of the world. Likewise the computer based empiricists who are seemingly expert at developing models and performing detailed computer analyses are often quite unable to generalize their results as useful theory. They too have been poor performers at the task of scientific discovery. Indeed, it is difficult to see how this theoretical and technical gap might be bridged especially as the various parties on either side of the divide often seem to belong to separate ontological universes, characterized by very different philosophies and research paradigms. Both sides seem to ignore the problems in what they advocate by adopting a philosophical stance that deems such concerns to be largely irrelevant; (Openshaw, 1991, 1992b).

Yet it is essential to consider how in the 1990s the computational geographer might start to bridge this conceptual chasm. This is important if only to avoid a rerun of the quantitative revolution disaster of the 1970s when many of the first generation of quantitative geographers wandered off into purely social theoretical domains never to return to spatial analytical concerns. Arguably one reason for this was their dissatisfaction with the conceptual poverty of the newly found quantitative analysis tools. Maybe also the justification involved a search for an improved social behaviouralist understanding of the world, following disillusionment about the shallowness of the spatial analytical paradigm at a time when computers were extremely slow, data were extremely poor, and the available analytical methods were often inappropriate. However, the computing and data situations are today vastly different from the 1960s and 1970s. Indeed the GIS revolution is generating vast amounts of spatial information and geography-relevant data. However, GIS is still extremely poor in terms of its representation of basic spatial concepts and its ability to generate geographical theory although very useful as an applications orientated tool.

It is with some justification that the critics of GIS regard the developments in the area as a re-birth of quantitative geography. Viewed in a narrow historic way it can be seen as suffering from similar limitations. Progress in basic digital map information handling technology has far outstripped the ability of geographers to do much that is new, creative or conceptually useful with the databases. Indeed, the current era of GIS induced data richness is emphasizing our inability to generalize pattern and process, and synthesize new concepts and discoveries from the minutiae of detail found in spatial databases. GIS can, therefore, be regarded as conceptually shallow and as a result it gives the impression of being not particularly intellectually satisfying in terms of either its ability to explain the observed spatial reality it can represent or as an aid to theoretical understanding. Additionally, its strongly applied focus sometimes fosters the misleading impression that is not even a particularly scholarly area; although, of course, it is essentially an enabling technology and not an end in itself. So to some extent, sometime in its future, GIS faces the spectre of a possible rerun of the great quantitative revolution disaster of the 1960s unless a much greater degree of development occurs to the spatial analytics theory generation, and scientific discovery side of GIS.

The challenge for the 1990s is to try and discover how this might be achieved and to identify ways by which the gap between the theoretical and empirical sides to geography and GIS can be narrowed, perhaps by the development of a new style and new forms of computer analysis. It is noted that the developments in computer hardware and computing methods that has stimulated the GIS revolution have also sparked off major, new, and often revolutionary developments in many other areas of computer science, pattern recognition, and artificial intelligence. In particular, there is an increasing set of practical AI based tools that can in principle handle soft information and abstract concepts. The implication is that it is no longer necessary to think only in terms of fairly crude (from a theoretical point of view) data-orientated, exploratory spatial analysis of the positivist sort dominated by number crunching applications and statistical methods dating mainly from the 1960s. Of course, there is no need to be exclusive and it is extremely important to continue the development of computational approaches to exploratory geographical analysis making use of new methodologies and massively parallel hardware as and when they become available during the 1990s. However, it is also noted that as it becomes possible to handle abstract information in the form of concepts and theories, that this may be regarded as providing a means of opening a second front in spatial analysis based on an inversion of the normal perspectives. This would be achieved by seeking to develop a new form of exploratory spatial analysis that is both concepts-rich and dedicated to theory generation and scientific discovery.

the development of a computer language, designed from first principles to handle the basic theoretical concepts of geography imbedded in a space-time framework (Oberg *et al.*, 1990). This is, however, likely to be a slow and lengthy process. The second strategy is to find a means of directly using existing theoretical and pseudo-theoretical concepts, thought to be relevant to geographical analysis by the development of new procedures that can take them into account when performing spatial analyses. A corollary is the corresponding requirement to try to rectify the surprising scarcity of relevant spatio-theoretical notions in geography. New ones would be sought by induction from the data rich GIS environments of the 1990s. The need for the creation of new knowledge by discovery is obvious; for example, it seems that our knowledge of the spatial patterning that characterizes our towns and cities has advanced little in the last 20–30 years despite massive improvements in data provision. Seemingly most towns exhibit strongly recurrent regularities in many key variables. We know these patterns exist but until recently we did not have access to the pattern recognition technologies needed to identify them in a formal scientific fashion.

The purpose of this article is to consider how to develop from scratch a concepts-rich form of spatial analysis by seeking to exploit developments in AI and computer vision in an era of massively parallel computing. The analysis process is designed to be based on concepts that derive from geographical theory; and where these are poorly developed, to discover new theory by exploratory data analysis. It is noted that a theory based concepts-rich style of spatial analysis is surprisingly undeveloped. Traditionally, this area has been almost entirely dependent on the application of inferential statistical methods, which are often too narrowly focused to be of any great value in geography. Additionally, there was always the lingering problem of what inference actually means in a spatial context. It should now be possible to do much better by the application of AI methods and by seeking geographical uses for recent developments in computer vision. In some senses also, the new opportunities concern how best to use the impending next generation of massively parallel computers. They are expected to be capable of sustained terraflop speeds by the late 1990s, and their use will make the most of the richness in spatial databases that GIS is providing. In some ways it might soon be possible to become smarter by becoming dumber in what spatial analysis tools are expected to produce. Why this might be considered useful and how in general terms it can be applied is the subject of this chapter.

9.2 Towards a concept based spatial analysis paradigm

9.2.1 Spatial autocorrelation as a concept

It is perhaps useful to start by focusing on a space-only form of spatial analysis. This is the traditional map metric of geography and is where GIS offers many manipulative and visualization tools. The extension into space–time is left for later study on the grounds that geographically speaking it is much less urgent.

Consider an example of an apparently simple spatial analysis task. For a long time it has been a widely practiced procedure to look for evidence of spatial autocorrelation in univariate mappable data. Various methods have been derived to test for its presence and to quantify its extent. However, a far more basic spatial question is not how much there is but where it might be found or not found. The so-called First Law of Geography (Bunge, 1966) is based on a very simple concept; that nearby areas are more likely to be similar than areas far apart. So why not operationalize this geographical concept as a concept rather than as a precise, assumption-dependent, statistical test of a hypothesis. The statistical test approach requires a much greater input of knowledge than is contained in the original theoretical notion; for example, a weights matrix and a null hypothesis to compare the observed result against. The weights matrix has to be chosen in advance of the analysis and reflects the form of spatial autocorrelation that is to be detected.

It is also sometimes necessary to provide a detailed specification of the alternative hypothesis; for instance that the value in one area is negatively correlated with neighbouring areas or that the spatial autocorrelation follows a first order Markovian process. Even when a significant level of spatial autocorrelation is found to exist, this result contains no information about what might have caused it. If the null hypothesis is not rejected, then there may still be spatial autocorrelation present but the test statistic may have simply failed to detect it - for instance, if the spatial autocorrelation is localized rather than widespread throughout the study region. In both instances, the geography of the problem is lost in the statistical analysis. It is not very interesting to have an answer that merely says that there is evidence of positive spatial autocorrelation throughout (or averaged over) an arbitrarily defined study region or even that it exists within x km of a specified point. The former is far too geographically vague to be of much assistance and the latter is too specific, too scale and aggregation dependent, and too parametric to be much help in an exploratory context where prior knowledge is both limited and uncertain. It is the non-stationarities or the spatial heterogeneities in the GIS database that often are of greatest interest. It should be relatively easy to spot spatial patterns that repeat throughout a study region. There is really little of interest in many such phenomena; mapping the data will usually identify them. Instead, it is the much more localized, perhaps surprisingly, neighbourhood based patterns or anomalies that are often of the greatest interest; and it is these that conventional methods miss almost completely. The geographical question is much vaguer than the statistical tests can handle. It can in essence be reduced to a series of linked questions: (a) is there any spatial autocorrelation? (b) where is it? and (c) what form does it take? Ideally, the answer should be visible on a map for visualization purposes. Question (b) is also the answer to questions (a) and (c). The hope is also expressed that the resulting map patterns may contain useful insights about underlying process.

9.2.2 Spatial association as a concept

Similar considerations apply to concepts of spatial relationships although here the problem is far more data rather than concept orientated as the idea of spatial association is usually an empirical notion that is application specific. Nevertheless, there is a whole plethora of statistical and mathematical modelling techniques that seek to identify, specify, and quantify spatial relationships. Again, they are all extremely precise; for example, a regression model requires a parameterized specification of the explicit mathematical form of the assumed relationship; for example a linear function or one specific nonlinear one. Of course, when this precise functional information is missing, then various exploratory methods can be used to help the researcher specify and test plausible or possible functional relationships. Unfortunately, once again when viewed from a concepts perspective, the questions that the statistical technology can answer are simultaneously too specific, too precise, too narrowly defined and too assumption dependent. The concept of spatial association is much vaguer; for instance, that there is a spatial relationship without necessarily being any more specific in the first instance. It is not necessary to start by assuming a functional form; for example, that it is linear. There is no idea of what form or shape or nature that the relationship should or must take in the concept of spatial association, only that there may either be one (or not) as the case may be. Testing a linear regression model only indicates whether a global linear association exists throughout the study region. This is extremely specific and only one of a myriad of different possible functional forms. It is also not very interesting because it contains so little theoretical information. Far more relevant would be the discovery that a spatial relationship exists but without necessarily knowing precisely how to specify or parameterize it, in the first instance. It would also be nice to know whereabouts on a map of the world or the region of interest such relationships exist and where they do not. Of course, still keeping things simple and general, the mathematical nature or shape of the relationship in terms of abstract functional forms might also be interesting to know; for example, do spatial variables X and Y increase together in some way, or is there an inverse relationship, or do they go up and down together, or do they merely fluctuate randomly in their values, or is there a small number of classes of different type of relationship. Again the aim is to categorize the map space by the general nature of the association that is found; free of having to be any more precise or being forced to assume that the spatial domain of the relationship is the entire map area. Modelling the relationship is a subsequent activity. Yet at present the ability to detect the presence of a relationship entirely depends on our ability to model it in some highly specific way before we can know what it is, and this restriction is quite unnecessary from a broader theoretical point of view.

9.2.3 Distance decay concepts

Another basic geographical concept is that of distance decay; for example, this is regarded as applicable to spatial interaction data and to urban population density patterns of cities around a central point. It is also implicit in the spatial autocorrelation concept. The classical quantitative geographic approach would involve trying to fit one or more *a priori*, supposedly globally defined, distance decay functions to a whole data set. Indeed there is a battery of mathematical models that could be used. However, the basic conceptual concern is whether or not there is any evidence of distance decay patterns in an abstract and highly generalized non-parametric sense. There is no need to worry about what might or might not be the precise nature of its functional specification, at least initially. Does it exist and if so, whereabouts. If the concept is in fact a valid generalization, then there should be evidence that it is applicable in many different parts of the world and also of some areas where it does not apply at all. By adopting only a statistical approach, there is a danger of being too precise in applications where the prevailing theoretical knowledge is general and far vaguer than the data or current knowledge about form of distance decay relationships. In short, spatial analysts have traditionally ended up testing highly specific bits of far more general theories, or of performing highly specific and detailed analyses in ways that make it difficult or impossible to subsequently express the results as generalizable concepts. No wonder the first quantitative revolution in geography failed.

9.2.4 Social pattern concepts

More complex and more abstract theoretical notions are also relevant here as a source of inspiration. Unfortunately, and once again, the classical Park and Burgess (1925) concentric ring model, Hoyt's (1939) sector model of theoretical urban social structure, the multiple nuclei model of Harris and Ullman (1945), and the Shevky-Bell (Shevky and Bell, 1955) models of multivariate urban structure are far removed from either the micro-scale, small area-based empirical results of the factorial ecologists (Davies, 1984) or, more latterly, the residential area taxonomists who analyse small area census data (Charlton *et al.* 1985). Traditionally, social geographers have attempted to relate their results in a vague and general way to these theories (Clark and Gleave, 1973; Johnston, 1979); but the task is extremely difficult. Somehow the researcher has to try and project the immensely spatially detailed and multivariate complex results of small area census analyses back into a much more abstract and generalized conceptual framework of theories about the structure of the urban mosaic (Robson, 1975). Seemingly this is an almost impossible task using conventional statistical approaches. The differences in scale and detail between general theory and the immensely detailed empirical analysis of census data are just too great to be easily reconciled, if at all. The microscopic nature of the data analysis emphasizes the unique, whereas the macroscopic and abstract nature of the theories of urban social structure emphasize generality, recurrency, and ignores the micro details. It is extremely difficult to relate one to the other in any sort of direct manner, and thus obtain useful and testable hypotheses or even a typology or library of recurrent urban area pattern types. It seems that all too often the uniqueness of place has combined with the micro detail of large scale analyses to hide the macroscopic generality from the spatial analyst. We know it exists but we have so far largely been unable to systematically seek it out and identify the precise nature of the recurrent patterns in a scientific manner. As a result, quantitative geography is rich in statistical and mathematical technology but poor in terms of its concepts. Again the challenge is to try and develop a different approach to spatial analysis that is based more on basic geographical concepts than on statistics and which directly addresses these concerns.

9.3 A pattern recognition approach

Consider again the Burgess concentric ring theory as a concepts-based model of a particular general form of spatial pattern. The question is which, if any, cities exhibit concentric ring patterns in their social structure? The traditional inferential statistical approach would be to test its existence in a data set by a statistical test of some kind. However, in practice it is difficult to devise a relevant null hypothesis to test. The statistical approach is far too precise for the abstract nature of the query given the scale-free nature of the original theoretical concept; for instance, what scale of data should be used, how wide should the rings be, around which point should the circles be formed. It is noted that many urban geographers in the 1960s and early 1970s did treat the concentric ring theory completely literally and did try and fit idealized rings to cities (Murdie, 1969). Additionally, it is possible to imagine situations where the concept applies but only partially, or only in certain directions or else has bits missing or added. No statistical method is going to be clever enough to handle these problems and as a result there are going to be problems with whatever null hypotheses are devised by the researcher. A conventional statistical test is also completely aspatial, in that it does not treat patterns in spatial data as a spatial entity. However, if it is treated as a pattern recognition problem, then the question is now simply whether or not a concentric ring type of spatial pattern can be found in urban social structure. The scale at which such a patterning exists is not known and has to be determined but need not be fixed. There is no need to specify the key dimensions as *a priori* information. Again the question is simply whether any kind of concentric structure can be found without worrying too much initially as to its precise morphology or to worry about parameterizing its size and shape. The latter are statistical concerns that are initially of secondary importance. If such patterns exist, then where do they exist and at what scale. If a concentric ring model is inappropriate then which of a small library of alternative urban spatial structuring concepts might be more appropriate. If different or new urban pattern types are found to be relevant, then they can be added to the library.

Another way of asking similar questions is to reformulate the task as an exploratory analysis problem. An even simpler question might be: do cities exhibit any kind of geometric or geographical pattern based regularities in their spatio-social structure at all? If so, what do they look like, at what scale, and how recurrent are they? Another example might be even vaguer. It has long been speculated that cultural and socio-economic differences between countries, and sometimes within them, have produced characteristic but different types of urban form, social spatial structure, and urban development. For example, the concepts of new towns or colonial cities or port towns or industrial towns abound in the old geographical literature. It would be possible using multivariate statistical methods to try and identify what these characteristics might be. However, such methods cannot handle any realistic representation of a town's spatial pattern. All the within-city geographical detail is lost once the city is represented as a list of values for M variables. Alternatively, you could identify from the literature a library of different city spatial pattern types and then use this as a reference list for template matching against particular cities. Note that obviously no two cities are going to be identical in their layout and structure for all sorts of unique reasons; for instance, there will often be effects caused by site, with preferences for development in different directions, differences in size, topological controls, etc. The pattern recognition question is whether their overall patterning is similar in an abstract, scale and rotation invariant, and size independent way. You could also ask the more focused question as to which cities are similar in a spatial pattern sense to Manchester or Newcastle or New York. In essence, the objective is once again to convert existing theoretical concepts into generalized abstract spatial patterns and then to search for their existence in a similarly abstract and generalized manner in GIS databases.

Pattern searching is not the same as hypothesis testing because there is no relevant null hypothesis. This point was lost on the original quantitative geographers. For instance, despite the superficially promising title *Patterns in Human Geography*, Smith (1975) in common with other statistical geographers, failed to develop a statistical theory of spatial analysis as distinct from providing examples of statistical methods being applied to spatial data in the search for largely aspatial patterns. The danger now is that the same mistake will be repeated 20 years later in the GIS era by a failure to appreciate that spatial patterns are themselves geographical objects that can be recognized and extracted from spatial databases. The question is does a theoretically derived abstract spatial pattern of urban social structure fit spatial reality as modelled in a GIS database? If it does, where does it, and at what scale. If it does not, are there any alternatives that might be more relevant? In this context then, the question is does Leeds show a similar spatial arrangement as Newcastle, or, are major British cities characterized by a small number of different theory-based abstract spatial pattern types and, if they are, at what spatial scales and levels of multivariate generalization are they applicable?

In essence the objective is to develop a concepts-rich form of spatial analysis by looking for the existence of abstract spatial patterns obtained initially from theoretical geography by searching for the existence of bits of them. In short, the gap between theoretical and empirical geographical analysis may be bridged by the development of a concepts-based pattern recognition approach to spatial analysis.

A final example concerns the concept of clustering in space. A test of spatial clustering is not very helpful at all. If the null hypothesis of some-kind of randomness is disproven, then we still know very little about the nature or extent or location or intensity of the clustering (Openshaw and Craft, 1991; Besag and Newell, 1991). It is also possible that the form of spatial randomness assumed by the null hypothesis is wrong; for example, the usual Poisson assumption is very much the assumption of last resort because we do not know any better. If the test of some kind of non-randomness is not disproven, then the data may still be clustered but all that may be happening is that the test may be too insensitive or the clustering too spatially localized to be detected. Disproving a null hypothesis of spatial randomness is likewise not very helpful because it provides no clues about what might be a more appropriate pattern. Additionally, the results can be affected by choice of study region and scale of data being analysed.

It would be much more useful to test or search for the existence of different types of clustering phenomenon, without expecting that the same type of clustering may be applicable to the entire map space. Some of these ideas are incorporated in the GAM-K cluster searcher of Openshaw and Craft (1991), although here only one type of cluster pattern could be detected (i.e. non-Poisson) in data viewed as excess cancer incidence; the effects of variations in population at risk, age-sex covariates, and spatial aggregation effects having been removed. There is no reason why the underlying principles cannot be developed further and the map space classified into regions of randomness and other areas where different types of scale and rotationally invariant clustering phenomenon might be more relevant. For instance, are there areas where the Seascale or Dounreay or Gateshead type of clustering phenomenon seem to be applicable. As the library of different types of clustering phenomenon is established, so the ability to recognize the locations of the different types becomes important and useful. However, in this context, there is no well established theory or concepts that can be applied off the shelf and it would be necessary to create a library of different theoretical clustering pattern types; for example, single centred with rapid distance decay or an elevated region of uniform clustering. The patterns of spatial clustering contained in such a library would provide the basis for further theoretical developments. In this way, it is also possible to develop a concepts-rich approach to exploratory spatial analysis and in the process develop appropriate theoretical generalizations.

There are a number of approaches that can be identified as a means of achieving these goals. Section 9.4 looks at production rule based approaches designed to identify generic features and section 9.5 considers neurocomputing methods. At present knowing more or less what you want to do is more important than how it is done. Nevertheless, a brief outline is useful to establish feasibility.

9.4 Rule based approaches

Clearly there is no single approach that can tackle all these issues and different techniques may be appropriate in different applications. The simplest and most direct approach is to encode the basic concepts as a general rule which can then be applied to a GIS database if this is possible, bearing in mind the requirement that the rules will have to be scale, rotation, and aggregation invariant. It is possible that this strategy will not always work; for instance, if the theory-based spatial patterns are highly complex. However, the GAM-K cluster searcher contains an example of this approach, in that there is a cluster recognizer that examines the spatial evidence for clustering and then looks for the presence of a certain type of clustering (viz. a peak value of excess with a distance decay structure). The rules for this recognition procedure are extremely simple (Openshaw, 1992a) but this particular approach may not always be generalizable to all types of spatial theory. It also assumes the pre-existence of idealized, theory- or concepts-based, notions of spatial pattern. An initial library of types could, however, be created by computer-based experimentation.

9.4.1 Identifying areas of positive spatial autocorrelation

A useful candidate for developing this rules-based approach is the concept of spatial autocorrelation. This may be encoded in a completely generalized fashion as follows. The spatial autocorrelation concept implies that the similarity of areas diminishes as the distance separating them increases. A simple algorithm to detect this type of spatial structuring is as follows.

Step 1. Select a free area, perhaps ranked in descending size order.
Step 2. Rank all other areas by distance from it.
Step 3. Identify those areas whose data values (or more generally, a measure of multivariate similarity) compared with the central case diminish with distance and remove these cases from the database.
Step 4. Link and plot these areas on a map.
Step 5. Return to *Step* 1.

It may also be useful to impose some continuity criteria on step 3; ie the distance increments should be less than some maximum. Also the inverse (i.e. negative spatial autocorrelation) can likewise be of defined; that is the values will increase with distance. The no spatial autocorrelation case would imply that there is little change in the values with distance. Distance can also be measured in different metrics. The use of a topological definition (i.e. first, second, third-, etc. order contiguities) might also be useful. In all cases, the results can be visualized as a map.

Note that there is no notion of scale or size. The method would be applied to basic spatial data and then to a sample of aggregation and scale changes. There is no notion in the theory of spatial autocorrelation that it should work at particular scales; maybe it will work at all scales, or maybe only at certain macroscales. An important task is to find out, where and when it works best by operationalizing the search process on a massively parallel computer. A general principle is that with terraflop parallel processors likely by the late 1990s, we can seek to become clever about our spatial analysis by global brute force searches that can exploit massively parallel hardware. It is useful to remember that map searching is an explicitly parallel process. Indeed computational geography is one of the few subject areas that has several problems that are explicitly suitable for parallel computation, and which cannot be adequately handled on serial machines.

Note also that in step 3, there is no attempt to quantify the distance decay effects. It would be possible, but is it necessary? It would (a) be far too specific, (b) require the definition of some globally applicable model or function (viz. a particular autoregressive function that is assumed to be applicable throughout the study region), and (c) assume a level of knowledge and statistical sophistication that is probably not necessary in the first instance in testing concepts whilst probably also being geographically inappropriate. It is argued that it should be possible to verify the reality of the concept of spatial autocorrelation in the same abstract fashion as it is expressed. To go further than this, represents a test of much more specific and less general concept. Finally, the relationship between the concepts of spatial autocorrelation and distance decay are clearly seen; viz. there is no real difference since the same rules would identify both.

9.4.2 Identifying areas where there is spatial association

Another rules-based approach could be devised to detect spatial association. Here the search is for areas of the map that can be categorized in the simple bivariate case as being: areas where variables X and Y increase together, areas where one or other decreases, and areas where they merely fluctuate in an uninteresting fashion. A suitable algorithm is given below.

Step 1. Rank unallocated data by Y values.
Step 2. Identify area i and flag all areas where X increases with increasing Y values. Compute a score for area i; viz. number of areas or map area of linked zones.
Step 3. Increment i and repeat until list of data exhausted.
Step 4. Display areas associated with largest score and remove from list.
Step 5. Repeat steps 1–4 until no further allocation possible.

In step 2 it may be useful to impose a minimum increase threshold (eps) to focus on strong rather than weak association. The procedure can also be run to detect areas of negative association. The unclassified areas are assumed to be regions of no or little association, i.e. less than eps . Note also that the procedure is scale and aggregation free. One of the objectives would be to search for map association that was indeed scale-specific; merely by re-running this procedure on a large sample of scales and aggregations. Alternatively, the effects of area size and varying levels of data precision could be removed by first engineering a zoning system that removed any of the relevant confounding factors; i.e. variations in zone size.

9.4.3 Identifying areas where Burgess and Park's concentric ring theory applies

This is a more complex search pattern. The following algorithm can be used to find areas where it may fit. It is assumed that a small number of diagnostic variables are defined for the smallest available data zones. The algorithm is described in its univariate form. The search pattern assumes a gradual increase in wealth with distance.

Step 1. Select a point x,y.
Step 2. Rank all data zones by distance from this point, out to a maximum window distance.
Step 3. It is assumed that social class increases within distance; so try and find a low value distance band by aggregating consecutive bands.
Step 4. Try and find a middle value band.
Step 5. Try and find a high value band terminated by a fall to background values.
Step 6. Score the resulting fit. If the score exceeds a critical threshold then save the details, else try somewhere else.

The search process is explicitly suitable for parallel computing. It could be systematic on a two dimensional grid or under the control of an intelligent search heuristic. Note that the scale at which any pattern is found is left undefined and is determined by interaction between the pattern rules and the data. Additionally, if there is more than one surrogate variable, perhaps with an assumed different pattern, (e.g. an increase, then a decay, or an inverse pattern to that defined here), then the goodness of pattern fit can be modified accordingly. Maps showing areas of relatively high fit or goodness of fits that are unusually good (in some senses) might then be useful as an exploratory analysis of the data within respect to the assumed theoretical pattern.

9.4.4 Searching for a multiple sector model pattern

This is more complex because the pattern consists of one or more separate pieces. A possible algorithm involves the following:

Step 1. Define a suitable multivariate data set.
Step 2. Define a point x,y and search window with side length z.

Step 3. Regionalize the data within this window using a multivariate, contiguity constrained grouping procedure.
*Step*4. Score the results by comparing the map pattern of the groups with a library of theoretical target map pattern.

Again the search process can be systematic or under the control of an intelligent method. The search need not be too smart provided it is dumb in a parallelizable fashion. Note that here the spatial arrangement of the groups may no longer be as relevant as previously. On the other hand, all the urban model theories do specify topological constraints; for instance, low-class residential is nearer the centre than high class and that there are social class gradients; for instance, low class adjoins medium class, medium class adjoins high class, and commuter zones may adjoin high class. These notions can also be built into the goodness of fit assessment. Additionally, the search rules will need to be adaptive. In the light of running with the simplest rules, new rules (perhaps embodying different concepts from those expressed in the original urban structure theories) may be needed e.g. to allow for differences between US cities for which the theories were originally designed and UK cities to which they are being applied. These modified rules will themselves contain extremely useful information.

9.5 Neurocomputing and template matching

9.5.1 Template matching

An alternative strategy to that of coding general rules is to search for defined target patterns using pattern recognition technology. This is far less sophisticated in that complex data invariant pattern recognition rules are no longer needed. On the other hand, once a library of spatial pattern forms are established, then there is a basis here for a far more general strategy. Spatial analysis now becomes a search of GIS data for previously identified types of spatial patterning that may be of some analytical, or theoretical, value. The problem now is how to achieve this goal in a completely general and spatial data invariant fashion. For instance, there is no value in searching for a sector model pattern with sectors of a fixed width and orientation; instead the essential characteristics of a sector feature need to be defined in some highly generalized and abstract manner that is scale, sector size and rotation invariant.

9.5.2 Black and white pattern search

In essence the pattern search process involves using a presence or absence template of the sort used in character recognition systems. Such a template would be of a fixed size. In a spatial context there would usually be several different templates depending on the nature of the search pattern; cf. an alphabet of letters. The search window would also have to be stretched and distorted in various ways to see whether it can be made to fit. Various methods exist for doing this (James, 1987; Hussain, 1991) and most are inherently suitable for parallel processors.

A search for positive spatial autocorrelation would now involve identifying areas where there is a pattern of high values embedded in an region of low values. Negative spatial autocorrelation is the converse. Zero or low spatial autocorrelation implies a different type of spatial pattern. The size of the areas of high or low value would be varied and would reflect the sensitivity of the pattern detector and the scale at which the analysis was to be performed. A search for a pattern that indicates the presence of a spatial relationship might also assume that areas which are high on one variable are either high (positive association) or low (negative association) on another. The black and white patterns can be coded accordingly. Multivariate equivalents might also be imagined.

A search for a concentric ring pattern, a multiple nuclei pattern or a sector pattern can likewise be coded as a series of black and white templates. This time there would be sets of target patterns, each of which forms an idealized spatial pattern alphabet. The data would be matched against a complete set.

A general search algorithm would be as follows:

Step 1. Define a search window centred around point x,y.
Step 2. Compare the data patterns found here against a pre-defined library of reference patterns.
Step 3. Compute a pattern goodness of fit.
Step 4. Keep if there is a good match, else either store the unmatched pattern in the spatial pattern library or discard.

The spatial search in step 1 could either be systematic (i.e. over a two-dimensional grid) or based on an intelligent search heuristic. The pattern goodness of fit in step 3 could be either a template match statistic or based on scale and rotation invariant measures derived from fast Fourier or Hough transforms. The step 4 stage offers the possibility of keeping patterns that appear to be present in the data but have not previously been specified in the spatial pattern library. This has the attraction of building up knowledge about unexpected spatial patterns, albeit at the cost of discovering how to discriminate between rubbish and interesting results.

9.5.3 Creating spatial pattern libraries

The discovery of apparently meaningful spatial patterns that do not fit theoretical expectations would be of great interest. They should be retained for detailed subsequent analysis and added to the spatial pattern library. Patterns that are frequently recurrent may well be of considerable theoretical interest because they clearly contain useful information; for instance, why else would they be so recurrent? Maybe they can be re-categorized as belonging to one or other of the hypothesized spatial pattern types; maybe they are indicative of other types of spatial pattern that have so far passed unnoticed. The general process outlined in Figure 9.1 would seem to be appropriate. This is clearly not a new idea; for instance, Wilson (1963) used a similar approach to generate hypotheses about bubble chamber events. Here the keep or reject decision would probably have to be based either on some Monte Carlo simulation to investigate whether the pattern was in some unusual (i.e. non-random) or on an arbitrary decision as to whether it possesses strong structural characteristics that make it interesting. The alternative criteria would be to store nearly all exceptions (based on very weak selection criteria) and then discard those which were either unique or had very low frequency counts later. Whether this is feasible depends on the speed of the matching process and the nature of the search. However, a limited sample of the study region could be scanned to establish an initial library of pattern types. Indeed it might well be conjectured that once the uniqueness of site, size, rotation, and local topologically-induced distortions are removed, there might well be only a limited number of different spatial pattern types of any particular class of phenomena in the world. The pattern recognition task would be to create a cumulative and comprehensive database of the more important ones. The analyst just needs to find, define, and generalize the most important ones on a global scale; the most important ones are those that occur most often!

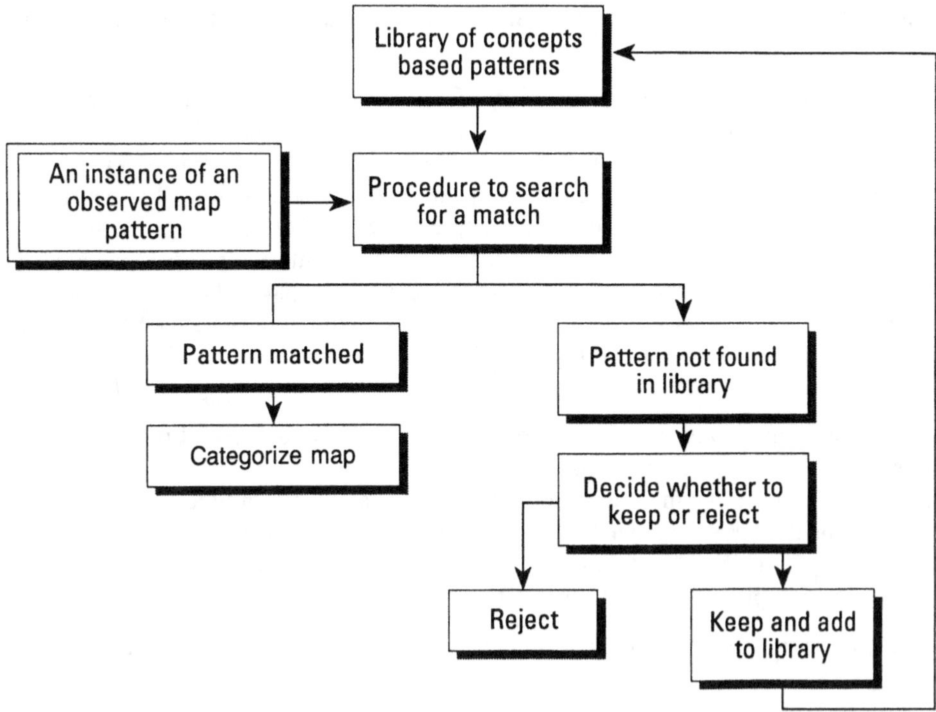

Figure 9.1 Processing a spatial pattern library.

9.5.4 Neurocomputing

The development of practical neurocomputing based tools is a feature of the late 1980s. Some of these developments concerning associative memory are of considerable relevancy here as a means of spatial pattern recognition. Of particular interest here is the associative memory nets; specifically Hopfield nets and Bidirectional Associative Memories (BAMs) (Kohonen, 1988; Kosko, 1992). These nets can handle high levels of noise and would be used to search for theory-based pattern keys again using multi resolution rasterized spatial data. Their main attraction is that the speed of recall is independent of the number of patterns stored, but they can also handle noisy data, and they are currently fashionable technology.

9.5.5 Machine learning and computer vision

Another set of tools comes from a machine learning and computer vision area. Exemplars, or idealized types of pattern template can be used as models against which data may be fitted via various deformations (Hill *et al.*, 1992; Lipson *et al.*, 1990). Genetic programming is another possible tool of considerable generality and potential usefulness here. It is noted that many of the pattern recognition procedures have been developed to assist in the task of battlefield target recognition at night. The problems of spatial analysis presented by GIS are probably considerably easier and to the extent that the problems overlap, there would appear to be a set of applicable tools just waiting geographical application (Daniel *et al.*, 1992). It is imagined that the civilianization of military technology may release many similar methods that might well be extremely useful here.

9.5.6 Searching for particular patterns or seeking to exhaustively classify entire map spaces

There are clearly two very different objectives of search that underlie both the rules-based and the pattern search. The first, is a spatially exhaustive systematic search for all occurrences of a particular spatial pattern or library of patterns. The alternative is to search for particularly good instances of one particular pattern type without necessarily performing a space exhausting search. This might well be a good way of determining whether it is worthwhile performing a systematic search for a particular pattern type. Additionally, it would also allow the analyst to speculate about the presence of particular forms of spatial pattern and then see whether they exist, perhaps subsequently modifying the idealized or hypothesized form in an interactive fashion. This suggests that there are three modes of operation:

1. categorizing a map in terms of a library of one or more model spatial patterns based on pre-existing concepts and theory;
2. theorizing by extending the library to include new spatial pattern types that are found to exist in the database or need to be added after the categorizing process is complete; and
3. speculating about what new types of idealized spatial pattern type may exist and then attempting to explain them in terms of theory.

The result is a seamless hypothetico-deductive-inductive spatial analysis system.

9.6 Conclusion

It would appear that the GIS modelled world may be full of recurrent and regular spatial patterns in whatever domains we care to look. Maybe they remain undetected because the technology for looking for them does not exist in the computer era. Maybe this is one function that the traditional, pre-computer geographer was able to perform with much greater efficiency when there was no compelling need to be statistical or computer-based. It is also clear that today's spatial patterns often need to be found. The patterns that matter are not self-evident and the limited sensory bandwidth provided by GIS makes it difficult to employ the full range of human cognitive skills. As a result the sensory-deprived geographers of the twenty-first century need advanced pattern recognition tools to make the most of the rich spatial databases available to them. It is also apparent that there is probably a tremendous amount of geographical knowledge just waiting to be discovered, and that given the right tools geographers are about to embark upon a new age of exploration and discovery.

Undoubtedly also some of the apparent patterns will have been produced by essentially random processes; i.e. chaos theory, but they may also be the product of underlying generalizable theories. Whatever the cause, it is essential to have spatial analysis methods that can try and make them visible and then identify what they are by reference to both existing theoretical knowledge and empirically-established spatial pattern libraries. A concepts-rich, spatial pattern recognition approach powered by massively parallel computing with libraries of pattern exemplars, is one way forward that can effectively combine the soft and hard forms of spatial analysis in a new approach to GIS-based exploratory descriptive geography. Clearly, there are some problems. Maybe this strategy assumes too much and attempts to be too clever. Perhaps the available spatial pattern-related concepts and theories are too crude to be of much practical value, whilst many contemporary theories are perhaps too complicated to be easily or readily expressed as simple models of idealized spatial patterns. Yet it is seemingly a valid challenge to return to basic geographic principles and then seek to link the theoretical spatial patterns with empirical reality. It is not a matter of coarsening the data or of seeking to convert abstract theories into explicit tests of hypotheses. Rather it is a matter of looking for evidence that the abstract theories fit the observed spatial data patterns in a suitably abstract manner. If they do not, then let us find some that do. This does not trivialize the analysis task; indeed, it greatly complicates it. Moreover it will only be practicable if the associated massive computational problems can be resolved. A start has been made, but much remains to be done.

References

Besag, J. and Newell, J., 1991, The detection of clusters in rare diseases. *J. Royal Stat Soc Series A*, **154**, 143–55

Bunge, W., 1966, *Theoretical Geography*, Lund Studies in Geography, Lund.

Charlton, M., Openshaw, S. and Wymer, C., 1985, Some new classifications of census enumeration districts in Britain: a poor man's ACORN. *J. of Econ. and Social Measurement*, **13**, 69–98.

Clark, B.D. and Gleave, M.B., 1973, *Social Patterns in Cities*, London: IBG

Daniel, C.E., Kemsley, D.H., Lincoln, W.P., Tackett, W.A. and Baraghimian, G.A., 1992, Artificial neural networks for automatic target recognition. *Optical Engineering*, **31**, 2521–31.

Davies, W.K.D., 1984, *Factorial Ecology*, Aldershot: Gower.

Harris, C.D. and Ullman, E.L., 1945, The nature of cities, *Annals of the American Academy of Political and Social Science*, **242**, 7–17.

Hill, A., Cootes, F.T. and Taylor, C.J., 1992, A Generic System for Image Interpretation Using Flexible Templates, in Hogg, D. and Boyle, R. (Eds), British Machine Vision Conference 1992, Springer-Verlag.

Hoyt, H., 1939, *The Structure and Growth of Residential Neighbourhoods in American Cities*, Washington, DC: US Federal Housing Administration.

Hussain, Z., 1991, *Digital Image Processing*, New York: Horwood.

James, M., 1987, *Pattern Recognition*, Oxford: BJP Professional Books.

Johnston, R. J., 1979, On the characterisation of urban social areas, *TESG*, **70**, 232–8.

Kohonen, T., 1988, *Self-organization and Associative Memory*, Heidelberg: Springer-Verlag.

Kosko, B., 1992, *Neural Networks and Fuzzy Systems*, New Jersey: Prentice-Hall.

Lipson, P., Yuille, A.L., O'Kavanaugh, D., Keefe, J., CaTaaffe, J., Rosenthal, D., 1990, Deformable Templates for Feature Extraction from Medial Images, Computer Vision - ECCV90, Faugeras, O. (Ed.), Springer-Verlag.

Murdie, R.A., 1969, Factorial Ecology of Metropolitan Toronto, 1951–61. University of Chicago, Geography Research Paper No 116.

Oberg, S., Holm, E., Goddard, J. B. and Openshaw, S., 1990, Towards a time geographical computer language for social science. Unpublished MS.

Openshaw, S., 1991, A view of the GIS crisis in geography, or, using GIS to put Humpty-Dumpty back together again. *Environment and Planning A*, **23**, 621–8.

Openshaw, S., 1992a, Some suggestions concerning the development of AI tools for spatial modelling and analysis in GIS. *Annals of Regional Science*, **26**, 35–51.

Openshaw, S., 1992b, Further thoughts on Geography and GIS: a reply. *Environment and Planning A*, **24**, 463–6.

Openshaw, S. and A. Craft, 1991, Using geographical analysis machines to search for evidence of cluster and clustering in Childhood Leukaemia and non-Hodgkin lymphomas in Britain, in Draper, G. (Ed.), *The Geographical Epidemiology of childhood leukaemia in GB, 1966*–83, OPCS Studies on Medical and Population Subjects, London: HMSO.

Park, R.E., Burgess, E.W. (Eds), 1925, *The City*, Chicago.

Robson, B.T., 1975, *Urban Social Areas*, London: Oxford University Press.

Shevky, E., and Bell, W., 1955, *Social Area Analysis*, Stanford University Press.

Smith, D.M., 1975, *Patterns in Human Geography*, London: Penguin.

Wilson, A. G., 1963, Hypothesis assigning in the testing of bubble chamber events, The National Institute for Research in Nuclear Science, Report 42. London: HMSO.

10

Exploratory spatial analysis in GIS: generating geographical hypotheses from spatial data

Richard J. Aspinall

This chapter presents a method of exploratory spatial analysis for geographic information systems that can be used to generate geographical hypotheses. The geographical hypotheses are developed using a Bayesian probability approach that addresses issues of integrating data from different basic spatial units and error analysis in GIS. It also provides a method for increasing the resolution of coarse (aggregated) spatial data. The method is illustrated using bioclimatic mapping for wildlife in Scotland.

10.1 Introduction

The ability to link automated methods for exploratory (inductive) spatial analysis with GIS offers great potential for analysis of spatial databases for large-scale, extensive geographic areas; this is especially true for development of data-based models and theories (Openshaw, 1989a; Openshaw *et al.*, 1987; Openshaw *et al.*, 1980). This approach is attractive for analysis of spatial data in applied land use, geographical and environmental sciences for three reasons: model relevance; the nature of the data available in databases; and the role of a spatial (geographical) perspective.

First, inductive methods develop models and theories using existing data and can usually be applied to wide geographic areas. This is particularly important in the current social and political context where the reasons for conducting a piece of research attract a great deal of attention (Openshaw, 1989a) and in which relevance to a particular problem or issue is demanded. These models contrast with more theoretically-based models that require data that are often impractical to collect and that have only limited geographic application beyond the original development/test site. Since much current work is directed towards understanding complex environmental problems and issues, this inductive approach gives results that build on available data and have a wide geographic relevance.

Second, environmental and socio-economic data in geographic and other databases are frequently collected as observations of the real world rather than as the result of experiment. In the absence of experiment, appropriate probability models against which to test and evaluate these data are not well defined (Haining, 1990) and analyses based on classical methods for inference testing are of lower relevance (Leamer, 1978). In this context, exploratory data analysis (EDA) (Tukey, 1977; Mostella and Tukey, 1977) has attracted attention (Haining, 1990), particularly for analysis of spatial data (Haslett *et al.*, 1990, 1991; Goodchild *et al.*, 1992b). EDA is concerned with hypothesis formulation from examination of data; its tools are resistent methods for describing structure and properties of data and graphical and pictorial presentation (Haining, 1990). For spatial data this has been enhanced by recent advances in computing, notably the facility to link windows dynamically in workstation and pc computing environments (Haslett *et al.*, 1990, 1991).

Third, a spatial perspective on data and analysis provides insights into data and relationships (Openshaw *et al.*, 1987; Goodchild *et al.*, 1992b; Goodchild, 1991) that can lead to identification of mechanisms and processes (Sebens, 1991); these may be in the form of testable hypotheses (Openshaw *et al.*, 1987, 1990). A spatial approach is also fundamental to investigations of global, continental, national, and regional environmental questions (Rhind, 1991) that are explicitly concerned with issues of geographical scale and inherently geographical questions asked by policy and decision makers.

This chapter presents a generic method of exploratory spatial analysis for GIS based on integrated application of two spatial analytical procedures: point pattern analysis (after Openshaw *et al.*, 1987) and an inductive learning process for pattern analysis that is based on use of Bayesian probability statistics (Aspinall, 1992a). The chapter concentrates on some of the background to the method, the assumptions made and tested in the analysis, and applications of the approach. Outputs from the analysis are hypotheses that define geographical and functional (ecological) relationships between a distribution being modelled (for example a plant species/community/biome) and predictor datasets (e.g. climate/soils/land cover). Bioclimate mapping is a fertile application area for this development since there is a wealth of data describing species distribution and also a clear need for information on climatic relations of species for environmental management purposes and to allow effects of climate change to be predicted (Climate Change Impacts Review Group, 1991) at different spatial and ecological scales. The context for developing and presenting the method is bioclimatic mapping in Scotland.

10.2 Bioclimatic mapping

The information needs of bioclimatic mapping, and its basis in ecological theory describing species distribution-climate relationships (Woodward, 1987), focus the development and application of the method by restricting both the range of relationships for which the analytical method need search and the data to be used. From the perspective of information outputs, bioclimatic (and other ecological) modelling methods must satisfy a number of criteria. They should:

- have a sound theoretical underpinning in ecology;
- be able to take account of the direct impact of climate on wildlife and habitat and the indirect effect of climate manifest through habitat; and
- be able to make use of the data resource already available describing wildlife distribution and abundance (Climate Change Impacts Review Group, 1991).

The approach to bioclimatic modelling used here is based on the concept of the ecological niche, which describes the relationship of a species with environmental conditions. It is defined in a mathematical space in which environmental variables form the abscissa and species response is the ordinate. In geographic space, ecological relationships are concealed in complex, geographically-related spatial variation associated with the geographic distribution of environmental resources and species. The method interprets patterns of variation in geographic space to generate hypotheses describing species response to environmental conditions in environmental (niche) space. Associations between species presence/absence and values of environmental variables are identified by comparing spatial datasets; relationships are mapped using an approach based on Bayes' theorem (Aspinall, 1992a). Since the factors limiting distribution can vary from one location to another, analysis is structured using point pattern analysis to identify geographic clusters in the data (Openshaw *et al.*, 1987). Clusters are each analysed independently to search for local/regional relationships within the overall (national) pattern.

Species distribution data are available as a series of national atlases (Sharrock, 1976; Heath *et al.*, 1984; Perring and Walters, 1962). These atlases typically are point data that are treated as a raster set with grid cell size of 10 × 10 km (based to the OS National Grid) as a basic spatial unit. The data describe presence (but not absence) within the grid square. This spatial unit is coarser than many environmental data available in digital form; for example than the climate data used here which have a spatial resolution of 1 km square. This is not unusual, for geographic data and questions related to the influence of spatial units on analysis are well known (Openshaw, 1989b). The difference in resolution offers analytical opportunities (Openshaw, 1989b) and the method here depends on the difference in resolution, both for managing uncertainty in the data and for generating the ecological and geographical hypotheses. Output has the spatial resolution of the fine dataset (1 km squares in the example here). The situation of coarse resolution-dependent data and fine resolution data that can be used as predictor datasets, is a generic situation in GIS and a variety of methods have been developed that begin to address this issue.

Digital climate maps for Scotland used in the bioclimate modelling are generated and managed using a GIS. Baseline climatic data, as 30-year means for the period 1951–1980, are calculated from maximum, minimum, and mean monthly temperature, and mean monthly rainfall data from meteorological stations in Scotland. Monthly potential evapotranspiration is calculated from temperature and rainfall using the Thornthwaite method. These point data are mapped digitally using a combination of trend surface analysis and point kriging (Aspinall and Matthews, in press; Aspinall and Miller, 1990) with topography as a covariate. The data are available as a series of raster surfaces with resolution of 1 km. Two surfaces describe each variable for each month. The first is the modelled value and the second is the uncertainty associated with the model estimate; the uncertainty surface provides input to Monte Carlo error analysis in the modelling procedure.

10.3 Modelling procedure

The basic modelling approach used has been described elsewhere (Aspinall, 1992a). It is based on calculating conditional probabilities from the relative frequency of association between attributes of a dataset to be modelled and attributes of a variety of predictor datasets (here, climate variables). The relative frequencies are calculated for random (spatial) subsets of the data to generate error bounds for the conditional probabilities. These allow uncertainties for both the input data and the model coefficients to be combined in analysis of error propagation (Aspinall, 1992a). The distribution to be modelled is also analysed using the Geographical Analysis Machine of Openshaw *et al.* (1987). This identifies spatial clusters in a point dataset through comparison of the distribution with a universe of spatial hypotheses. Clusters are identified in the atlas data used for bioclimatic mapping using square search regions of increasing size and a universe of point distributions generated from a Monte Carlo procedure.

Operation of the method is based on the difference in resolution of the basic spatial units beteen the dependent and independent data. The coarse resolution units of the dependent data allow multiple stochastic realizations of species presence/absence to be generated at the finer spatial resolution of the independent data. This allows a set of spatial relationships between the datasets to be calculated and error/uncertainty estimates to be derived for model parameters. In the implemented procedure, a FORTRAN program incorporating functions from the NAG and UNIRAS libraries, the coarse resolution units are used as a sampling frame to generate multiple sets of presence/absence data with 1 km resolution by random selection of a 1 km square from within each of the 10 km squares in the sets for presence and absence. Presence and absence initially are assigned as for the 10 km squares. The assumption of uniform presence (and optionally absence) throughout 10 km squares is tested during the modelling process (below).

This process of within-coarse-unit random sampling allows multiple estimations of relationships of presence/absence with climate variables; relationships are established as relative frequencies and rescaled as conditional probabilities. The set of individual estimates are used to calculate the mean and variance of conditional probabilities for both presence and absence for particular values of each climate variable (Haining, 1990). These summary statistics are used to describe ecological relationships. The relative frequencies satisfy the probability axioms (Learner, 1978) and are used in mapping through application of Bayes' theorem. The variance of conditional probability estimates provide input to Monte Carlo error analysis. Climatic conditions that consistently are associated with either presence or absence have a relatively low variability in conditional probabilities between realizations.

The method described here considerably extends the basic approach (Aspinall, 1992a) in six ways, four analytical and two concerned with error testing.

1. Inductive cluster analysis based on the Geographical Analysis Machine of Openshaw *et al.* (1987) is used as a form of pre-processing for the dataset being modelled. This is described above. In the context of bioclimatic mapping this allows local (regional) controls on distribution to be modelled from spatial subsets (clusters) of the full dataset. Any regional clusters of distribution identified within the national (Scotland-wide) distribution dataset are analysed independently.
2. Selection of a dataset describing 'absence' is automatic and aims to maximize discrimination of environmental relationships for presence and absence. The 'absence' dataset provides an inductively generated comparative probability distribution describing the available environmental resources in the study area; it is used as a probability distribution against which to assess association between species presence and environmental variation. Three strategies for selection of this 'absence' dataset are available:
   (a) random sampling from the entire geographical area;
   (b) random sampling from a remote geographic area; and
   (c) random sampling from a proximate geographical area.

Random sampling from the entire geographic area can lead to weakened discrimination between environmental relations for presence and absence through different frequencies of sampling in geographic areas of presence and absence; for widespread species this sampling method may be appropriate. Random sampling from a remote geographic area may fail to distinguish sufficient detail in the presence/absence data for rare or restricted species and result in discrimination between presence and absence being over-generalized. The third strategy uses random sampling from within a geographically restricted area constrained to be within a specified distance of squares for presence. Analysis is thus targeted on the basis of the observed distribution and since strong spatial dependence is inherent in the independent climate datasets, this method focuses analysis on subtle differences between climate conditions for presence and absence. This leads to improved discrimination of environmental relationships at the limits of the distribution; it is the basic method for analysis of regional/local clusters.

3. Since the climate data are represented as surfaces of continuous variation they are simplified during analysis by applying class intervals. For each dataset the class interval is increased gradually and the narrowest interval that produces a significant association is identified; this retains the maximum information content of the climate map. Conditional probabilities for presence and absence for all datasets with a significant association are recorded and graphed. Values of variables where conditional probabilities for presence and absence are equal represent critical limits for the species. A maximum of one (monthly) dataset from each set of climate variables is selected to produce the output map model; the programme which implements the procedure can constrain analysis to particular datasets (climatic variables and/or months). The dynamics of climatic variables (e.g. change in temperature during the year) can be incorporated by generating bioclimcatic indicator variables, as, for example, described by Lindemayer *et al.* (1991).
4. Datasets for use in modelling are selected on the basis of chi-square analysis of the table of association between presence/absence and categories on each individual climate map. Haining (1990) discusses the use of inferential significance testing in exploratory data analysis applied to observational data. For all but very rare species, a significance level of $p \geq 0{\cdot}95$ is used as a cutoff criterion; this is possibly over-rigorous and can be relaxed for very rare species that occur in only a few squares.

The applied context, inductive approach, variability (and low values in some parts of the variable range) of conditional probabilities for presence and absence, mathematics of rescaling in Bayes' theorem (error propagation), assumptions made in order to generate the multiple representations of conditional probability, and uncertainty in the input data all combine to highlight the need for adequate error analyses to accompany the main outputs from this method. Error analysis is consequently an important element of the method and integral to the derivation, interpretation and use of the results of modelling. Two (automatic) error tests are included which together indicate the sensitivity of the outputs to possible sources of error:

1. An iterative procedure is used to test the assumption of species presence in all parts of the 10 km grid squares. This is part of the model generation process. It is tested by repeating the entire analytical procedure but allocating 1 km square realizations of 10 km presence to either presence or absence classes based on the results of the first model. Reallocation of absence is not usually carried out although available as an option. The rationale for this is that a species need not occur ubiquitously or uniformly throughout any 10 km square in which it has been recorded, but only somewhere within it. Each 10 km square can therefore, contain both suitable and unsuitable conditions for the species. Using the output from an initial model to allow areas within 10 km squares for presence to be used to contribute to the absence category is a logical approach to relaxing this assumption. It provides an estimate of the influence of data assumptions on analysis and improves the discrimination of presence and absence in the model; if it does not do this the original model is not good and the dependent data probably cannot be modelled adequately with the independent data being used. Testing the assumption of presence also enhances definition of the ecological hypotheses generated. The probability value that determines occupation of the 1 km square within the 10 km squares for presence can be identified in a number of ways (Aspinall, 1992b); it may also be allowed to vary and subject to Monte Carlo simulation to identify its effect on the hypotheses and model specification although this is not currently implemented due to the computing requirements!
2. Monte Carlo simulation is used to determine the sensitivity of the model to uncertainty

in the input datasets and conditional probability estimates. Combined effects of these uncertainties are generated for the mathematical operations involved in the use of Bayes' theorem. This approach contrasts with the original implementation (Aspinall, 1992a) which used estimates of variability in conditional probabilities to derive error estimates through equations describing mathematical error propogation and local geographic variability in model outputs to indicate data uncertainty. The Monte Carlo approach is mathematically less complex and more readily implemented with multiple data layers, although demanding of processor time (Openshaw, 1989b). Output can be interpreted as:

1. the variance of the probability output for each location;
2. the number of times that a particular threshold probability value defining presence is exceeded;
3. through generalization using an error model for categorical data (Goodchild *et al.*, 1992a); and
4. as entropy (Cressie, 1991; Press *et al.*, 1992).

Combinations of these approaches might also be used.

10.4 Bioclimate maps – geographical hypotheses

The selected climatic datasets (with associated class interval) are combined with Bayes' theorem using the calculated conditional probabilities. This provides a map at the full spatial resolution of the climate data, which shows the probability of occurence for the species. Many such maps can be generated using a variety of combinations of independent data; these maps combine with graphs and tables of relationships to provide the geographical hypotheses. Use of conditional probabilities and Bayes theorem provides a framework for mapping which is independent of sampling frequency. Some values will occur relatively infrequently in the sampling distribution for a map (extreme values of the independent variable in a standard statistical analysis). In most methods of data combination (e.g. regression) these values are peripheral to the mass of the data and have a different relationship with the model. In contrast, the implementation of Bayes theorem used here produces estimates of presence from the ratio, rather than magnitude, of the conditional probabilities. Although low conditional probability values for presence and absence may be associated with particular values of climatic variables because of their relative infrequency in the data sample, use of Bayes theorem to recode the datasets provides output which is independent of sampling frequency and reduces the influence of these low values.

Bayes' theorem is a consequence of the probability axioms (Leamer, 1978) and, when used to calculate one conditional probability from another is not controversial. Use of a Bayesian approach for inference is, however, more controversial (Berger, 1985) mainly because of the question of selection of prior probabilities to represent the initial subjective estimation of likelihood for presence and absence. This has been discussed by many authors (Leamer, 1978; Cressie, 1991). In inference, prior probabilities represent subjective information and are modified by the conditional probabilities generated from data analysis. In image analysis, maximum entropy image restoration is based on Bayes theorem and it has been suggested that negentropy be used to establish prior probabilities (Press *et al.*, 1992). In general, when estimating a small number of parameters from a large amount of data, the prior probabilities are trivialized since their effect is lost by the weight of evidence contained in the conditional probabilities. This will frequently be the situation in GIS applications. An alternative, constructive view of prior probabilities is that they are an important facility provided by a Bayesian approach; allowing prior probabilities to be set for each location by each model constructed allows different (*a priori*) hypotheses held by the GIS user to be evaluated and tested. This encourages an interaction between the user and the GIS database that allows the user to be more closely part of the data exploration process. In the method here, prior probabilities are set equal at 0.5 (to indicate no *a priori* preference for a site on the basis of location); this prior probability is commonly used in image classification (in the absence of estimation of negentropy). It can be altered if the user wishes.

10.5 Discussion

The output map and the species–climate relationships for each month's climate data (where identified as significant) provide testable hypotheses concerning the ecology of the species being modelled. The graphical and tabular (probability) results represent hypotheses concerning ecological relationships between the species and its climatic regime, the maps visualize this information and the associated error sensitivities. Use of Red/Green/Blue display in an image analysis system or GIS allows the model output and error analyses to be presented in (visual) combination. More traditional map output conveys the pattern described by the model without indicating the sensitivity to error (Figure 10.1). Graphs show the relationships between presence and absence for each climatic variable.

For bioclimatic analyses, output maps represent the climatic regime in which a species might be expected to occur (and the sensitivity of the map model presented to error). They provide hypotheses that can be tested through experiment using targeted survey although testing probability surfaces presents its own problems. Testing has proved possible for some species that have been subject to independent and comprehensive field survey; for these species the models have been shown to be of practical use.

The approach has wide application in environmental and socio-economic research. The bioclimatic implementation provides useful information for a range of applications including assessing impacts of climate change, developing understanding of species ecology at a range of spatial scales, and for habitat evaluation in conservation biology. Bioclimatic relationships identified also provide a framework for interpreting likely effects of habitat and land use changes and defining relative importance of climate and habitat factors in controlling species distribution. For example, habitat management for species conservation or land use changes in areas outwith the climatic regime of the species are unlikely to have an impact on the species concerned. The information output for individual species can also be used to investigate communities, community composition and functional relationships between species that adds to the range of interconnected hypotheses that can be constructed for conservation biology. This method has also been used to model farm distribution by linking the coarse spatial resolution maps of Scottish Office Agriculture and Fisheries Department farm types (used to categorize the June agricultural census) with more detailed environmental data (Aspinall, 1993). This, in turn, allows integrated analysis of socio-economic and biophysical land use issues and changes by enhancing the resolution of coarse spatial resolution socio-economic census units with results of environmental models operating in concert with the detailed environmental data of the GIS.

Conclusion

The analytical method presented offers considerable potential for analysis of existing data to produce useful and relevant information. The information generated using this approach

Figure 10.1 Bioclimatic map output for Hen Harrier in Scotland.

offers an opportunity to generate and investigate geographical hypotheses in a logical and scientific manner, the geographic link being particularly important for policy and planning issues and for effective use of GIS databases.

Acknowledgment

Climate mapping was carried out as part of an integrated research programme into the possible impacts of climate change on land use in Scotland; this involved Gordon Hudson, Keith Matthews, Diane Pearson and RJA. The analytical procedure and its application to bioclimate mapping were developed as part of a spatial analysis research programme that is part of a programme investigating wildlife – environment relationships using GIS for application in wide area conservation evaluation. Both projects were funded by the Scottish Office Agriculture and Fisheries Department.

References

Aspinall, R.J., 1992a, An inductive modelling procedure based on Bayes' Theorem for analysis of pattern in spatial data. *International Journal of Geographical Information Systems*, **6** 105–21.

Aspinall, R.J., 1992b, Spatial analysis of wildlife distribution and habitat in a GIS. *Proceedings of the 5th International Symposium on Spatial Data Handling*, **2**, 444–553.

Aspinall, R.J., 1993, Exploratory spatial data analysis in GIS: generating geographical hypotheses describing farm types from spatial census data. *Proceedings of the Fourth European Conference on GIS*, **1**, 517–26.

Aspinall, R.J. and Matthews, K., In press, Climate Change Impact on Distribution and Abundance of Wildlife: An analytical approach using GIS. *Environmental Pollution*.

Aspinall, R.J. and Miller, D.R., 1990, Mixing climate change models with remotely-sensed data using raster based GIS, in Coulson, M.G. (Ed.), *Remote Sensing and Gobal Change*. Proceedings of the 16th Annual Conference of the Remote Sensing, 1–11.

Berger, J.O., 1985, *Statistical Decision Theory and Bayesian Analysis*, 2nd Edn. New York: Springer-Verlag.

Climate Change Impacts Review Group, 1991, *The Potential Effects of Climate Change in the United Kingdom*. First Report. London: HMSO.

Cressie, N., 1991, *Statistics for Spatial Data*, New York: Wiley.

Goodchild, M.F., 1991, Spatial Analysis with GIS: Problems and Prospects. *GIS/LIS 1991*, Vol. 1, 40–8.

Goodchild, M.F., Guoqing, S. and Shiren, Y., 1992a, Development and test of an error model for categorical data. *International Journal of Geographical Information Systems*, **6**, 87–103.

Goodchild, M.F., Haining, R. and Wise, S. *et al.*, 1992b, Integrating GIS spatial data analysis: problems and possibilities. *International Journal of Geographical Information Systems*, **6**, 407–23.

Haining, R., 1990, *Spatial Data Analysis in the Social and Environmental Sciences*. Cambridge: Cambridge University Press.

Haslett, J., Bradley, R., Craig, P.S., Wills, G. and Unwin, A.R., 1991, Dynamic graphics for exploring spatial data, with application to locating global and local anomalies, *American Statistician*, **45**, 234–42.

Haslett, J., Wills, G. and Unwin, A.R., 1990, SPIDER - an interactive statistical tool for the analysis of spatially distributed data. *International Journal of Geographical Information Systems*, **4**, 285–96.

Heath, J., Pollard, E. and Thomas, J.A., 1984, *Atlas of Butterflies in Britain and Ireland*. New York: Viking.

Leamer, E.E., 1978, *Specification Searches: Ad hoc Inference with Non-experimental Data*. New York: Wiley.

Lindemayer, D.B., Nix, H.A., McMahon, J.P., Hutchinson, M.F. and Tanton, M.T., 1991, The conservation of Leadbeater's possum *Gymnobelideus leadbeateri* (McCoy): a case study of use of bioclimatic modelling. *Journal of Biogeography*, **18**, 371–83.

Mosteller, F. and Tukey, W., 1997, *Data Analysis and Regression*. Addison-Wesley.

Openshaw, S., 1989a, Computer modelling in human geography, in MacMillan, B. (Ed.), *Remodelling Geography*. Oxford: Blackwells, 70–88.

Openshaw, S., 1989b, Learning to live with errors in spatial databases, in Goodchild M.F. and Gopal, S. (Eds), *Accuracy of Spatial Databases. London: Taylor & Francis, 263*–76.

Openshaw, S., Charlton, M., Wymer, C. and Craft, A., 1987, A mark 1 geographical analysis machine for the automated analysis of point data sets. *International Journal of Geographical Information Systems*, **1**, 335–58.

Openshaw, S., Cross, A. and Charlton, M., 1990, Building a prototype geographical correlates machine. *International Journal of Geographical Information Systems*, **4** 297–312.

Perring, F.H. and Walters, S.M., 1962, *Atlas of the British Flora*, BSBI, Nelson.

Press, W.H., Teukolsky, S.A., Vettering, W.T. and Flannery, B.P., 1992, *Numerical Recipies*, CUP.

Rhind, D., 1991, Geographical information systems and environmental problems. *International Social Science Journal*, **XLIII**(4), 649–68.

Sebens, K.P., 1991, Habitat structure and community dynamics in marine benthic systems, in Bell, S.S., McCoy, E.D. and Mushinsky, H.R. (Eds), *Habitat Structure: The Physical Arrangement of Objects in Space*. London: Chapman & Hall, 211–34.

Sharrock, J.T.R., 1976, *The Atlas of Breeding Birds in Britain and Ireland*. Calton: T. & A.D. Poyser.

Tukey, J.W., 1977, *Exploratory Data Analysis*. Addison-Wesley.

Woodward, F.I., 1987, *Climate and Plant Distribution*. Cambridge: Cambridge University Press.

11

Visualizing spatial association in area-value data

Jason A. Dykes

Classifying area-value data sets with a spatial characteristic as the primary consideration has potential for communicating patterns and revealing data trends. The capacity for classifying in this manner is explored by producing a series of maps with various classification schemes and calculating a coefficient of spatial association for each. Moran's *I*, a fundamental geographical statistic is used, and found to be variable irrespective of zone regularity. However, the coefficient only equates with perceived simplicity when zones are regular. This leads to questions regarding the appropriateness of visual analysis, or that based on a single metric distance statistic, from irregular area value data. An alternative representation which can be analysed visually is briefly introduced.

11.1 Introduction

11.1.1 Visualization and cartography

Innovations in digital technology are driving changes in the way in which data are collected and displayed. More data are available, and detailed visual representations of data, which can be rendered and interrogated in real time, are contributing to analysis. Visualization is being hailed as the answer to many of the problems associated with analysing and exploring masses of data (McCormick *et al.*, 1987). This is occurring across the sciences with relatively little thought for the implications of relying on the human visual system as a scientific tool. Whilst the presentation of data in visual form would seem to encompass the realm of the cartographer, the visualization community has made little of the long cartographic history of involvement with representing complex multi-variate data in restricted dimensions. Yet cartographic techniques have been developed to facilitate the communication of specific geographic themes, and gain knowledge and insight from an overwhelming mass of data. As visualization becomes an established mode of scientific enquiry, cartographic research into the ways in which data are displayed becomes increasingly important.

The research offered here relates to the presentation of valued area data, and initially addresses the multi-nomial choropleth classification question (Evans, 1977). The issue has a long tradition, but ample scope for potentially novel insights. Whilst the classified choropleth typically resides at the communication end of the visualization 'user continuum' (Bishop *et al.*, 1994) the findings outlined here point to ways in which more abstract representations of enumerated data can utilize the 'visualization for data exploration' strategy for ideation.

11.1.2 The classed choropleth map

Area-value data are commonly represented visually as a choropleth map, particularly since the widespread adoption of computer assisted cartography for which they are the most easily implemented scheme. In such a representation the land surface is sub-divided into mutually exclusive polygons, each of which is associated with an item of information, usually relating to a single theme. The map comprises zone outlines with symbolism inserted in a manner consistent with the information content of the polygons. The popularity is a function of ease of production, spatial data agglomeration at collection time, and a perceived simplicity in map interpretation.

It can be argued that the classed choropleth has been superseded by powerful systems which provide real time interaction with area-value data, or unclassed maps where symbolism relates directly to value. However the static classed map retains several distinct roles. First classification decreases the communicative bandwidth and imposes standards on the data, reducing the potential for subjective (mis)interpretation. Second, the static map provides a model of the data set which may illustrate an aspect of the data or reveal an important pattern. In addition, communication of spatial information may be achieved more easily with a classed static map. Muehrcke (1990, p.9) summarizes by stating that 'it is abstraction, not realism that give maps their unique power'. Furthermore, interactive systems cannot be distributed to a mass audience, and even if this becomes technologically feasible, not everyone has the capacity to process all of the data that they are presented with. A static map can, in the same manner as a graph or chart, communicate relevant aspects quickly.

11.1.3 Classification

The classification of data into internally contiguous and externally distinct classes has historically posed problems throughout the sciences. The variety of methods available to the choropleth cartographer is reviewed by Evans (1977). Traditional classification techniques employ any of a number of algorithms to delimit classes in a single dimension. Maps have been the basis for inventory and visual spatial analysis and so the spatial data set has been tabulated by adhering to the statistical values as closely as possible (Jenks and Caspall, 1971). Digital data storage and computer mapping have removed the need for maps to portray raw data values spatially. Alternative aspects of the data can now be mapped easily with particular objectives. This allows other factors to be considered when classifying the choropleth. As geographers are concerned with space, and the distribution of phenomena within it, the visualization of spatial association would seem to be an important aspect.

In this chapter it is initially argued that choropleth maps can be used most successfully to emphasize spatial association at the expense of other functions. Classification which takes spatial variables into account is proposed. This proposition has two objectives. First the map may be communicated to an audience more effectively if patterns are simple. Complex map structures and patterns are less likely to be interpreted successfully or remembered. Map 'communication effectiveness' has been linked to complexity previously (Muehrcke, 1973; Monmonier, 1974; MacEachran, 1982). Secondly an implicitly spatial model of the data is created if maps are classified to highlight patterns. This may reveal spatial associations that would not be apparent under a statistical classification mechanism.

11.1.4 Proposal

Considering the amount of research into choroplethic communication, it is astonishing that so little work has been attempted concerning the effects of class interval selection on spatial aspects of maps. Previous studies have been limited by avoiding measurements of attribute similarity (Olson, 1974), using crude measures of spatial association (Olson, 1974), ignoring irregular areas (Monmonier, 1972). The relationships between spatial association and classification and the potential for using a coefficient of spatial association in classification can be investigated by combining a range of data sets with the methods of interval selection as defined by Evans (1977). By varying map parameters such as zone shapes, underlying spatial distributions and number of classes, as well as classification scheme, a large number of maps can be created. By calculating a rigorous coefficient of spatial association for each data set the variability of the coefficient through classification can be investigated. Preliminary conclusions as to the potential for inferring communication effectiveness from such coefficients and the relationship between spatial association and perceived pattern can then be drawn.

11.2 Methods

11.2.1 Coefficients

A classification procedure based on the presence of a simple pattern requires a measurable coefficient. Various map pattern coefficients exist in the literature (Olson, 1974; Muehrcke, 1973; Monmonier, 1974). The most rigorous which accommodate multi-nomial ratio data are indices of spatial autocorrelation (Goodchild, 1986). Autocorrelation was developed in economics where a one-dimensional, unidirectional dependency in variables was observed between neighbours in time. When extended to a map model (two dimensions and multiple directions) the interdependency can be used to quantify Tobler's first law of geography: everything is related to everything else, but near things are more related than distant things. If features which are near display similar attributes, then positive spatial autocorrelation exists. Negative autocorrelation occurs where features which are close together in space tend to be more dissimilar in terms of their attributes than those which are further apart. Zero spatial autocorrelation implies that attribute values are independent of location. An autocorrelation function can be described as in (11.1)

$$I=\sum_{i=1}^{n}\sum_{j=1}^{n}c_{ij}w_{ij} \qquad (11.1)$$

where c_{ij} is a measure of attribute similarity and w_{ij} is a measure of spatial proximity. The ways in which c_{ij} and w_{ij} are measured depend on the type of data and object. A popular method of recording ordinal or ratio attribute similarity is to count the number of joins, but this results in a loss of information regarding the degree of dissimilarity between non-contiguous neighbours. Moran (1948) suggested an attribute similarity index I as shown in (11.2).

$$c_{ij}=(z_i-\bar{z})(z_j-\bar{z}) \qquad (11.2)$$

where $\bar{z}$ is the mean of all z values. His index is thus based on the cross products of the deviations of the z_i from the mean of z, the covariance. I is restricted to a fixed range by dividing by the weightings and variance as shown in (11.3). The extreme values of

1·0 and −1·0 (under normal circumstances) match intuitive notions of positive and negative spatial autocorrelation.

$$I = \sum_{i=1}^{n} \sum_{j=1}^{n} \frac{w_{ij}(z_i - \bar{z})(z_j - \bar{z})}{s^2 \sum_{i=1}^{n} \sum_{j=1}^{n} w_{ij}} \tag{11.3}$$

where s^2=sample variance, and

$$s^2 = \sum_{i=1}^{n} \frac{(z_i - \bar{z})}{(n-1)}$$

A great number of spatial proximity or distance weighting functions are possible. A binary function relating to zone contiguity is one option, but it omits much of the spatial nature of the attribute data. A '$1/d^2$' function (where d is the Euclidean distance between adjacent centroids) incorporates spatial proximity data and remains popular in the literature. This function was used here as it is virtually standard despite the hinderance of a reduction from area to point data, difficulties concerning the objective definition of centroid location and apprehension over the use of Euclidean distance as an interaction index from some authors (Gatrell, 1983).

11.2.2 Experimentation.

Data distributions were derived, ranging from positively autocorrelated to random, by adding a random element to a distance function and imposing a grid zone structure over the data. Negatively autocorrelated surfaces were not analysed initially as their contravention of 'the first law of geography' makes them rare. A 10 by 10 grid was used as n=100 was regarded as sufficiently large to respond to subtle variations in class limits and yet small enough to keep files at a manageable size and make visual interpretation possible. Irregular zones were required to test the autocorrelation present in naturally occurring spatial phenomena. The electoral wards of Leicestershire were chosen as an example of a typical set of irregular zones. Data for the county were readily available, the number of zones (166) was manageable and they exhibited considerable size and shape variation and so constituted a genuine alternative to the regular grid. Values were derived for the zones as above, and population data were included to provide an authentic element to the study.

Classification methods were chosen from those recognized by Evans (1977). Exogenous and arbitrary values were not regarded as important here as maps which use such schemes illustrate a pre-defined relationship and so are not generally open to change. Of the idiographic techniques, a multi-modal option was deemed appropriate (classification at natural breaks) although contiguity biased limits over-lapped with the notion under investigation and correlation biased limits were regarded as irrelevant due to their dependency upon external data. Zone number (percentiles) and area (equal area division) quantiles were included in the investigation. The nested means method was not due to its requiring 2^n classes. Of the serial options, equal intervals and geometric and arithmetic progressions were incorporated, although normal percentiles; standard deviation and curvilinear classes were considered to be too obscure at this stage due to their assumptions concerning data being statistically distributed in a specific manner and so they were not included in the preliminary analysis. This gave a total of 16 different methods of selection. The data were classed into three, five, seven and nine classes following the pattern of Mersey (1990) which encompasses 'the range used on the majority of choropleth maps'.

The use of 16 classification schemes, four sets of classes and 11 data sets would have produced 704 maps. In order to reduce this figure to a more manageable level two of the data sets and the progression classification schemes that changed at other than constant rates were dropped. This resulted in 140 grid maps, and 192 maps of Leicestershire.

11.3 Results

11.3.1 Graphical Analysis

The classified values of Moran's I were plotted against classification scheme for each of the four class numbers to investigate the effects of classification on I. The line graph format was regarded as the most satisfactory method as although the x axis accounts for no numerical series, the vital differences in I values could be appreciated more successfully than from other representations, and by plotting each classed data set the variation associated with the number of classes was highlighted particularly well (see Figure 11.1).

Several trends were observed in the gridded data, from which provisional conclusions can be drawn. First, positive autocorrelation is rarely increased through classification, and the trend is certainly toward a reduction in autocorrelation coefficient with the regular data. Second, the number of classes has a dominant effect, with the three-class map regularly returning the lowest autocorrelation coefficient. This trend of autocorrelation increasing with number of classes continues up the range, but dependencies are less noticeable as class numbers increase; Figure 11.1 gives an example with one of the regular data sets.

The data surface being classified provides the most significant effect on the potential for variation of I. Values associated with the extremes of the range of autocorrelation (positive autocorrelation and zero autocorrelation) vary little with any scheme, and show only slight signs of the inverse relationship with class number. Consequently, where autocorrelated or random data are concerned, class limit variation will reveal fewer changes to the spatial structure of the mapped variable. Most geographic data fall in a range between these extremes, with local variation occurring over a regional trend, and so manipulation of I is conceivable.

The irregular data were less conclusive regarding the dependence of I upon the number of classes than were the grid data. Whilst the graphical analysis illustrated potential for varying classification to alter autocorrelation, obvious trends and effects were not forthcoming.

11.3.2 Visual Analysis

Visual examination of the mapped patterns and the coefficients prompted several questions. Where the grids were concerned it proved relatively easy to associate visual pattern with values of I, and highly autocorrelated surfaces are indeed more representative of a perceived spatial trend (Figure 11.2 provides some examples). However, the coefficients returned for the data sets involving irregular zones are much more difficult to estimate by eye. Similar looking maps returned very different values, and even extremes of I were often hard to differentiate (Figure 11.3). The apparent reason is that a combination of small neighbouring zones, whose influence on the statistic depends on '$1/d^2$', combined with skewed data, places great emphasis on a few individual values. The statistic is extremely sensitive to these small zones, which, due again to their modest size, do not provide obvious visual effects. In essence autocorrelation measures are sensitive to small scale variation, and visual analysis initially detects large scale features. The problem is particularly acute

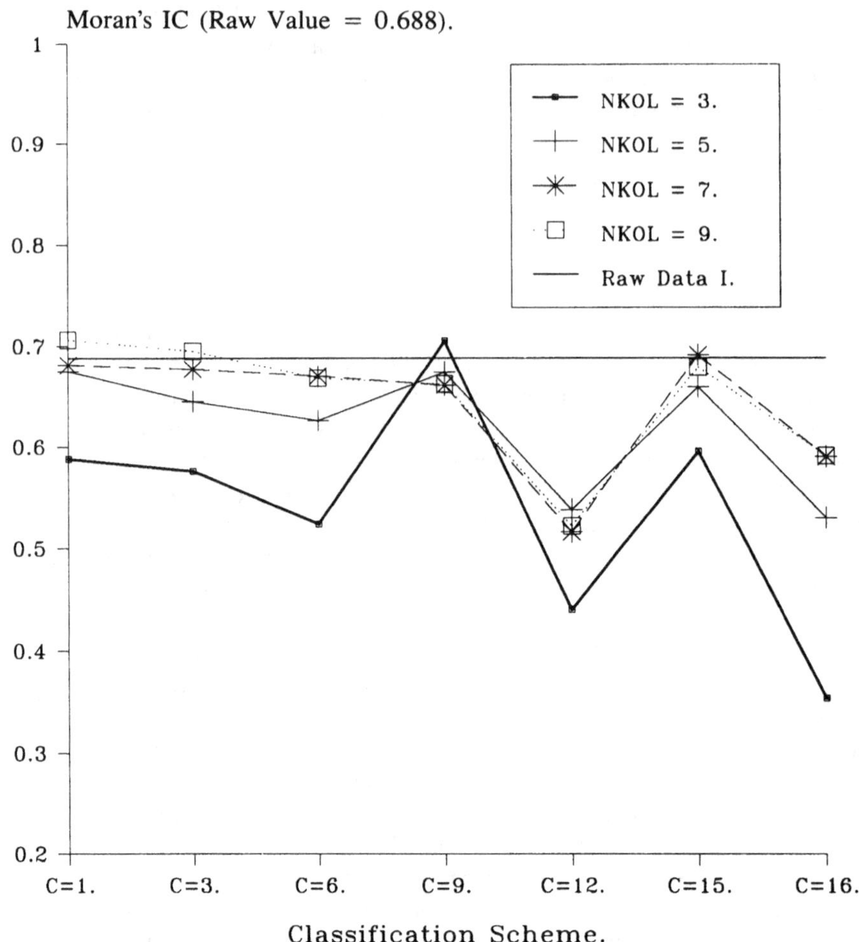

Figure 11.1 Classed Moran's I and classification scheme for a gridded data set. NKOL refers to the number of classes. The associated map for the C1 scheme is shown in Figure 11.2.

where population statistics are concerned as extreme values often occur in small zones (poverty indices, for example, may well equate with space per individual, which relates directly to the way in which enumeration districts are organized to equalize population totals per zone). The occurrence of a class limit between two adjacent zones with extreme values compared to the rest of the surface (and thus data mean) can greatly reduce *I* irrespective of the contiguity between larger, visually dominant zones. In short, where zone sizes are variable, and particularly if data are skewed, autocorrelation coefficients are unpredictable, and extremely sensitive to classification (for further details see Dykes, 1991).

11.4 Conclusions and discussion

These findings suggest that there is ample scope for varying spatial association as measured from geographical distributions through class delimitation. Where regular areas are

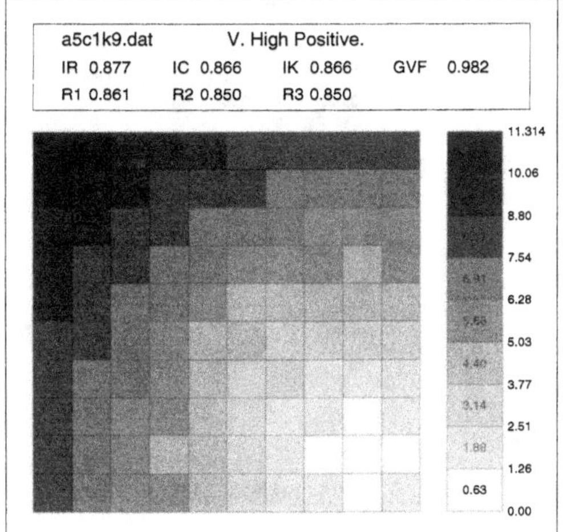

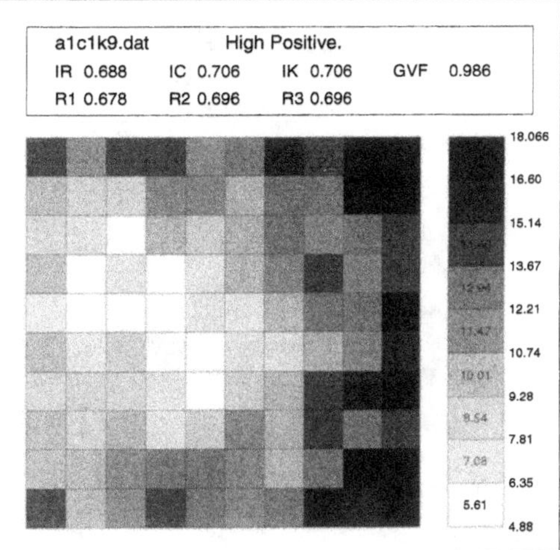

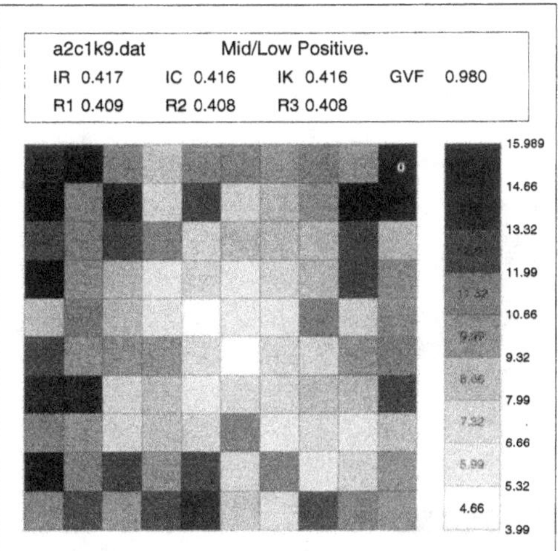

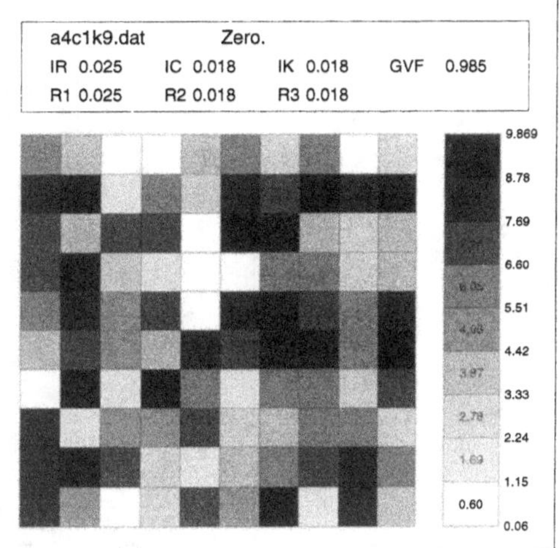

Figure 11.2 Gridded data sets, classified through equal intervals.
The classified values of I (IC) reflect perceived spatial trends. These examples cover a considerable range of values of I, but even relatively minor variation in I can be detected visually. Moran's I is not varied substantially from the raw data value (IR) through this equal interval subdivision into 9 classes.

concerned, Moran's *I* can provide a basis for classifying to enhance spatial trends. Despite this, autocorrelation coefficients as described here do not identify pattern parameters that can be used successfully for measuring map complexity or amplifying data trends when classifying irregularly zoned areal data sets. Moreover, and perhaps more importantly, the research suggests that a fundamental geographical statistic cannot be easily inferred from a prevalent mapping technique.

Four conclusions arise from this work and are the basis for continuing research:

1. If sophisticated autocorrelation indices do not reflect visually perceived trends, then map readers involved in ideation will find it difficult to interpret the complex spatial data presented by the irregular choropleth. This has an important bearing upon the

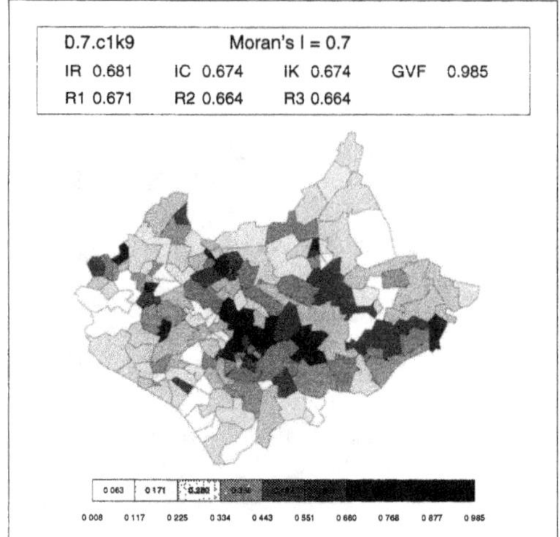

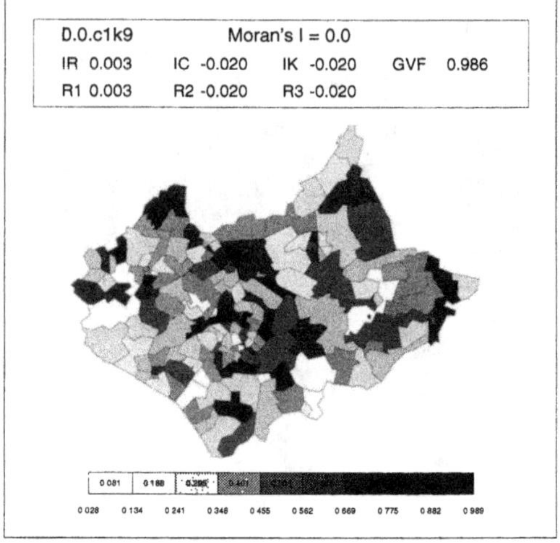

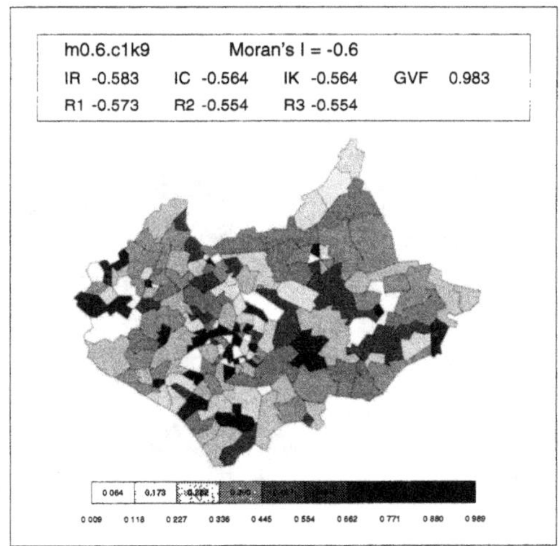

Figure 11.3 Data sets comprising irregular zones.
These data display classified values of Moran's I (IC) across a substantial range. The data are taken from a rectangular (not skewed) distribution and classified by an equal interval scheme. It is difficult to interpret the autocorrelation structure visually. Skewed data, where extreme values correspond with small zones are even more difficult to interpret.

visualization of irregular data, as the results of statistical and visual spatial analysis will differ. This is a problem basic to the tenets of the visualization methodology and is notable as many of the maps used previously to measure complexity involve fewer zones of substantially less size variation than those used here (Muller, 1976; Lloyd and Steinke, 1976; Jenks and Caspall, 1971). It is likely that the influence of the zone size variation, skewed data and subjective equation parameters may not have been fully appreciated by studying irregular zones that approach a regular lattice.

2. A regular grid presents data in a structure from which spatial associations may be interpreted successfully.

3. There is a need for a statistic or function that *does* relate to perceived map complexity from irregular maps, so that a spatial component can be intrinsic to map classification.
4. The reduction of an irregular mass of spatial data into a single statistic is not satisfactory in terms of ideation. The subjective input regarding the choice of w_{ij} and c_{ij} when using

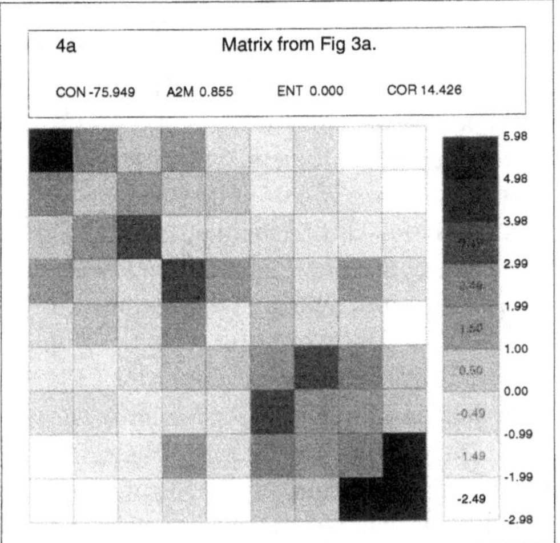

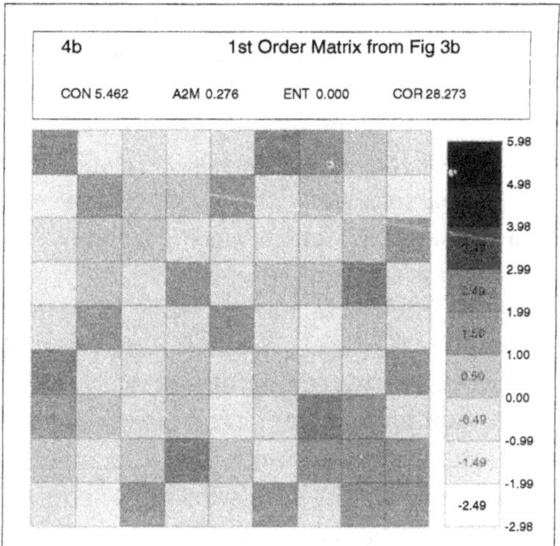

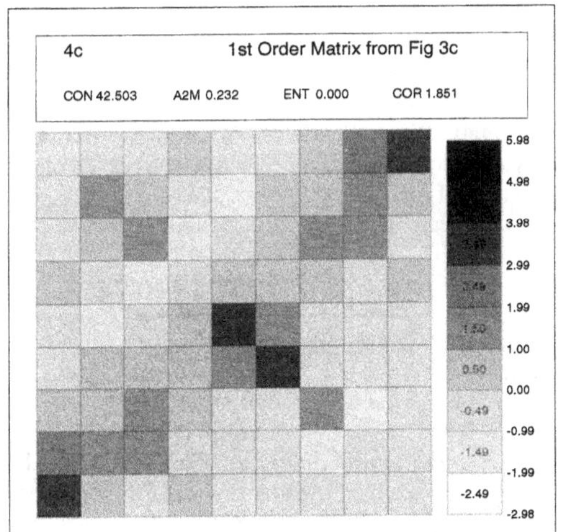

Figure 11.4 First order co-occurrence matrices from maps 3a, 3b and 3c.
The rows and columns of these matrices represent class numbers from the maps in Figure 11.3 (11.3a, 11.3b and 11.3c are ordered from left to right). The values in each cell relate to the number of times the two classes indicated by the row/column location are adjacent, hence the symmetry about the line row/column. The counts are standardized under the assumption of non-free sampling. Matrices can be created for lagged adjacencies of orders from one to the diameter of the map to illustrate the variation of spatial association over distance.

Matrix 11.4a indicates autocorrelation, especially amongst high classes, as concentration occurs around the row/column axis. Matrix 11.4b is less distinct. There is a tendency for high classes to co-occur but spatial association is not apparent. Matrix 11.4c shows that classes 5 and 6 cluster, but the reverse diagonal indicates that the trend is for spatial association between extreme values, indicating negative autocorrelation.
The associated values of Moran's I for maps 3a, 3b and 3c are 0·674, −0·020 and −0·564.

a statistic, along with the assumptions required to construct a Euclidean distance coefficient detract from the value of the statistic. Furthermore, the wealth of information available concerning relationships over spatial lags is ignored.

The way in which spatial association can be used to extract information from a series of values when zone structure is regular, the suffocating loss of spatial information that results from reducing a complex map to a single statistic, and the restrictions of metric measurement and distance function calculation point towards the use of abstract rather than metric spatial representations for visualizing spatial association in area-value data sets. This can be achieved by mapping irregular zones as regular shapes in an equal area cartogram. By using adjacency to define spatial relationships autocorrelation coefficients will correspond with visual interpretations of association. An alternative is to map counts of adjacent classes in a space whose dimensions are referenced by class numbers. This technique is equivalent to Haralick's (Haralick *et al.*, 1973) production of co-occurrence matrices from raster images. The usefulness of such representations is currently under consideration and examples of matrices for the maps from Figure 11.3 are shown in Figure 11.4. While this work provides an indication of the way in which a fundamental geographical statistic and a familiar map interact, it also shows that visual data analysis is not a straightforward task where irregular enumerated data are concerned. It is the author's belief that visualization may best be employed when simple, non-metric and regular cartographic representations of data are created and used to interrogate more complex data sets traditionally looked at, and perhaps overlooked rather than interrogated, by the geographer.

References

Dykes, J.A., 1991, An investigation into the effects of classification upon spatial autocorrelation in an attempt to maximise the 'communication effectiveness' of the choropleth map. Unpublished masters thesis, University of Leicester.

Evans, I.S., 1977, The selection of class intervals. *Transactions of the Institute of British Geographers*, **2**(1), 98–124.

Gatrell, A.C., 1983, *Distance and Space: A Geographical Perspective*, Oxford: Clarendon.

Goodchild, M.F., 1986, *Spatial Autocorrelation, Concepts And Techniques in Modern Geography* (CATMOG) No 47, Norwich: Geo Books.

Haralick, R.M., Shanmugam, K. and Dinstein, I., 1973, Textural features for image classification. *IEEE Transactions on Systems, Man and Cybernetics*, **3**(6), 610–21.

Jenks, G.F. and Caspall, F.C., 1971, Error on choroplethic maps: definition, measurement and reduction. *Annals of the Association of American Geographers*, **61**, 217–44.

Lloyd, R. and Steinke, T., 1976, The decision making process for judging the similarity of choropleth maps. *The American Cartographer*, **3**(2), 177–84.

MacEachran, A.M., 1982, The role of complexity and symbolization method in thematic map effectiveness. *Annals of the Association of American Geographers*, **47**(4), 495–513.

MacEachran, A.M., Gatrell, A.C., Dykes, J.A., Dorling, D.F.L. and Bishop, I., 1944, Advances in visualizing spatial data: Introduction, in Unwin, D.J. and Hearnshaw, H. (Eds). *Visualization in GIS*, Bellhaven Press.

McCormick, B.H., DeFanti, T.A. and Brown, M.D., 1987, Visualization in scientific computing. *ACM SIGGRAPH - Computer Graphics*, **21**(6).

Mersey, J.E., 1990, Colour and thematic map design: the role of colour scheme and map complexity in choropleth map communication. *Cartographica*, **27**(3), Monograph 41.

Monmonier, M.S., 1972, Contiguity-biased class-interval selection: a method for simplifying patterns on statistical maps. *The Geographical Review*, **62**, 203–28.

Monmonier, M.S., 1974, Measures of pattern complexity for choroplethic maps. *The American Cartographer*, **1**(2), 159–69.

Moran, P.A.P., 1948, The interpretation of statistical maps. *Journal of the Royal Statistical Society*, Series B, **10**, 243–51.

Muehrcke, P.C., 1973, The influence of spatial autocorrelation and cross correlation on visual map comparison. *Proceedings of the American Congress on Mapping and Surveying*, 315–25.

Muehrcke, P.C., 1990, quoted from Buttenfield, B.P and MacKanness, W.A., Visualization, in Maguire, D.J., Goodchild, M.F. and Rhind, D.W. (Eds) (1992). *Geographical Information Systems: Principles and Applications*. New York: John Wiley & Sons.

Muller, J.-C., 1976, Number of classes and choropleth pattern characteristics. *The American Cartographer*, **3**(2), 169–75.

Olson, J.M., 1974, Autocorrelation and visual map complexity. *Annals of the Association of American Geographers*, **65**(2), 189–204.

12

Probable and fuzzy models of the viewshed operation

Peter F. Fisher

Various researchers in the GIS community show confusion between the concepts of fuzzy and probable models of uncertainty and particularly how these may be applied to modelling spatial phenomena. This chapter will use the viewshed function, familiar to many users of both raster and vector GIS, to exemplify the distinction. At the same time, a mathematical formalization is presented for some of the empirical methods employed by landscape architects in landscape evaluation and aesthetics, and results of significance to visual impact assessment are outlined.

12.1 Introduction

It is possible to document considerable confusion between the models of probability and fuzzy membership from past research. This confusion is perhaps understandable, because of the similarities of the two measurement scales (both range from 0–1) and the relatively late arrival of fuzzy set theory (Zadeh, 1965). It does, however, ignore fundamental differences in the concepts of the two.

To illustrate the point here, two quotes will suffice. Howard and Barr (1991) in their discussion of the British countryside change database, state that (p. 218):

> ITE is investigating the application of fuzzy logic (Chang and Burrough, 1987) to present such boundaries. Although ARC/INFO is designed to present crisp boundaries, it is possible to include probablistic information to present parcels as part of a continuum (Wang *et al.*, 1990).

Similarly, Peuquet (1988) states that (p. 390):

> One mechanism that has been investigated is fuzzy set theory... It is based on the assumption that the imprecision in natural languages is probablistic in nature (Robinson *et al.*, 1986).

It is sad to note that both these writers are actually mis-quoting their sources, who do not equate fuzziness with probability. The present writer, indeed, is not innocent of disseminating confusion, and, in particular, he has used the term 'fuzzy viewshed' incorrectly in the past (Fisher, 1991). This confusion is not restricted to geographers, however, and Zadeh (1980) himself has come to the defence of others in attempting to clarify the differences between fuzzy sets and probability. A motivation for this chapter is to rectify the author's own misuse of the term, but in so doing it is hoped to clarify the distinctions for the broader community.

In the spatial database literature, Robinson and Frank (1985) set out the differences between the fuzzy and probability models of uncertainty, but their work has not received the attention it deserves. On the other hand, Robinson (1988) unfortunately fails to provide as clear a distinction, and much other geographical literature on spatial applications of fuzzy sets fails to draw these distinctions. For example, Leung (1988) in an otherwise very thorough book on the use of fuzzy sets in spatial analysis, gives no clear spatial examples of the distinction. Leung *et al.* (1992) furthermore obfuscate the distinction, by describing a system for visualizing either fuzzy or probable images. They use the terms correctly, but fail to give a thorough distinction.

The aim of this chapter is to clarify the difference between the concepts by reference to a specific GIS function, namely the viewshed. The next section outlines the underlying concepts of fuzzy and probable uncertainty. Then the methods of deriving the probable and fuzzy viewsheds are discussed separately. Finally, methods for combining such viewsheds are reviewed.

12.2 Concepts

The primary distinction between fuzzy and probable models is conceptual, and confusion seems to arise primarily because they are both measured on a scale of 0–1. The probable model of uncertainty assumes that there is a Boolean phenomena which can be said to either exist or not; in the GIS context it is, of course, said to exist or not at a location. The line-of-sight determines whether one point is visible from another, and is therefore a Boolean process. When applied exhaustively over the area of a digital elevation model (DEM) within a GIS, as the viewshed function, the set of locations visible from a particular point is identified. Probability is the method we use to determine the accuracy with which we can state that the Boolean event of being visible occurs. Thus we may state that there is perhaps a 95 per cent chance that location x is visible from the viewing point, while location y has only a 10 per cent chance of being visible. Exactly how the probable viewshed can be determined and some of its justification is discussed below.

The fuzzy viewshed is conceptually different and should be compared with the Boolean viewshed, not the probable viewshed, whatever the probability of the set of points being visible. The fuzzy concept is used to determine the degree to which an object is a member of a set. If a set of points is visible from one location with some level of confidence, then an object at one location in that set may not be distinguishable, while at another location, the same object may be recognized. There is in fact a continuum of conditions of the clarity of the object between being distinguishable and being indistinguishable. This continuity is the degree to which objects are clear, and depends on a host of variables which might include: the size and colour of the object, the colour of the surrounding vegetation, the eyesight of the observer and their knowledge of the class of object, and the weather conditions at the time (sunglare, haze or fog). The degree of clarity is then the primary property of the fuzzy viewshed. This is an experiential phenomenon, and so closer to the human condition than the probable viewshed which simply reports whether a location is visible.

In essence the probable model of error allows us to determine the area which should be visible from a particular viewing point, while the fuzzy models tells us how distinct any object might be: the visible and the distinguishable locations.

12.3 Study site

Any study site could have been selected to illustrate the concepts discussed in this chapter. A 100×100 cell subset of the Ordnance Survey 20×20 km tile SO64 which includes the dramatic scenery of the Malvern Hills in Worcestershire is used (Figure 12.1). This subset includes the highest points of the ridge of hills. The two viewpoints analysed here are the highest points of the Worcestershire Beacon and North Hill.

The Boolean viewshed found by the point-to-point algorithm (Fisher, 1993), is shown in Figure 12.2. This is the standard version of the viewshed which can be derived by functions in many different GIS packages.

12.4 The probable viewshed

The digital elevation model (DEM) itself is laced with errors - the USGS reports up to 15 m root mean square error, and the OS state 2 to 3 as typical values. Thus there is indeed a level of uncertainty whether a particular location is actually visible from another even if the line-of-sight calculation reports that it is (Fisher, 1991). This error, of course, effects both the viewing location, the target and all locations between. Felleman and Griffin (1990) and Fisher (1993) have shown that different algorithms yield vastly different viewsheds (up to 300 per cent is reported), and that alternative commercial implementations, while not documenting the algorithms used also give differing results. Various reasons therefore exist for the use of a probable model of the visible area.

There are a number of methods for determining the probable viewshed. Fisher (1992) used Monte Carlo simulation of error in the DEM to generate alternative versions of the DEM, and by summing the viewsheds found in each version of the DEM, it is possible

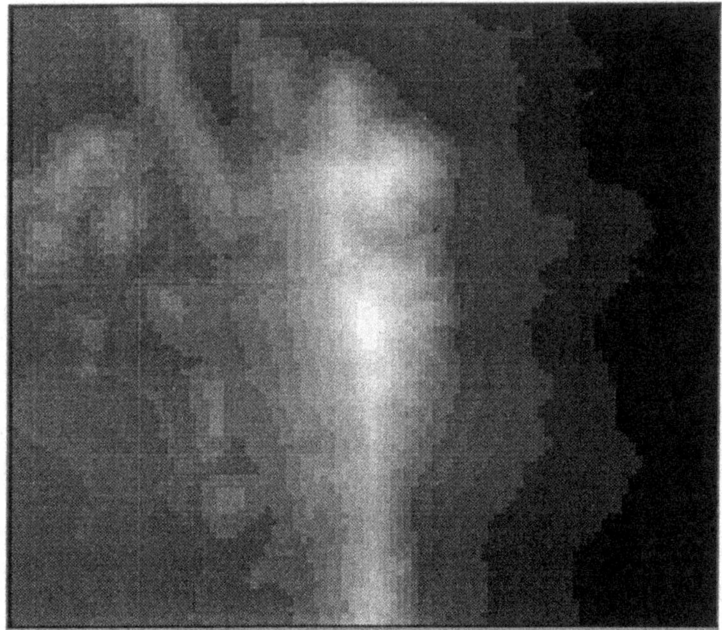

Figure 12.1 Digital elevation model of the Malvern Hills. A 100×100 cell subset of the Ordnance Survey's SO64 DEM is shown, and analysed. Crown copyright reserved.

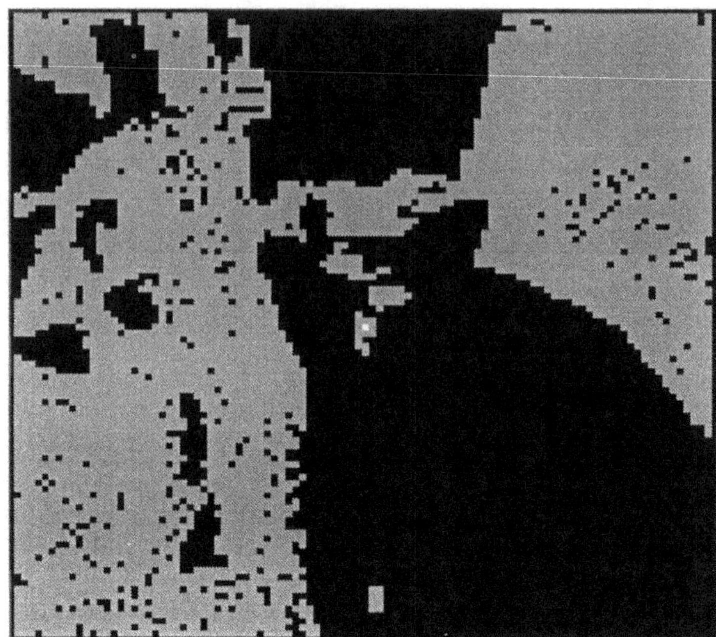

Figure 12.2 Boolean viewshed from Worcester Beacon for which alternative analyses are shown in subsequent figures.

to derive the probability of any point being visible using equation (12.1); see also Felleman and Griffin (1990). In Fisher's (1992) work the phenomenon derived was incorrectly called the fuzzy viewshed.

$$p(x_{ij}) = \frac{\sum_{k=1}^{n} x_{ijk}}{n} \qquad (12.1)$$

where $p(x_{ij})$ is the probability of a cell at row i and column j in the raster image being visible, and x_{ijk} is the value at the cell of the binary-coded viewshed in realization k such that k takes values 1 to n.

The result is shown in Figure 12.3, which, when compared with Figure 12.2, shows clearly an increased information content. Other approaches which may be used to derive the probable viewshed include randomization of the elevation at the viewing location, and systematic variation, again finding averages of the multiple Boolean viewsheds in (12.1). All these may involve excessive computational overheads due to the number of times the Boolean viewshed needs to be found.

It is also possible to find the probability of the line-of-sight being higher than the surface, given the RMSE of the DEM, and solving for the z value. Heuristic methods are also under development, but tend to place locations in a limited number of probability classes (Teng and Davis, 1992).

12.5 The fuzzy viewshed

It is within all our experience that objects visible from a particular location can be seen with different clarity, and, furthermore, that what one person can see does not necessarily

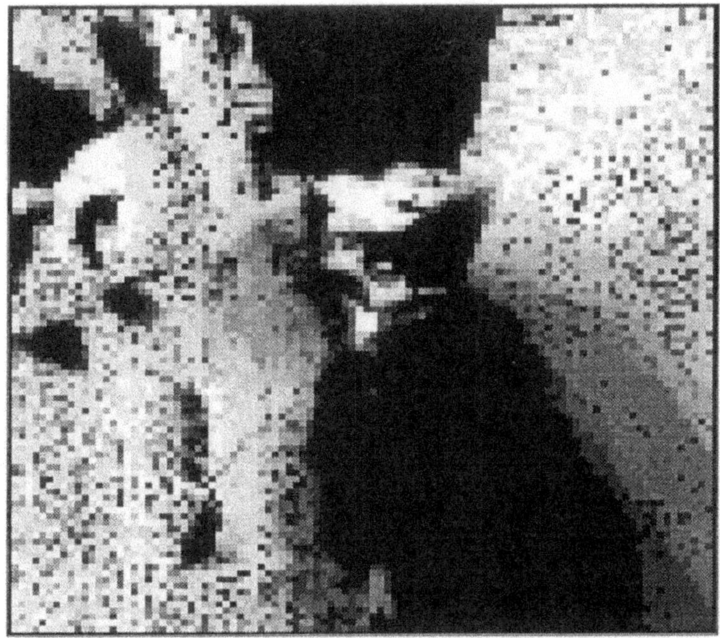

Figure 12.3 Probable viewshed from Worcester Beacon derived from 20 simulations (i.e. $n=20$ in equation (12.1) and root mean square error=2 m.

correspond to what another can see or to what the same person can see at another time. Definition of the fuzzy viewshed is concerned to identify this clarity.

The simplest form of the fuzzy viewshed is as a distance decay function away from the viewing location. The form of the decay should be from 1 at and near to the viewing point, and asymptotic to 0 at infinity. Such a curve is generated by the formula given by Kandel (1986), which is used by Burrough *et al.* (1992), and presented here as (12.2). Other formula could also be contemplated with similar properties, and it is far from certain that this is the best curve, particularly since it shows a very steep decline from a membership of 1. The resulting viewshed is shown in Figure 12.4.

$$\mu(x_{ij}) = \begin{cases} 1 & \text{for } d_{vp \to ij} \leq b_1 \\ \dfrac{1}{\left(1+\left(\dfrac{d_{vp \to ij}-b_1}{b_2}\right)^2\right)} & \text{for } d_{vp \to ij} > b_1 \end{cases} \tag{12.2}$$

where (x_{ij}) is the fuzzy membership at the cell at row i, column j, $d_{vp \to ij}$ is the distance from the viewpoint to row i, column j; b_1 is the radius of the zone around the viewpoint where the clarity is perfect, and the target object can be seen at the defined level of detail, and b_2 is the distance from b_1 to fuzzy membership$=0 \cdot 5$, sometimes called the cross-over point.

There is normally a region around the viewing point where clarity can be considered to be perfect, and this is accommodated by b_1. If the object is a very small, b_1 will be correspondingly small, but normally if a person, for example, is the object to be observed, the region is about 1 km from the viewing point, and is referred to by landscape architects as the foreground. The middle ground is where the clarity decays, and is presumably a

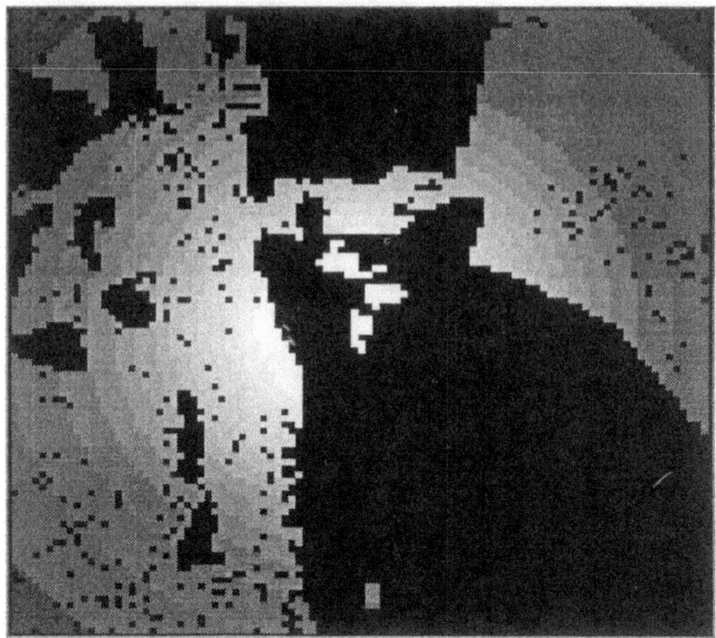

Figure 12.4 Fuzzy viewshed from the Worcester Beacon, using equation (12.2) and giving a spatially symmetric distance decay function.

little in excess of b_2 in width. Some people will be able to see objects with considerable clarity, even at the further edge of the middle ground, while others will have a harder time. The background is essentially a zone of little clarity, where only very broad landscape features can be distinguished.

In certain weather conditions the form of the decay will vary. With fog or mist, for example, the extent of the foreground will be curtailed to some extent, and the decay in the middle ground will be abrupt, to no clarity in the background. Equation (12.3) gives a trigonometric function which yields this form of decay, which is illustrated in Figure 12.5.

$$\mu(x_{ij}) = \begin{cases} 1 & \text{for } d_{vp\to ij} \le b_1 \\ 0 & \text{for } d_{vp\to ij} > b_1 + 2 \times b_2 \\ \sin\left(\left(\dfrac{d_{vp\to ij} - b_1}{2 \times b_2}\right) \times 90^\circ\right) & \text{for } d_{vp\to ij} > b_1 \end{cases} \tag{12.3}$$

A more complex situation occurs when solar glare is introduced and the viewer may only see a limited distance into the sun, but a larger distance away from it. A much more complex trigonometric function given in (12.4) is required to describe this situation.

$$\mu(x_{ij}) = \begin{cases} 1 & \text{for } d_{vp\to ij} \le b_1' \\ \dfrac{1}{\left(1 + \left(\dfrac{d_{vp\to ij} - b_1'}{b_2'}\right)^2\right)} & \text{for } d_{vp\to ij} > b_1' \end{cases} \tag{12.4}$$

where

$$b_1' = b_{1_{min}} + ((b_1 - b_{1_{min}}) \times c) \tag{12.4a}$$

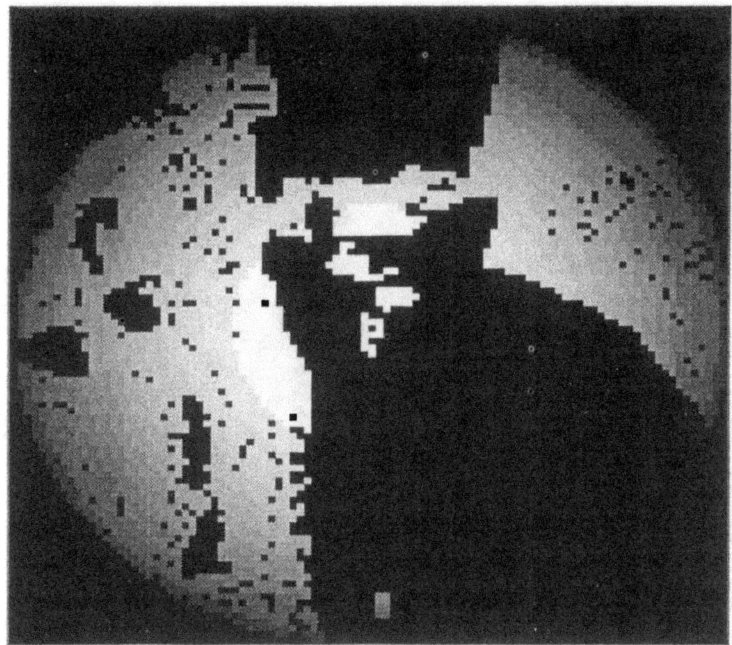

Figure 12.5 Fuzzy viewshed from the Worcester Beacon with fog or mist simulated from equation (12.3).

$$b_2' = b_{2_{min}} + ((b_2 - b_{2_{min}}) \times c) \tag{12.4b}$$

$$c = \begin{cases} \sin \psi & \text{for } \left|\sin\left(\frac{\theta}{2}\right)\right| < \sin \psi \\ \left|\sin\left(\frac{\theta}{2}\right)\right| & \text{for } \left|\sin\left(\frac{\theta}{2}\right)\right| \geq \sin \psi \end{cases} \tag{12.4c}$$

and θ is the angle of the sun, such that the position of the sun is transformed to 0, ψ is the azimuth of the sun above the horizon, and $b_{1_{min}}$, $b_{2_{min}}$ are minimum values for b_1 and b_2 respectively.

Although manipulation of (12.3) can give an approximation to conditions of atmospheric haze, it is entirely a distance decay function. This is not necessarily an accurate model of the situation, because haze commonly stays close to the ground. Thus if the line-of-sight passes close to the ground surface it will suffer degradation of clarity, while if it does not then clarity may be maintained, and so objects in the background may have higher clarity than those in the foreground. Under these circumstances, the clarity increases with distance of the line-of-sight above the ground. This is a condition of equations (12.5) and (12.5a).

$$\mu(x_{ij}) = \begin{cases} 0 & \text{if } ij \text{ is out-of-view} \\ \min\ [\mu(m_1), \mu(m_1), \ldots, \mu(m_n)] & \text{if } ij \text{ is in-view} \end{cases} \tag{12.5}$$

where (m) is the fuzzy membership of an object being viewed through the haze at each of the n locations along a line-of-sight where the elevations are compared, and is derived from (12.5a).

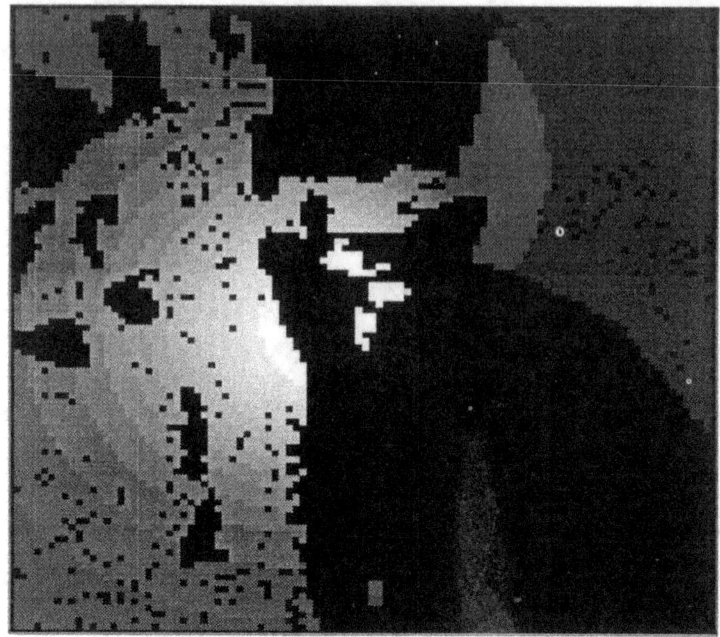

Figure 12.6 Fuzzy viewshed from the Worcester Beacon at sunrise, derived from equation (12.4) using parameters given in the text.

$$\mu(m) = 1 - \frac{d}{(d+h)} \tag{12.5a}$$

Here the line-of-sight is h metres above the terrain at that location, and d is the height of significant haze (taken to be the cross-over point of $(m)=0\cdot5$). A fuzzy viewshed effected by haze is shown in Figure 12.7.

The major problem in using equations (12.2)–(12.5) is defining values for the different parameters, such as b_1, b_2,$b_{l_{min}}$, $b_{2_{min}}$, and h. The magnitude of all will depend on the object being observed, as well as factors mentioned above. In landscape applications, and when the individual tree (as an element of the landscape) is the primary target, $b=1$ km and $b_2=3$ km might be appropriate, meaning that individual trees would be losing clarity at distances much over 4 km. The question asked by the observer viewing the tree is, however, important; for example, does the observer wish to tell the species of tree or simply note the existence of the tree.

Numerous other factors which might be considered in defining the fuzzy viewshed can be conceived and have been discussed in general terms in the landscape architecture literature (Hardian *et al.*, 1988), but space precludes more than summarizing them here: target contrast with background; low cloud; observer's personal knowledge of the target; landscape aesthetics; observer's personal preference; vegetation screening.

12.6 *The probablity of a fuzzy event*

The probability of a particular fuzzy viewshed being true may be assessed in a number of ways. If the weather conditions are important, then the probability of the associated fuzzy viewshed may relate to the return period of those weather conditions. From the

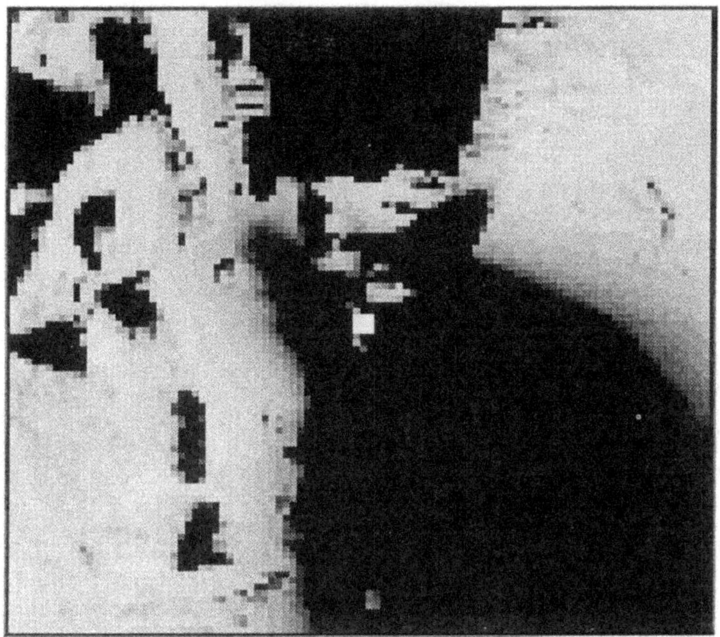

Figure 12.7 Fuzzy viewshed from the Worcester Beacon with 2 m haze over the whole area.

point of view of the observer, the frequency of a particular quality of eyesight in a human population may define the probability of another fuzzy viewshed.

12.7 Combining viewsheds

12.7.1 Combining binary viewsheds

Binary viewsheds from more than one location may be combined. Given that individual viewsheds are coded as 0 or 1, indicating invisible and visible locations, the most effective method for combination is to derive the 'times seen' image known to landscape architects. For every cell it is simply the sum of all binary viewsheds (12.6) where m is the number of binary viewsheds to be combined.

$$y_{ij} = \sum_{c=1}^{m} x_{ijc} \tag{12.6}$$

The resulting image actually encodes a variety of information, as is illustrated in the image in Figure 12.8 where m=2. Any cell where $y_{ij}>0$ is part of the union of all the viewsheds, i.e. they are the locations visible from either viewpoint, and if isolated in a single image would be the result of the logical OR operation (the union). That is to say that if one observer was at each viewing point, this is the total area they would be able to see, although many parts would be invisible to one or the other. Where a cell has $y_{ij}= m$, it is part of the intersect of the two viewsheds, the result of a logical AND operation, and the sites visible to both observers, such that they can compare notes on what they can see.

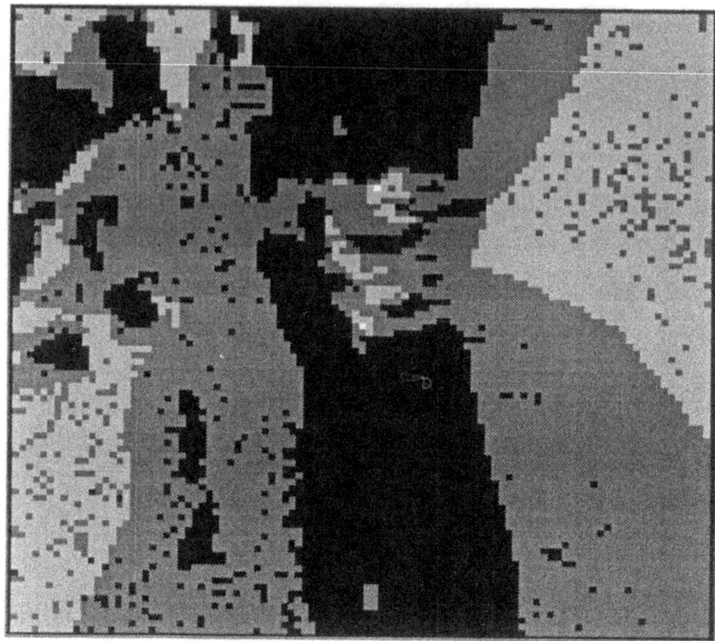

Figure 12.8 Sum of binary viewsheds from Worcester Beacon and North Hill. The light grey shows the intersection, and both light and dark grey areas show the union.

12.7.2 Combining probable viewsheds

It is also possible to find the union and intersect of multiple probable viewsheds. Equation (12.7) is the standard formula for the probability of an object belonging to the union of sets A and B from the independent probabilities of belonging to A and to B. Figure 12.9 shows, therefore, the probability of any cell being visible to observers on either the Worcester Beacon or North Hill. The multiplication rule (12.8) gives the intersection of the two viewsheds, showing the probability of a cell being visible by both observers.

$$p(x_{A\cup B})_{ij}=p(x_A)_{ij}+p(x_B)_{ij}-(p(x_A)_{ij}\times p(x_B)_{ij}) \qquad (12.7)$$

$$(px_{A\cap B})_{ij}=p(x_A)_{ij}\times p(x_B)_{ij} \qquad (12.8)$$

These images are very useful, in their own rights, and give considerable advantage over the binary equivalents, but it is not the purpose of this chapter to dwell on interpretations. It is clear that the combination can be viewed as a Bayesian process if the viewpoints are sufficiently close together, and particularly if they lie along a route, whether a road or path. The problem is that as m increases, so the probability derived by Bayes theorem tends to zero, because of the combination of probabilities by multiplication. Similarly the result of (12.8) will tend to zero for all locations as m increases. This result is counter-intuitive, and is a problem with these methods which has contributed to the move by researchers and developers in artificial intelligence to find alternative methods for handling uncertainty.

12.7.3 Combining fuzzy viewsheds

Fuzzy viewsheds may meaningfully be combined in a number of ways. Figure 12.11 shows the result of fuzzy intersection (12.9) of two fuzzy viewsheds from the same viewpoint,

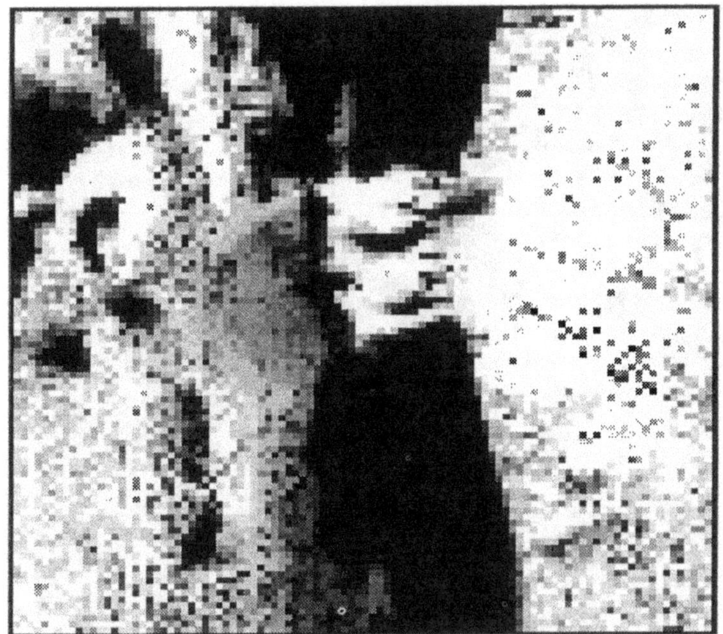

Figure 12.9 Union of probable viewsheds from Worcester Beacon and North Hill, being the probability of cells being visible from both locations.

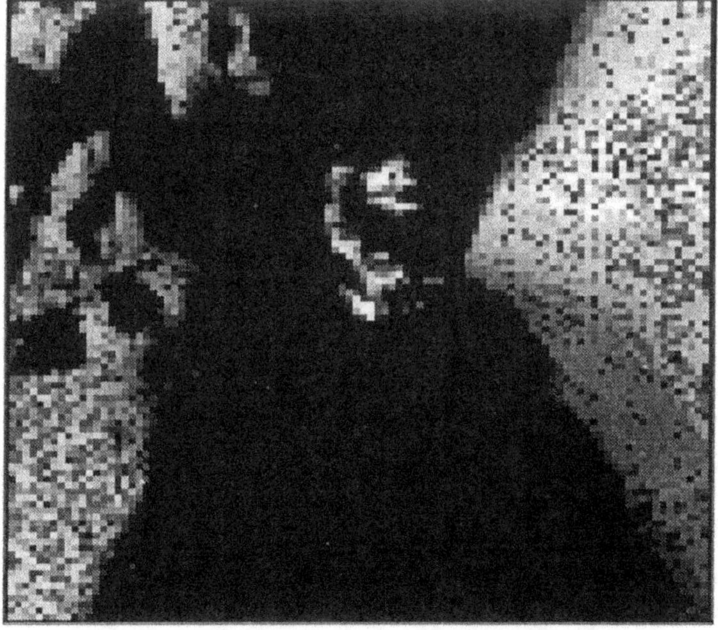

Figure 12.10 Intersection of probable viewsheds from Worcester Beacon and North Hill, being the probability of being visible from both locations.

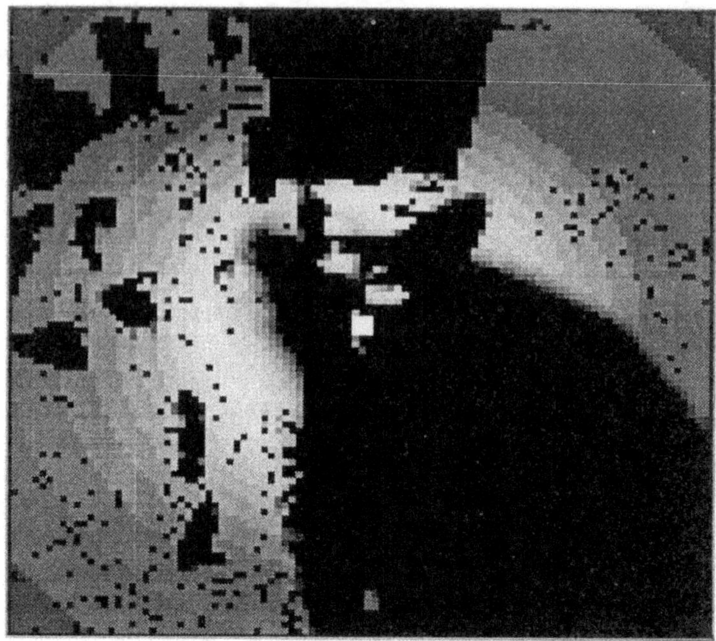

Figure 12.11 Compound fuzzy viewshed from Worcester Beacon from intersection of symmetric and 2 m haze.

with symmetric distance decay, and with 2 m haze (Figures 12.4 and 12.7). It shows the effects on clarity of both processes formulated in the associated membership functions. The minimizing operation of the intersect ensures that the least clarity is carried over for any location.

The intersection of the fuzzy viewsheds from two (or more) viewpoints may also be determined. Using the minimizing function in (12.9) again, yields the worst clarity of locations when viewed from both locations (Figure 12.12). On the other hand, the fuzzy union is given in (12.10), and is shown in Figure 12.13. Here the maximum is recorded at each location, and it reports the clarity with which any point could possibly be seen by observers at both locations.

$$\mu_{(A\cap B)}(x) = \min\ [\mu_A(x),\ \mu_B(x)] \tag{12.9}$$

$$\mu_{(A\cup B)}(x) = \max\ [\mu_A(x),\ \mu_B(x)] \tag{12.10}$$

12.7.4 Combining fuzzy and probable viewsheds

Finally, probable and fuzzy viewsheds may be combined. The probable viewshed is simply a statement of the confidence with which it is possible to state that a location is visible. Whatever level of $p(x)=t$ is set by the user, it is possible to define an associated binary viewshed, by thresholding the probable viewshed, and declaring any location with $p(x)>t$ in view (coded as 1), and any locations with $p(x)>t$ out-of-view (coded as 0). The largest possible binary viewshed would be derived by setting t to the lowest value of $p(x)$ which is not zero. Once a binary viewshed is defined, the associated fuzzy viewshed may be derived by solving the membership function for any location which is in view. This can be impossible when elevation data is required in the membership function (12.5 a). This problem arises because blocking land may be lowered in the randomization algorithms,

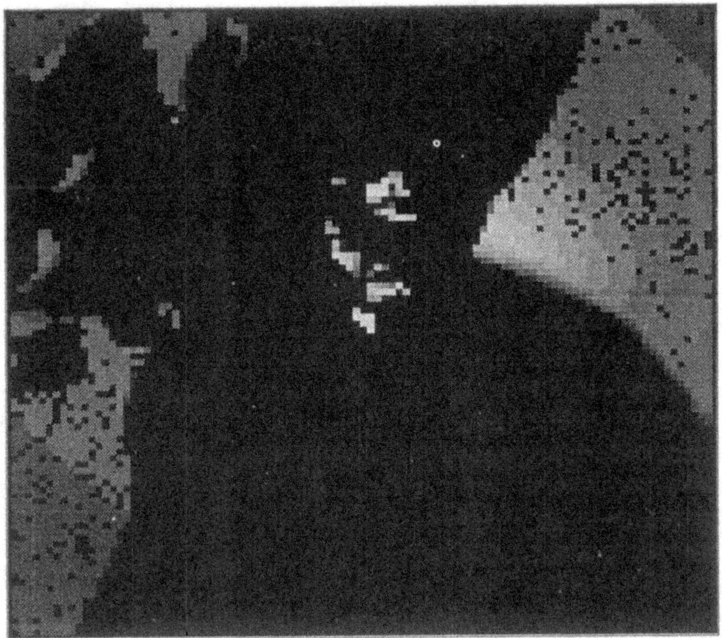

Figure 12.12 Intersection of compound viewsheds from Worcester Beacon and North Hill, using equation (12.8).

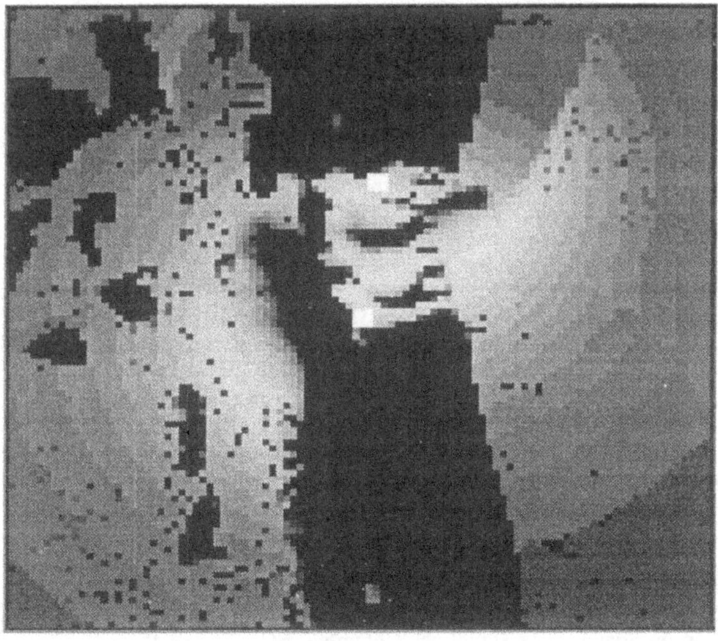

Figure 12.13 Union of compound fuzzy viewsheds from Worcester Beacon and North Hill, using equation (12.9).

or not be judged as significant in others when the probable viewshed is derived. A solution is to find the fuzzy membership at the same time as the probable viewshed, and at each randomization the union of preceding fuzzy memberships is recorded.

12.8 Conclusion

The possible applications for both the probable and the fuzzy viewshed over the simplistic Boolean are myriad. In landscape planning, it will be possible to assess whether new constructions will actually have a gross impact on the view, even if they are visible, and it will be possible to determine the probable consequences of the construction. In planning walking and other trails or roads, it will be possible to derive both snapshots and total experience assessments of the landscape viewed by the hikers. Landscape planners attempt this already most commonly with the times-seen image, but also with extensions of the viewshed along the lines specified here (Hardian *et al.*, 1988). With probable and fuzzy viewsheds there is perhaps a greater clarity in the meanings, and there is a firmer mathematical basis giving improved analytic tools which should greatly enhance the analyses.

In this chapter the necessary building blocks are introduced. Further work will extend and show how to parameterize the different models of the fuzzy viewshed, as well as exploring how multiple probable and fuzzy viewsheds should be combined. Another approach to deriving membership functions, which will be persued, is to model radiation-atmosphere interaction as light transmitted to the viewpoint. The meaning of all such analyses will be crucial, as will a re-examination of the influence of vegetation on the character of the viewshed.

References

Burrough, P.A., MacMillan, R.A. and van Deursen, W., 1992, Fuzzy classification methods for determining land suitability from soil profile observations and topography. *Journal of Soil Science*, **43**(1), 193–210.

Chang, L. and Burrough, P.A., 1987, Fuzzy reasoning: A new quantitative aid for land evaluation. *Soil Survey and Land Evaluation*, **7**(2), 69–80.

Felleman, J. and Griffin, C., 1990, *The role of error in GIS-based viewshed determination: a problem analysis*. Technical report EIPP-90-2, IEPP, SUNY-Syracuse.

Fisher, P.F., 1991, First experiments in viewshed uncertainty: the accuracy of the viewable area. *Photogrammetric Engineering and Remote Sensing*, **57**, 1321–7.

Fisher, P.F., 1992, First experiments in viewshed uncertainty: simulating the fuzzy viewshed. *Photogrammetric Engineering and Remote Sensing*, **58**, *345–52*.

Fisher, P.F., 1993, Algorithm and implementation uncertainty in the viewshed function. *International Journal of Geographical Information Systems*, **7**, 331–47.

Hardian, D.R., Bishop, I.D. and Mitchelltree, R., 1988, Automated mapping of visual impacts in utility corridors. *Landscape and Urban Planning*, **16**, 261–82.

Howard, D.C. and Barr, C.J., 1991, Sampling the countryside of Great Britain: GIS for the detection and prediction of rural change, in Heit, M. and Shortreid, A. (Eds), *GIS Applications in Natural Resources*, GIS World, Fort Collins, Co., 217–21.

Kandel, A., 1986, *Fuzzy Mathematical Techniques with Applications*. Reading, MA: Addison-Wesley

Leung, Y., 1988, *Spatial Analysis and Planning under Imprecision*. Amsterdam: North-Holland.

Leung, Y., Goodchild, M.F. and Lin, C.-C., 1992, Visualization of fuzzy scenes and probability fields. *Proceedings of the 5th International Symposium on Spatial Data Handling*, IGU Commission on Spatial Data Handling, Columbia, SC., **2**, 480–9.

Peuquet, D., 1988, Representations of geographic space: toward a conceptual synthesis. *Annals of the Association of American Geographers*, **78**(3), 375–94.

Robinson, V.B., 1988, Some implications of fuzzy set theory applied to geographic databases. *Computers, Environment and Urban Systems*, **12**, 89–97.

Robinson, V.B. and Frank, A.U., 1985, About different kinds of uncertainty in collections of spatial data. *Proceedings of Auto Carto 7*, American Congress on Surveying and Mapping and American Society for Photogrammetry and Remote Sensing, Falls Church, VA, 440–9.

Robinson, V.B., Thongs, D. and Blaze, M., 1986, Man-machine interaction for acquisition of spatial relations as natural language concepts, in Opitz, B. (Ed.), *Geographic Information Systems in Government*, Hampton, Va.: A. Deerpak Publishing, 433–53.

Teng, Y.A. and Davis, L.S, 1992, Visibility Analysis on Digital Terrain Models and its parallel implementation. Technical report 625, Center for Automation Research, University of Maryland.

Wang, F., Hall, G.B. and Subaryono, 1990, Fuzzy information representation and processing in conventional GIS software: database design and application. *International Journal of Geographical Information Systems*, **4**(3), 261–83.

Zadeh, L.A., 1965, Fuzzy sets. *Information and Control*, **8**, 338–53.

Zadeh, L.A., 1980, Fuzzy sets versus probablity. *Proceedings of the IEEE*, **68**(3), 421.

13

Error simulation in vector GIS using neural computing methods

Chris Brunsdon and Stan Openshaw

13.1 Introduction

The problem of error in GIS has long been recognized. The power of GIS derives from its ability to combine data from a variety of sources, such as digitized paper maps, aerial photography and satellite imagery. Each of these information gathering processes is subject to different levels of error. A major technical and scientific weakness in GIS is the lack of any adequate methods for simulating the effects of data error on the results of any application. There are two separate problems. The first involves devising pragmatic methods for estimating the effects of error propagation within GIS environments. It would seem that this problem has already been solved in that various error handlers now exist; Openshaw (1990). These include partial derivative based methods (Burroughs *et al.*, 1989), Monte Carlo simulation (Openshaw *et al.*, 1991), and probabalistic techniques (Brunsdon *et al.*, 1990). The second problem is more difficult and involves the modelling and prediction of the amounts of localized error in spatial databases. This task is considered hard because nearly all the world's digital map data has been stored without any error information and, there is by no means any standard methodology for measuring localized rather than global error in vector GIS. Furthermore, so far very little basic research has been performed on how to predict error for spatial data stored without it; or even if the errors are known how to measure, represent, and model it. This neglect or error in GIS is considered to be a major scientific problem and one that needs immediate attention.

This chapter presents a methodology for modelling and predicting error in vector GIS databases. The research is focused on the problems of modelling digitization error. It is recognized that this is just one source of error and uncertainty in GIS but it is also an obvious starting point.

13.2 Measuring local line complexity

13.2.1 Local versus global line complexity

The modelling challenge involves estimating error levels in vector data given that the only information available relates to the geometry of the line feature for which the error estimate is required. The scale of the original source document may also be known if this is relevant to the information source concerned or if data from different scales are being combined. The assumption here is that the levels of error in the positional accuracy of the line reflects its geometric complexity but it is recognized at the outset that the relationship need not be linear nor is it a simple linear function of map scale. Indeed, it is possible that operators take more care and work more slowly with highly irregular line segments than they do with simple ones. Likewise long straight lines are highly dependent on the end points being correctly located with very little opportunity for what might be termed measurement redundancy. Other factors that might be important concern the level of cartographic generalization that was applied to the original map source (this might well be map specific, reflecting purpose and design aspects), the thickness and nature of the map lines (i.e. contours may be different from roads), as well as map rotation and subsequent rectification induced distortions. Not all these latter variables can be handled explicitly although the methodology described here will attempt to include them.

In digital cartography, the concept of numerical indices of geometric complexity for cartographic lines has been widely discussed. These indices are used to provide a single real-valued number to summarize the shape of a given line, or to measure difference in shape of a pair of lines. The latter has been applied to assess the effect of line simplification procedures (McMaster, 1986; Jasinski, 1990). Algorithms to achieve this comparison may vary greatly in complexity, from simply measuring the ratio of lengths for a pair of lines to using their estimated fractal dimensionality as a statistical measure of complexity (Muller, 1987). A common property of all of these indices is that they represent a global attribute of the lines in question.

For example, consider the line shown in Figure 13.1. Any of the indices documented in the articles cited above would yield some averaged measure for the line as a whole, not reflecting local complexity in any of the distinct regions. Given that the index is a single dimensional variable, this is inevitable. To describe reasonably the complexity of the line in Figure 13.1 would require at least three indices, one for each distinct region of the line.

Despite this problem, the measurement of line complexity plays an important role in the analysis of map-related data. Complexity measures offer a means of quantitatively analysing the effect of generalization, of line type identification, and of assessing likely error in digitizing. However, in many cases the varying nature of shape within a single geographical feature suggests that the calculation of a single measure may lead to an averaged quantity which lies somewhere between the true nature of region of the line whilst failing to represent any of them adequately.

Poulton (1974) finds experimentally that human subjects tracking a two dimensional signal are prone to an average error that is functionally related to the frequency content of that signal. In more complicated line sections, the error encountered will differ from that expected in simpler ones. Thus, even within a single line feature, the appropriate

Figure 13.1

model for digitizing error (which is closely related to signal tracking error) may be subject to change. This situation could be improved upon if a method for subdividing digitized lines into regions of similar complexity were devised. In this case, a separate complexity index or error estimate could be applied to each section, and a separate set of results could be attributed of each identifiable entity. Taking for example the coastline of the UK, it can be seen that the north-eastern part of England has a mostly straight outline, whilst the western coast of Scotland exhibits much greater angular complexity. This suggests that, in terms of a human digitizer operator, the task of digitizing one of these sections of coastline will require very different physical actions than would digitizing the other. Because of this, it seems reasonable to assume that the degree of error in digitizing each of these regions will be different. It is therefore a fundamental task to identify regions of digitized lines having distinct different characteristics before attempting to estimate the degree of error.

13.2.2 Haar functions: frequency content as a measure of complexity

The desired descriptive indices should measure the local shape of the line as it is traversed. One way of achieving this goal is to use a Haar function. Strong clues to the degree of digitizing error may be found by examining the complexity of the lines being digitized. The notion of feature complexity will most likely exist in the intuition of many GIS users, or indeed of anyone who has ever examined a map. It is less likely that this notion will be envisioned in a mathematical form, or that complexity will be thought of as a numerical quantity. However, numerical information is the stock in trade of computer algorithms, and in order to automate the analysis of digitizing error such a conceptual leap must be made. How can this be done? In this section and the next, methods of measuring line complexity will be considered, together with ways of dividing cartographic lines into distinct sections, or features, having distinct characteristics. This second problem follows on from the first. Having gained an understanding (in quantitative terms) of the complexity or 'wiggliness' of a large map feature, say a river, the next task is to attempt to separate this feature into distinct parts, and assess the digitizing error attributable to these parts (see elsewhere in this chapter).

Although line shape, or complexity can be perceived in many ways, one particularly pertinent approach is to consider the degree of winding that is observed in a particular line. For example, a rugged coastline, a meandering river or a straight expanse of motorway would generally be considered to differ in complexity when viewed on a map, and this difference could largely be attributed to the frequency of turning for each of these objects. It is this frequency that is the key to the quantitative analysis described here. Given a mathematical function, there are various ways of decomposing it into different frequency components. Thus, a function showing a high degree of oscillation (Figure 13.2a) would have a large content at higher frequencies, whereas the function in Figure 13.2b would have a no content at these frequencies, but a large contribution at lower frequencies. However, frequency contribution need not be confined to one region within the spectrum. In Figure 13.3, the function illustrated has high components at both of the frequencies seen in Figures 13.2a and 13.2b. The mathematical task of computing the levels of each frequency in quantitative terms is often referred to as a 'transformation to frequency space'. On a map, line features may be represented of in terms of mathematical functions, and so they may be transformed into frequency space. By examining the outcome of the transformation, estimates of the likely digitizing error may be obtained.

An advantage of this method is that, rather than summarizing the complexity of a line with a single parameter, the feature is transformed into an entire space. Thus, if the error

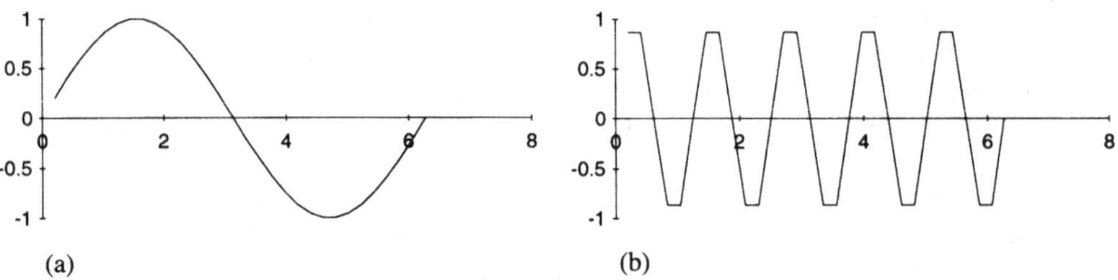

Figure 13.2

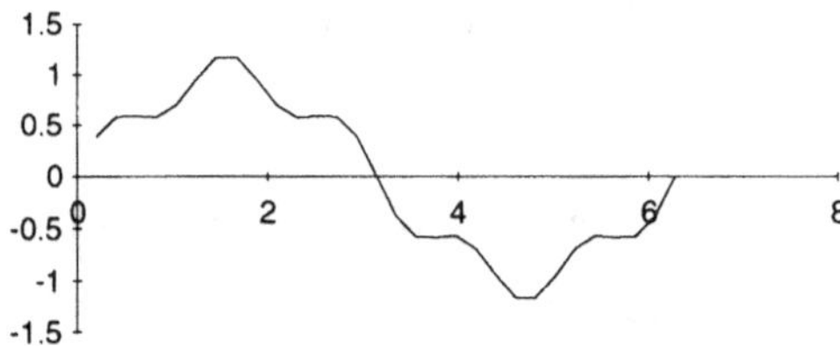

Figure 13.3

is associated with higher frequency components, then both Figures 13.2 a and 13.3 may be identified as being similarly error prone, since despite having different shapes, they would have similar attributes in the relevant part of frequency space. Clearly this could not be achieved if a single quantity had been used to summarize *all* of the characteristics of line complexity.

13.2.3 The Fourier transform

The most commonly encountered transformation into frequency space is the *Fourier transform*. Given a function $f(t)$, the Fourier transform derives a new function, $f^*(s)$ defined as:

$$f^*(s) = \int_{-\infty}^{\infty} e^{isx} f(x)\, dx$$

The newly derived function in s gives information about the frequency component at frequency s. Values of s giving high values of $|f^*(s)|$ correspond to frequencies having a high contribution to f. Thus, by drawing graphs of the Fourier transform of a function, it becomes possible to see which frequencies contribute to the composition of that function. This is often referred to as the frequency spectrum of the function. The frequency spectrum of the function in Figure 13.3 is shown in Figure 13.4. The two high points correspond to the two main frequency components of this curve. It is also worth noting that although the frequency spectrum is depicted here as a continuous curve, this need not always be the case. For example, when $f(x)$ is a periodic function, then the spectrum would be zero for all non-integer values of s. More importantly, when f is not continuous itself, but is tabulated for n points, then the spectrum for s also consists of just n points.

13.2.4 Transforming map lines to frequency space

At this stage it may be seen how to examine the frequency content of a mathematical function, but cartographic lines, even as stored in a GIS in vector form, are not of the

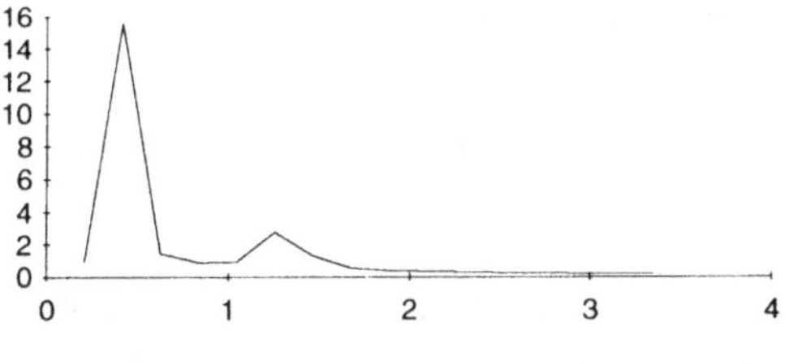

Figure 13.4

$f(x)$ form discussed above. For any given argument, f may only have one value, so that a coastline that doubled onto itself, or the north and south banks of a wide river running east/west could not be represented. A simple way of overcoming this problem is to represent cartographic lines not as a single function, but as a function pair $(x(d), y(d))$. In this case, x and y are functions of t, which could be the distance travelled along a line feature from some point. A typical digitized line in a vector GIS would then be represented as two piecewise linear functions x and y. The joins of the linear sections of the functions would correspond to the digitized points. These points would occur at the same values of t on both of the x and y functions. Since t is a value that always increases as the line is traversed, both x and y are well-defined functions.

In order to investigate the Fourier spectrum of the cartographic line, then the Fourier transforms of both functions in the pair need to be computed. This may be done as before. However, at this stage two frequency spectra will be presented, for both x and y functions. This in itself is not satisfactory; for a single map line, a single frequency spectrum should suffice. How may these be combined to give an overview of the spectrum? One possibility, also suggested elsewhere in this chapter, is to express the digitized coordinates in terms of complex numbers. Here, instead of a coordinate pair, the digitized coordinates could be expressed as the number $x(t)+iy(t)$. The Fourier transform of this complex-valued function of a real number is simply

$$\int_{-\infty}^{\infty} e^{ist}(x)t) + iy(t)\ \mathrm{d}t$$

As before, this gives a function $f^*(s)$ which describes the spectrum of the observed line, when its modulus is taken. An important property of this transform is that, at least in modulus form, it is invariant under rotation. This can be seen more clearly if it is considered that rotation here can be shown to be equivalent to multiplying $x(t)+iy(t)$ by a complex constant k, such that $|k|=1$. This constant may be taken outside of the integral sign, and if the modulus of the transform is taken, then the spectrum of the rotated map feature is just $|k|\,|f^*(x)|$. Since $|k|$ is 1, this is just the same as the spectrum of the unrotated feature.

This property may appear to be a mathematical nicety, but is in fact of much more practical value. Maps are sometimes drawn at unusual angles, and are seldom fixed to digitizing tablets perfectly parallel to the grain of the electronic censors below the tablets surface. Thus, the angle of the map, the angle of the digitizer and the orientation of the real-world feature under examination are seldom in perfect agreement. Any means of representing or measuring line shape that did not have this property would yield a measure of line complexity that would be liable to change whenever there was a change in the angle of the map on the digitizer, or an unusual map was selected. Clearly, such a measure would be unsatisfactory, and this mathematical property has relevance to practical problems of error estimation in GIS.

13.2.5 Global vs. local characteristics

In the previous sections, a means of measuring the frequency spectrum for cartographic line features has been presented. However, as discussed elsewhere in this chapter, there is one major shortcoming that has been overlooked. The spectra derived here are global so that they reflect aggregate properties of the cartographic features as a whole. This is reasonable when dealing with distinct features, but is often not helpful when presented with a large number of linear features, such as a network of rivers of varying complexity. Some of the features in this collection may have one characteristic, but others may differ. Sometimes, more importantly, characteristics will alter within a single digitized line feature. In these cases, the use of Fourier transforms will be particularly unhelpful, because the spectra will give information about aggregate properties, not of individual regions within the lines. To overcome this, local characteristics of lines, rather than Fourier's global frequency space transforms should be adopted.

13.2.6 The Haar transform

At this point it is useful to consider another spectral transformation, the Haar transform. In the form considered here, the Haar transform works in a similar manner to the discrete version of the Fourier transform. A list of values (effectively an n-dimensional vector) is transformed into another n-dimensional vector, which presents spectral information about the function being transformed. The Haar transform may be computed as a series expansion, which is also the case for the Fourier transform. However, while the Fourier transform expands functions in terms of trigonometric series, the Haar transform expands in terms of a series of Haar functions:

$$f_i = \sum a_{k,j} h_j^k(i)$$

Here, $h_j^k(i)$ is a Haar function of the integer value i. This is defined as follows:

$$\begin{aligned} h_j^k(i) &= 1 \quad \text{if } 2^j < i \leq 2^{j+k} \\ &= -1 \quad \text{if } 2^{j+k} < i \leq 2^{j+2k} \\ &= 0 \quad \text{otherwise} \end{aligned}$$

Thus, Haar functions may be thought of as a form of square waves, but that each wave only exists on a small section of the domain of the function. Thus, the coefficients of the Haar function do correspond to frequency components as k varies, but j also gives information about the position along the function to which this frequency corresponds. Since the function values are zero outside of the section of domain mentioned above, the value of $a_{j,k}$ gives information about the frequency content within this window, but in no other place. This appears to have the desirable property that local spectra, not based on the entire functional domain may be computed, unlike the Fourier transform. It must be pointed out, however, that this is not without cost. The frequencies for which a coefficient may be computed are now restricted to integer powers of two, whereas for the Fourier transform applied to a discrete function frequencies may be calculated at all integer frequencies. For the problem in hand, it seems that the restriction on frequency components for the Haar transform is the lesser of the two equals.

As before, at this stage the discussion has considered mathematical functions rather than cartographic lines (as stored within a GIS). Fortunately, the approach may be transferred to a cartographic measuring procedure by expressing lists of digitized coordinates as a complex function, $x(t)+iy(t)$, as before. Again, by considering the modulus of coefficients of a particular Haar function, a spectrum of the cartographic line expressed as two functions may be obtained. This operation also shares the rotational invariance property of the cartographic Fourier transformation. Thus, the Haar transform provides a rotationally invariate method of measuring the frequency components of mapped lines, with the useful ability to identify local changes in frequency content. This may be considered as a basis for identifying distinct sections within cartographic lines, alongside the neural network techniques discussed later in the chapter.

13.2.7 Smoothed length-angle indices

One problem with Haar functions is that they relate to a complete line segment. Line segments are arbitrary entities, they need have to natural meaning, and are modifiable. It is the wrong object for analysis even if it provides a good summary description of localized line complexity. Something much more localized at the subsequent level is needed. It would be best if this index was calculated at every point along the line and reflects local changes between points. An analogy with a time series springs to mind with the ordered (x, y) pairs forming a point line series. What is required is a type of data transform capable of enhancing the differences between points where coordinates lie in different regions of complexity regardless of the density of points in the neighbourhood. A good choice of transform will help sharpen any distinct patterns that occur in the line data and thus improve the performance of any pattern recognition technique.

There are a number of possible line complexity indices that could be used. After some experimentation it was decided that a length angle pair of indices provides a good representation of line complexity, as well as having some relevancy to the digitization of the nature of the line; highly irregular segments will tend to have small inter-vertex distances. However, this measure is by itself not sufficient; for instance, in stream digitizing there is a constant distance option. Also the coordinates may have been generalized, which affects point density. Inter-vertex length is not the only variable that may well be related to digitizer error. The windiness of the lines is also likely to effect the amount of concentration and digitizer error. A good measure of local windiness is the angle between successive line segments subtended at each vertex. The more wiggly a line is near to a digitized point, the greater the angle turned through. As a means of examining complexity, it also has the desirable property of invariance under rotation. This is an intuitively sensible property for a complexity indicator, since a pair of lines identical in all aspects except orientation do not differ in frequency of oscillation, camber, angularity or any other aspect of shape. Indeed, the transform could be viewed as a filter which removes the orientation information from the data, leaving mainly information about shape.

For the digitized line of Figure 13.1, a scatter plot of length against angle turned is shown in Figure 13.5. Three distinct clouds are visible, with a large density cluster inside the second. The first corresponds to the lower part of the line, where generally greater changes in angularity occur. This also requires that shorter distances occur between digitized points on the line, in order to capture the closely-packed windings. The second and third clouds correspond to the lower section of the feature. Here, the complexity or windiness is not as great, so there is less of a range in the angles turned, and the distance between digitized points is reduced accordingly. The higher cloud of these two appears to correspond to the jutting feature of coastline in the middle of this section. Finally there is a densely-packed cluster of points above these. This is near the axis corresponding to the straight section at the end of the line. In order to capture the slight non-linearity of the feature, digitized points are not much farther part than in the second section, although the angles between line segments are obviously smaller.

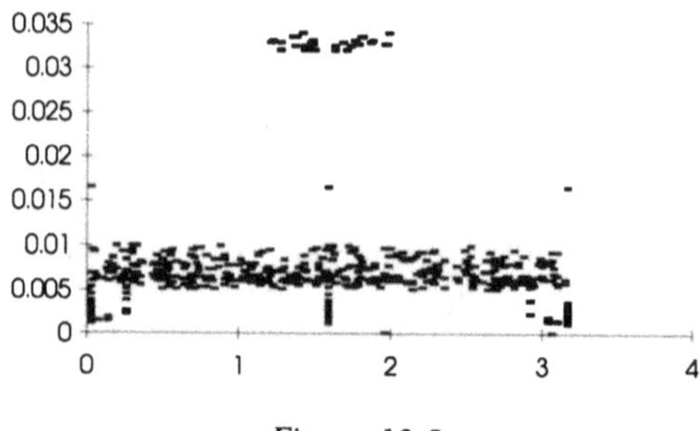

Figure 13.5

Thus, the length and angle plot helps to identify local shape changes within the digitized lines. However, there are still some shortcomings. The dense cluster of points associated with the straight section is encapsulated within the more diffuse cluster relating to part of the other section. It is therefore difficult to decide which cloud a point in the region of the denser part of the plot belongs to, since both will have points in this area. The data needs further processing in some way to attempt to overcome this difficulty. The main problem is that the distributions of angular change in both segments have ranges of angles in common. This is only to be expected, particularly in stream digitizing situations, where the position of the digitizing cursor is automatically recorded at regular intervals. If these intervals occur when the map line is not on a turning point, a near zero angle will result. This could occur on any map line in some instances, regardless of shape. For example, a line having alternatively long and short segments or a spike that occurs at every five or six vertices would have a very distinctive shape but local indices based only on measurements taken at each vertex have oscillatory behaviour. There is an interesting dilemma here between a global view which loses too much detail and a vertex-orientated view which retains too much.

Fortunately, this problem is avoidable. Although the angles at individual points may fall in the same range for different shaped line segments, the mean angular change will differ for each segment. A coastal region having a large number of spurs and peninsulars would have a higher count of large angles than a relatively straight region, and this in turn would lead to a higher average angularity. At this point, though, there is some danger of a circular argument. If the distinct regions can be identified, their mean angularities can be calculated; but the reason for calculating mean angularities is to identify distinct regions! A way of avoiding this paradox is to work with moving averages. If an average is associated with each point, taking into account itself and a small number of points on either side then a mean value of angularity in the neighbourhood of each point will result. This will have the effect of smoothing the angularities in the direction of their local regional average. Thus, the near zero values encountered previously are less likely to occur unless the line segment is very smooth.

It would seem that applying moving average windows to the angle/length data may enhance the ability of plots of the sort shown in Figure 13.5 to discriminate between different shapes of line. There is, however, one final problem of the method to be overcome. One of the effects that a moving average has when applied to an ordered sequence of data is to smooth out sudden changes. Thus, as the digitized points on a line move from one different region to another, there may be a relatively slow ramp in angularity when computed as a moving average. The angles from the previous section may still be affecting the mean values well into the current section. To some extent this is unavoidable. If there were no smoothing the previous problems would result, but if there is smoothing, the response to sudden change is damped. However, the effects can be limited by the use of weighted moving averages. In this case, more weight is given to the angle associated with the points closer to the centre of the moving window. Thus, while some degree of smoothing is applied, the effect of the more distant points is downweighted. It is, of course, these distant points that are most likely to come from a different shaped line segment. A possible weighting scheme is illustrated in Figure 13.6. The values roughly follow a Gaussian curve, over a window of 10 points in each direction. While no claim is made for the optimality of these values, they have been found to work well in practice.

The scatter plot of the lengths and angles after they have been smoothed as suggested above is shown in Figure 13.7. The clouds corresponding to each of the different shaped line segments are now clearly discernible, and occupy distinct regions of the scatter plot. The ambiguity of Figure 13.5, when one cluster was located within another, has now been removed. It is now possible to subdivide the plotting space into distinct regions, and assign a shape classifier to each point in the plot according to which of the regions it lies within. This is a relatively simple task for a human being but for use in a GIS some kind of automatic pattern recognition procedure is needed.

13.3 Dis-assembling line segments using a neural net

13.3.1 Design

The averaged length and angle measures provide a two dimensional descriptor of line shape for any cartographic line. The next task is to classify the line sections into chunks that share a common local line complexity. Indeed in Figures 13.2 and 13.4, a number of points were seen to be in different regions of the scatter plot, and by identifying regions of different shape within, digitized lines could be found. The process of labelling points according to

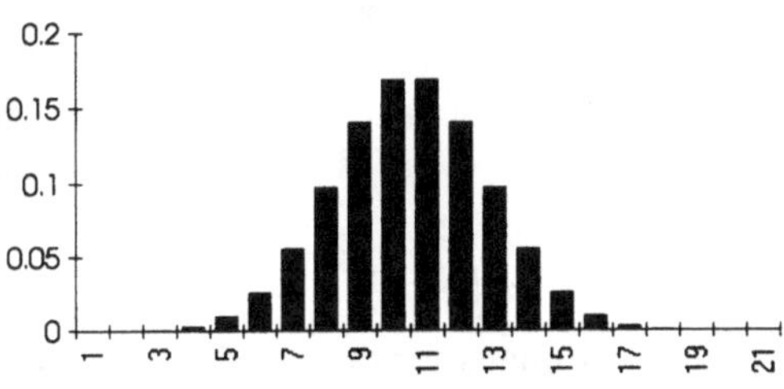

Figure 13.6

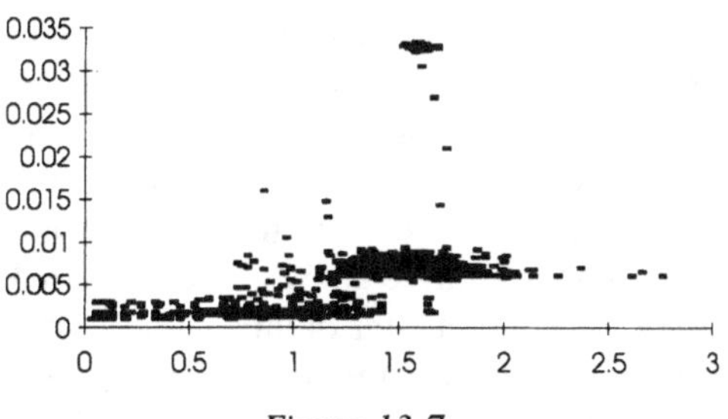

Figure 13.7

which grouping they belong to may be thought of as a form of digital quantization (Gray, 1984; Nasrabadi and King, 1988). The point clouds in the figures could be caricatured by a smaller number of points (say one to three for each cluster to suggest their shapes). If it is necessary to decide which cluster a given original points belongs to, this could be done on the basis of finding which of the caricature points is its closest neighbour. If this approach is to be taken, the problem of automatic assignment of each point to a particular category could be split into two stages. First, find a small set of points which caricature the scatter plot reasonably well, and then classify the data in the scatter plot according to these.

13.3.2 An unsupervized neural net

The first of these problems, the caricaturing or quantization stage, has been the subject of recent study in the field of neural computation (Kohonen, 1989; Ahalt *et al.*, 1990). Each of the points in the caricature set may be thought of as a neuron and its position coordinates as weights. The set of all of these points is a neural net. The general problem in neural computation is to train the neural nets to respond to input data in some desirable way. The output of a neural net comes from subset of the neurons, which fire (i.e. yield some positive output) if their input stimulus is sufficiently large. The weights control the neurons sensitivity to the inputs, and it is be progressively adjusting these weights that the nets may be trained to respond in the required manner. In this case, the input corresponds to one of the points in the cloud plots. The output will come from a single neuron, one of the caricature points, chosen to be nearest to the input point. Note that only one output unit ever fires in a single event and that there is competition to decide which unit is to fire, so the units are often referred to as winner-take-all units. It is also worth noting that in this example there is no prior information about cluster labelling. Given simply the smoothed length and angle values, the net must decide for itself what clusters there are in the data. There is no calibration data set with known cluster classifications. The technique is therefore referred to as unsupervized competitive learning.

Initially the weights will be set in some random way. The training process will consist of randomly selecting points from the input data set to be fed into the net. Alternatively, they could be fed in turn. When a point X is fed in, one of the units, i, with weights W_i will fire, where

$$[W_i - X] < [W_j - X] \quad \text{for all } j$$

The learning step is to update the weights for the winning unit only, by making it closer to the current input X. This will make this unit more likely to win in the future when the same input or one similar to it occurs. One method for doing this, the standard competitive learning rule, is to update weights in the winning unit by W^* where

$$W^* = (1-a)W_i + aX \quad \text{with } 0 < a < 1$$

The value chosen for a controls the level of response of the network. If a is close to zero, the weights will adapt very slowly to the input. If it is close to one, the weights may over-respond. This may be particularly problematic when the weights are already reasonably well positioned, since the oversensitivity may cause them to jump about within the area that they are classifying, as different points from that area are input. It would seem that initially larger values of a are more useful, to allow the network to position weights rapidly in a roughly representative way, but that later on this value should be reduced, allowing the weights to settle. This suggests that some form of reduction schedule for a should be incorporated into the learning algorithm. A simple possibility would be to reduce a

linearly (and fairly slowly) and stop the algorithm when it reaches zero. This method has been found to work well in practice.

A six-element network was used to analyse the data points shown in Figure 13.7. The initial weightings of the units were chosen to be uniformly distributed random points within the length and angle space shown in figures 5 and 7. The units were fed random selections from the data set, and their response was reduced linearly, as suggested in the last section. The positions of each of the units is shown in Figure 13.8. With six nodes, two are used to indicate the shape of the upper elliptical cluster, two the lower cluster, and a final one indicates the smaller and more densely packed collection of points. This leaves one node which has not fired sufficiently to be close to any of the clusters. This is quite a common phenomenon. In some cases, adding further output nodes corresponding to more output classifications than are necessary to reflect the structure of the input information may result in these nodes becoming redundant (Openshaw and Wymer, 1990).

The classification of each point depends on its position in the length and angle plot, and to which of the neurons it is closest. The structure of the classification provided by the analysis may be viewed as a Voronoi diagram constructed from the neural points. This provides a partition of the space in terms of the neurons for which each area is nearest. This is shown in Figure 13.5 by dotted lines. It can be seen that one of the regions contains no points from the data set, due to redundancy, but that the dense cluster is entirely within one region, and the other two regions are contained by the unions of pairs of regions. Thus, the network has yielded a classification scheme which sub-divides the space to contain unique groupings, in some cases more than one neuron, and hence more than one region has been used to provide classification for a single 'cloud'. The joining together of regions which represent the same grouping gives a reasonable geometrically adaptive means of partitioning the space.

There is a final problem before the classification is complete. Since the neurons are initially positioned at random, and no classification information is given during the training stage, there is no direct way of identifying cases where more than one neuron is used to classify the same cluster. Thus, joining together regions will give good classifications, but it is not immediately clear from the network output alone which of the regions should be combined. This problem can be overcome by considering the sequential nature of the data. The points in the plot can be ordered by their sequence in the original map line. When considering this order, it is expected that the earliest points will all lie in one particular cluster, corresponding to the local shape at the beginning of the line, and then move on to the next cluster for the next shape change and so on. Thus, rather than skipping at random between all clusters, the points will tend to jump from cluster to cluster only a small number of occasions. However, within the cluster such strong patterns of behaviour would not be expected, and while within a cluster it may not be unreasonable to expect fairly random

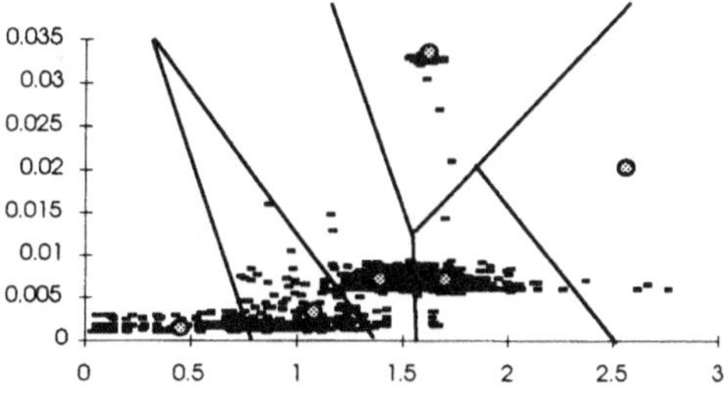

Figure 13.8

transitions between subregions. For example, in Figure 13.8, there may be several crossings between some regions (which together cover a cloud), but only one between other disjoint regions. Thus, by examining the regions occupied by each point in sequence some useful information about multiple region group labelling may be obtained.

In Table 13.1, a frequency count of regional transitions is given. As was expected, most of the transitions occur in the leading diagonal, suggesting that jumps between groups are relatively uncommon. If the distribution of frequencies of crossings in all by the leading diagonal are considered, two pairs of outliers emerge. Firstly, there is a very large number of crossings (47) between regions 2 and 3. There are also 5 crossings between regions 4 and 5, and although this is much less than between the other pair, it is still unusually high. All other edge crossings have frequencies of either zero or one. This suggests that the edges between regions 4 and 5, and 2 and 3 should be dissolved. If the redundant node is excluded from the analysis, there are now three distinct labels that could be attached to each point. Applying these labels to the points on the original digitized line, the classification shown in Figure 13.6 is obtained. From this it is clear that three distinctly shaped subsections of the line have been selected. In Table 13.2, one measure of line complexity, the average angle between segments is calculated firstly for the entire line, and then for each of the three subsections.

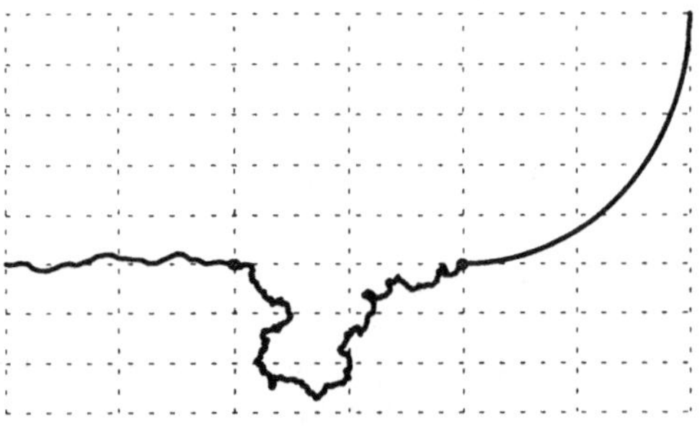

Figure 13.9

Table 13.1 Sequential transference between classification groups

To group	From group 1	2	3	4	5	6
1	0	0	0	0	0	0
2	0	608	0	0	1	47
3	0	0	118	0	0	0
4	0	0	0	39	5	0
5	0	0	1	5	160	0
6	0	47	0	0	0	333

Table 13.2 Local vs. global line complexity

Average angle (radians)	Region
0·55504	Entire line
0·66053	Section 1
0·36962	Section 2
0·03672	Section 3

From this it is clear that the global indicator does not reflect the three distinct values associated with the three identified line subsections, lying between the values for sections 1 and 2. Thus, the method has succeeded in identifying regions having distinct shapes within the supplied line segment, which may be used for local tagging of complexity.

13.3.3 Implementation

The neural net based line dis-assembler needs to be built off-line. There are only a finite number of possible vertex length and angle values; most of which can never occur. The neural classification would be built using a reasonable complex digital map data-base. Provided there was sufficient diversity then the resulting net would provide a very good nearest neighbour classification. The resulting trained net is easily implemented in a GIS and can dynamically dis-assemble line segments input to it. Accuracy could be improved by broadening the range of training data used in the classification, although the nearest neighbour nature of the results would probably be sufficient to provide an adequate generalization capability. This net-based approach offers the following advantages: speed of application; sensitivity of the probability density function of the training data; simplicity; fuzziness is retained. The classification can be made self-monitoring, line segments that belong to more than one line segment type could be given hybird values; and a monitoring file built of poorly handled exceptions so that performance increases through time.

13.4 Making estimates of error using a neural net

13.4.1 Openshaw's conjecture

The line dis-assembler will classify any line segment in terms of local complexity with some regard for context. It only now remains to establish plausible error margins for each line type. These may also be map scale dependent and they could even be map source specific. Valid error distributions could be established by experimentations since idealized representative line segments belonging to each line type could be generated. Error and uncertainty information could then be associated as attributes of data by a simple look-up operation.

The conjecture is that there will only be a relatively small number of line segment types with different error characterizations. The resulting error database once established could be loaded with a GIS, and revised as experience increases.

13.4.2 Brunsdon's conjecture

The alternative model is that of an infinite range of error characteristics that are source specific, scale specific, and not easily generalized. A supervized neural net would be needed to predict error levels from various attributes measuring the characteristics of the disessembled line segments. This is a non-parametric regression approach that involves additional levels of complexity. In particular, a back propagation net would have to be trained to map the function that relates predictor variables based on line features with experimental estimates of positional error.

According to Brunsdon and Openshaw (1993) this strategy would work best in an interactive situation, in which as part of the data capture process, error modelling is also being performed. It is possible to imagine that this would become an integral part of the digitization process, whether manual or automatic. It is clearly less useful as an off-line process.

13.4.3 A hybrid approach

There is no reason why both conjectures cannot be combined. If need be, or warranted by the nature of the application, the line type error information could be represented as a trained supervized neural model. There are two possibilities:

1. The error information used for a particular application could be stored not in table form but as a neural model. Key aspects of the application (i.e. map scale, qualitative indicators of quality of data, etc.) could be used to generate a look-up table.
2. The global look-up table would itself have been generated off-line by a neural net trained to represent the function relationship between line geometry and error.

Clearly, the first option is the ideal long-term goal. However, both need considerable quantities of experimental data from which to model error. The data needs to be broadly representative of the whole range of digital databases likely to be used in GISs. This will take several years to establish and requires a corporate and co-operative effort by the GIS community as a whole. It also assumes that established error handlers exist which can exploit the information. So maybe option zero is to start with a simple fixed table, perhaps even of guessed values, with a view to improving precision as the error handling technologies develop and the necessary error modelling data bases and libraries are established.

13.5 Computing techniques required

The approaches advocated here require the application of several new techniques, many of which were devised by the authors themselves. Thought must be given to the development of new algorithms, and the environment in which this is to take place. First, we will consider the latter problem. Although generally neural network algorithms exist, it is neccessary to place these in a cartographic context. The choice of a programming language and software development system that can handle cartographic data with ease could provide a significant contribution to the success of the algorithms used. If a less suitable setup were to be chosen, a considerable amount of unnecessary effort would be spent coercing the language to handle this kind of information. Thus, before embarking on any analysis, some consideration of the requirements of the software used could pay for itself many times over in terms of intellectual effort. Taking heed of this a 'wish list' of desirable properties has been drawn up for this research. These properties are: the ability to handle two-dimensional data effectively; the ability to handle lists of data and vectors effectively; interactive graphics support; reasonable software debugging and diagnostic utilities.

The authors chose the XLISP-STAT (Tierney, 1991) software package, as meeting all of these criteria to a good degree. Throughout this section each of these characteristics will be dealt with in detail, illustrating how XLISP-STAT can use them to advantage in tyical examples.

The other, more fundamental issue to address is the choice of the algorithms themselves. Considering the two general approaches outlined earlier in this chapter, there are several main algorithms to be provided:

1. a taxonomical tool, which can assign a general category to a set of line shape parameters;
2. a feature identifier, capable of splitting a cartographic line into distinct features;
3. a non-parametric regression tool, modelling the relationship between several 'predictor variables' and one 'response variable.'

In addition to these, there are several interstitial techniques that will need to be combined with these to achieve a full working system. As suggested earlier, the taxonomical work can take place using an unsupervised neural network approach (Hertz *et al.*, 1991); non-parametric regression can be dealt with using several techniques (Tapia and Thompson, 1990) for a discussion), but in this case the supervised neural network is proposed. The final procedure mentioned above is the identification of distinct line features. This is perhaps the most distinctly geographical problem. This will require the handling and manipulation of cartographic lines as a data type, and since it involves classification, will also have to call on the taxonomical procedure. An outline of the approaches to these problems using XLISP-STAT will also be contained in this section. Thus, the development of the algorithms will be considered in two parts; first a discussion of how various aspects of the XLISP-STAT package are used to handle geographical information and implement the neural network algorithms, and then a discussion of the implementation of the algorithms themselves.

13.5.1 The XLISP-STAT package

At this point, it may be helpful to provide a brief description of the XLISP-STAT package, since it will become an integral part of discussions on algorithms and implementation in this section. XLISP-STAT is a graphics-based implementation of the common Lisp (Touretsky, 1989) language. It is capable of running on IBM PC compatible micros, under Microsoft Windows, can be compiled to run under X-Windows or on Macintosh computers. The version of common Lisp, XLISP, was developed by Betz (1988) who placed it in the public domain. As the title of the package suggests, statistical functionality has also been added to the basic Common Lisp standard, by Tierney (1991). Another important feature of the implementation its graphics capability. It is extremely easy to create spin-plots (Tukey, 1977), linked plots and histograms, and various other sophisticated forms of interactive statistical graphics. In addition to this, it has the full functionality of the LISP language, which includes list processing, functional programming and manipulation of functional information using lambda notation (see later), and a complex number data type. Also, the language is interpreted and controlled from a command line (in a particular window) so that Lisp expressions can be entered and tested interactively, and the values of all variables inspected allowing the development of algorithms to take place dynamically, without the compile–link–run cycle of compiled languages. Some of these features will now be considered in relation to the geographical information processing techniques required for this problem.

13.5.2 Graphical aspects

When working with map-based data the ability to see the geographical features rather than just the numerical information representing them is invaluable. For example, in terms of feature identification, some visual indication as to which distinct features have been detected on a particular line is the most fundamental, and perhaps most powerful diagnostic technique available. Clearly, the ability to turn geographical information quickly and interactively into some graphical representation gives insight into the functionality of any algorithms that are applied the data. In XLISP-STAT, this can be achieved reasonably easily, using the 'plot-lines' function. At the command line, entering

```
(plot-lines x y)
```

will create a window and plot a graphical line whose x-coordinates (or eastings) are stored in the list *x* and whose *y*-coordinates are stored in the variable *y*. A reference for this window can be stored in another variable (called 'map') for example, and subsequently further lines (or points) can be added to it, or other changes made. In this way, to test a feature detection algorithm, it is possible to draw parts of the line seen as distinct features in different colours, or to plot points on the line delimiting the end points of each characteristically different line segment.

As well as investigating line segments, graphical windows can also provide diagnostic tools in cluster analysis and taxonomy. Recall from earlier that unsupervised neural networks may be used to provide a classification for a set of multidimensional data points. If the points are in only two dimensions (which have, as stated earlier, provided satisfactory results) then a scatter plot showing all points, and their classification (perhaps by colour coding) demonstrates whether a classification algorithm has performed in a satisfactory manner. Again, if the two-dimensional space consists of two local line shape indices, say 'param1' and 'param2', then the XLISP-STAT command

```
(plot-points param1 param2)
```

would provide the appropriate scatter plot in a new window. Again, the reference to this window could be stored in a variable, and it would be reasonably simple to change the colour (or plot symbol) of points according to their classification. If more than two parameters were supplied, some visual techniques are still possible. In particular, in three-dimensions, a spin-plot is possible. This is effectively a three-dimensional plot of the data, with the ability to be rotated about each of its axes interactively. This allows the detection of clusters in the data, and again, the effectiveness of a classification can be evaluated by colour coding the points.

A final useful feature of the graph-based diagnostics is the ability to link several of these graph windows. If the graph window references for plots showing the map line and its local shape parameters are stored in variables called 'map-space' and 'param-space' respectively, then by entering

```
(link-views map-space param-space)
```

the two windows become 'linked'. This implies that if a point, or set of points is selected in one window using a mouse on the PC or workstation, as well as being highlighted in that window they will also be highlighted in the other. Thus, a particular cluster of points in the local shape parameter space can be linked to the corresponding points on the digitized map line. In this way, different pairs of shape description parameters can be experimented with, and their effectiveness at detecting features can be assessed.

The main utility of all of these techniques is the ability to apply graphical diagnostics to algorithms undergoing construction. Although many of the graphs could be produced outside of the development software (with the possible exception of the linked plots) they could not be done so with this degree of immediacy or interaction. Taking advantage of the interpreted nature of the XLISP language, if an algorithm were to fail unexpectedly, not only could one inspect suspicious variables, but could actually draw them in their state at the time of failure. This kind of debugging is a vital step forward when handling geographical data, which is essentially 'drawable'.

A final point to be made here, illustrating the importance of choice of development environment, is that it is possible to produce linked plots in just three lines of code, due to the features already provided. To implement a similar feature in a C or FORTRAN compiler might require several thousand lines of code. Not only does this system provide a graphical debugging feature, but it also provides easy access to it. This is perhaps the one most powerful attribute of the system for geographical software development.

13.5.3 A list-based approach

The last subsection dealt with the issue of debugging and diagnostics of any software developed, but clearly we must also deal with the develpment of the algorithms themselves. Under the next few headings, aspects of the Lisp language that enable efficient implementation and development of geographical data handling will be considered. The first, and most obvious of these is the use of list handling techniques. The name Lisp is a compaction of the phrase 'List Processor', and as this might suggest, the language has inherent abilities to handle lists of information. Why is this valuable to geographical information handlers? A list is an ordered set of items, of any size. A digitized map line can be described in this way. It is a set of (x,y) pairs, stored in the order that these pairs appear on the cartographic line. A language providing features for handling lists can therefore apply them this sort of geographical data. The sort of features that are offered are the ability to pick out the *n*th item in a list, to join two lists together (for example two sections of coastline) and to apply a mathematical function to each element of the list in turn, yielding a new transformed list.

The power of lisp is that the syntax for such actions is considerably more compact than in many other languages. To represent a list of items, each item is placed in order, separated by spaces between a pair of brackets. Thus

```
(1 2 3 4)
```

is a valid list. To apply a function to each element of the list in turn, one would enter

```
(mapcar #'square '(1 2 3 4))
```

which would return the transformed list

```
(1 4 9 16)
```

therefore, to apply what is effectively a loop structure to a list of numbers can be achieved in only one line. Many operations come ready 'vectorized' in XLISP-STAT so that

```
(square '(1 2 3 4))
```

would yield the same result. This vectorizing also applies to dyadic operators such as +, −, * and /. These will be applied element by element to a pair of lists of the same length, giving a new list of each result in order. Again, these operations would normally require a loop in many languages. The relevence of the above to this research is that mathematical operations of this sort can constitute a large part of the computation of the local shape description indices, and of both the supervised and unsupervised neural network algorithms. This means that alogorithms can be specified more efficiently, and also altered more rapidly if improvements are to be made.

A final, and rather striking, advantage of the Lisp language is the fact that a program (or more correctly a function definition) in Lisp is itself a list. It may be thought of as a sequence of actions to be performed on some input data. A list in Lisp that specifies such a sequence of actions is referred to as a lambda expression. Since Lisp functions may also be thought of as lists, it is possible to pass them as parameters to other Lisp functions. At first this may sound esoteric, but turns out to be of great value here. For example, the use of window-based methods (such as a windowed mean) passed over a set of data can give a reasonable indication of local trends in a list. This could be applied here to, say, angles between line sections in a digitized boundary. However, there may be some doubt as to which operation should be applied to the moving window. At times, a mean is applied, but a median may also be of use. Similarly at other times, a measure of texture, such as a windowed variance, may be useful. In all of these cases, the general moving window idea is apparent, but in each case the function to be evaluated changes. Using a lambda expression as one argument, a general moving window operator could be programmed, applying a generic function across the window, and the actual windowing operator (which could vary of course) could be supplied at each individual call of the function.

13.5.4 The use of complex numbers

It is a surprising fact that despite the inherent two dimensional nature of much geographical data, there has been very little use made of complex numbers to represent and statistically analyse cartographic points or lines. In this particular study, the algebraic and algorithmic treatment of cartographic lines has been greatly enhanced by their use. A major advantage of complex numbers is that they may be directly added and subtracted, this being equivalent to measuring spatial displacement. The modulus of the difference of two complex numbers is the distance between them, and the argument of this is the bearing of one position from the other. Also, multiplying a complex number by another with unit modulus is equivalent to a rotation. Multiplying by a real number simply re-scales the coordinate. Thus, when working with complex numbers, simple operations such as multiplication and subtraction cover a wide range of mapping operations.

The utility of complex numbers has value added when working with lists also. A cartographic line can be thought of quite simply as a list of complex numbers. To rotate a cartographic line, then, it is necessary only to specify a multiplier (see above) and multiply the list representing the line by this quantity. As with much of the list-based advantages, the utility of adopting a complex number-based approach is in the brevity of notation. Suppose x contains a list of complex numbers representing a particular cartographic line. Applying the 'delta' function to the list returns a list of first differences, that is, a set of two dimensional pairs (in the guise of a complex number) representing each digitized line section. To find the length of these sections, the modulus function 'abs' is applied to the list. This function has been vectorized, so that the entire operation can be expressed in lisp as

```
(abs (delta x))
```

which is clearly more compact than in many languages. Here, it is the combined utility of complex notation and list processing working together that is of value. An expression of this sort could then be fed into a moving window function, as described above, to give a local shape measure based on a moving average of line segments. A measure based on angles of line segments would not be much more complicated. Here, the argument, rather than the absolute value, is of interest. This can be obtained by applying the 'phase' function. However, this in itself only gives the bearings of the line sections. It is the change in these that are of interest, as these measure the angles between lines. These can be measured by applying the 'delta' function once more:

```
(delta (phase (delta x)))
```

Now, as before, the resultant list of angles can be used as input to a moving window function, and measurements of the 'windiness' or texture of the line become possible.

The use of complex numbers is also of value in the algorithms related to neural networks. Although these are generally more lengthy than the shape index expressions, it is possible to express many of the operations considerably more succinctly when working with lists of complex numbers than would otherwise be possible. It will be seen, for example, that an important stage in the unsupervised neural network algorithm is the discovery of the nearest point from one set of points (the neurons) to another single data point. Again using vectorized operations, and lists of complex numbers, it is relatively easy to find which point in the first list is nearest to the data point. Generally, although these algorithms are generally more complex, they may be broken down into a sequence of operations of this sort, and therefore expressed in a reasonably simple manner in Lisp. In the next section, specific issues of implementation will be discussed.

13.5.5 Neural algorithms: unsupervised neural networks

As discussed previously, both supervised and unsupervised neural algorithms will be utilized at various points in the project. First, the unsupervised case will be considered. As suggested earlier, these are essentially classification algorithms. The points to be classified are, at their most general, at set of points in n-dimensional Euclidean space. Here, n is generally two, and the points are sometimes thought of as lying in the complex plane. Each neuron contains n weights, also thought of as representing a point in the same space as the data. Each data point is input to the network of neurons, and the neuron closest to the point is said to fire, as suggested earlier. When training is considered to be complete, a new data point is input to the network, and its classification is determined by the identity of the neuron that is closest to it in Euclidean space. In Lisp, both the set of points being classified and the list of neurons can be represented as lists of n-tuples. A further Lisp function can provide a classification for each point in the data set, based on the neuron list. Using some of the linked plotting techniques suggested earlier, it is possible to identify members of each class by plotting different coloured symbols, and to link these back to the original digitized nodes on the map. This allows an intuitive form of visual assessment to take place, when feature detection algorithms are under development.

13.5.6 Neural algorithms: supervised neural networks

Supervised networks may be thought of as performing the task of non-parametric regression. The training set this time consists of inputs and desired outputs, and the network is trained to respond to inputs by producing outputs as close as possible to the desired outputs. After training has been completed, it is hoped that when new inputs are presented to the system, the outputs produced will be good predictors of the true values of the output variables. In our case, line complexity indicators are to be input, and the output will be confidence bands that may be placed around digitized map lines.

The technique used here is the back-propagation method (Rumelhart and McLelland, 1986). The network here can be thought of as several layers of neurons, the output of one layer feeding into the input of the next. The first layer takes the data as its input, and the final layer gives predicted values as output; for more detailed discussion, see Hertz *et al.* (1991). Each layer consists of an array of neurons, whose output is a function of a linear combination of its inputs. The weights in this case are the coefficients for this linear combination. The concept of back propagation is that the weights of the output layer are adjusted so as to make the actual outputs closer to the required outputs. However, not only may the weights be adjusted, but also the input values. Recall that these are the output values from the previous layer. Thus, to adjust these, the weights for this layer must be altered. The requirements for adjustment, based on a steepest descent optimization algorithm, are propagated back from the output layer to the previous one. This back propagation may be repeated through the network backwards, until the input layer is reached.

A back propagation network is a considerably more complicated entity than an unsupervised network. As well as values required for weights and outputs of the neuron, methods (effectively functions assigned to an object) are supplied to carry out the forward propagation (for prediction and training) and back propagation (for training) actions. In this case, entire layers of neurons are supplied as a subordinate object, and a network object is a list of these with some extra methods. In this case, it is possible to exploit the use of lists, vectorized functions and also of matrix operations, to simplify the algorithms used.

It should be noted that the price that is paid for using an interpreted language is speed of execution, and that supervised back propagation networks require much more training than the unsupervised examples discussed above. In practice, it is the case that the use of Lisp for this type of application is only of value in the software development stage, where relatively small test data sets are used.

13.5.7 Feature identification

Having addressed the implementation of both supervised and unsupervised networks, the final major problem is that of feature identification. In simple terms, this is the task of partitioning the cartographic line into distinct sections, so that each section has a clearly identifiable shape characteristic. In previous parts of the chapter, the notion of obtaining shape indices from a cartographic line segment has been discussed, as has the concept of classifying these indices using unsupervised neural networks. These will be the building blocks of feature identification. Starting with the line segment, first the shape index functions transform the set of digitized points defining the line onto a set of points in the index space. In this space, the classification techniques take place, and a category may eventually be attached to each point in the data set. The final part of the task is then to transfer these classifications back from the index space and onto the map line. In this way, a shape classification is attached to each point on the digitized line.

Once this has been done, points whose neighbours on the line have a difference classification to themselves can be identified. Using list-based techniques this is relatively simple. One possible way to achieve this is to take a delta function of the classification list and identify its non-zero elements. If classes are stored in the variable class, then the lisp expression following achieves this:

```
(/=0 (delta class))
```

If a cumulative sum is calculated for this variable, then a new list of numbers is formed, which will start at zero, and increase by one each time a division is encountered. Finally, selecting elements from the original map line whose corresponding elements on this new list are equal to a particular integer will give the relevent line feature. From a set of classifications and the original line, individual distinct geographical features may now be selected.

This gives the basic line feature detection algorithm. Various forms of improvement can be made to this, however. One criticism of this method is that the feature detection could be affected by the positioning of the two end points of the cartographic line. If, in a geographical database, part of a coastline is covered by two digitized lines then the breakpoint between the two lines may occur in the middle of a feature. This could be problematic, as the feature may not be easily detected by the algorithm when only viewed from one of its two parts. This can be combatted if, along with the information about the line itself, some data relation to the adjacency of all of the lines in a database is provided. If this is the case, the end features of adjacent lines could be analyses together and testing applied to see whether the two features should actually be classed as one. This could be done by examining shape indices for the two sections, and for both sections joined together as a single section. If the two distinct indices differed sufficiently from the aggregate one, the features would not be amalgamated, but otherwise they would be. A further adaptation might be to extend this method to cases where three lines met at a point, such as the meeting of two county boundaries or a railway junction. In this case, the feature extension could be tested over all possible combinations of line junctions, and the most likely classification decided on similar grounds.

13.5.8 Error modelling

Having identified the distinct line features, it only remains to tag each of these with some estimate of error. The methodology for doing this has been covered earlier, in terms of supervised neural nets, or alternatively by applying a grand classification to the feature (see section 4) in a universal taxonomy of line features, and applying a corresponding error tag. Once a line feature has been identified, using the methods outlined above, global shape indices may be computed. In this case the use of global rather than local indices are justified, as the line feature is chosen to have distinct, and presumably reasonably uniform characteristics. These are the predictor variables which may then be either fed into a backpropagation network, or into a classification scheme. At this stage, the error value (in the form of either an rms figure, or a confidence band) may be attached to the feature.

13.6 From design to a working system

Much of the emphasis of this research has been on experimentation, and the building of prototype systems. However, it is clear that this would be wasted effort if some form of practical application could not also be achieved. Although some of the earlier work in this research has been implemented as a set of Arc/Info macros (Carver, 1991), work on the more recent advances has yet to take place. It is hoped, however, that this will be the subject of future research, and towards this end this section will be concluded with a discussion of some issues relevant to this problem.

Clearly, to achieve this, the algorithms developed here will need to be modified to run in a working GIS environmentment. Two problems are identified here. First, although XLISP-STAT provides an excellent development environment, the interpreted nature of the language means that algorithms will take longer to execute than when using a compiled system. The second problem is the issue of integrating these tools into a GIS environment.

13.6.1 Speeding up the algorithms

As stated previously, XLISP-STAT provides an excellent environment for experimenting with various algorithms, for quick modification, and graphical diagnostic techniques that is unlikely to be rivalled by the more traditional compile–link–run–debug cycle of most compiled languages. However, once a suitable set of algorithms has been identified, there will be problems with the continued use of this package. As stated earlier, the most notable of these is speed, but it is also more difficult to integrate the XLISP-STAT environment with that of a GIS. At this stage, an implementation of the techniques in C or FORTRAN, which will of course be compiled is required.

This is not considered to be a major problem. It is the authors' belief that the effort saved and the utility of Lisp as an algorithm development tool more than pays for the minor inconvienience of translating the code into C once the algorithms have been developed. In fact, this could be managed incrementally, so that whenever a new or updated algorithm is developed in the XLISP-STAT environment, it could be translated into C and the working C code updated on a module-by-module basis. In fact, even the ask of converting the Lisp code could be automated, as there are lisp-to-C converters commercially available.

The quantum leap from interpreted to compile code is likely to bring about great improvements in the performance of the algorithms, but this could be further improved, if resources were available, by hand optimization of the C code. If the algorithms were profiled, and the most commonly executed statements identified, this optimization could be targeted to 'bottlenecks' in the code. Another interesting possibility here stems from the fact that the original Lisp code makes use of 'vectorized' operations. These are not neccessarily executed in parallel, the intention here being syntactical convenience rather than hardware parallelization, but it could be argued that this approach does in fact lend itself to this kind of parallel processing, and that massively parallel systems running GIS software may, at some point in the future, take advantage of this facet of the algorithm design.

13.6.2 Linkage to GIS

The final issue to be discussed is the linkage of the above software to GIS systems. The main problems here are the transference of information between the GIS and this software, the representation of the output from the error analysis software in the GIS, and the seamless integration of the software into the GIS system. Initially, it is intended to transfer information between the GIS and the software using a text-based standardized format. Although slower than subroutines which directly read the binary files, it is considerably less system-dependant, so that new sets of subroutines will have to be used each time the error handling module is to be married to a new GIS. In essence, the final implementation of these algorithms will expect some text-based representation of lists of point pairs, corresponding to line segments, and will output a new set of line segments in the same format together with a corresponding text file listing the error bounds.

The representation of the output from our error tagging algorithm is difficult on early versions of vector-based systems, since although map lines or arcs may have a data attribute related to the entire line, the method here detects features within those arcs. The only way this can be overcome here is for the software to produce a new set of smaller arcs, and then require the system to rebuild the topology from these new, feature-based arcs. On the other hand, more recent systems offering dynamic segmentation allow attribute values to be attached to subsections of lines, so that rebuilding of topolgies will not be neccessary each time the software is used. This is particularly relevent when very large map coverages, such as the railway network for the entire UK, are being processed. Finally, the issue of integration should be raised. If, as suggested, the issue of error in GIS and the computation of error margins is to be ignored at the policy makers' peril, then the means of handling should be made as accessible as possible. Ideally, seamless integration of this technique into the rest of the GIS should be achieved, so that the user finds the use of this algorithm as commonplace as overlay and buffering. In various systems, this may be achieved through a macro language or modelling language, provided this gives links to external software. At the core of the macro will be the central program to carry out the error tagging, and around this will be the operations in the GIS needed to perform the data exporting, from the GIS to the program, and the importing of the ouput from the program, exploiting the dynamic segmentation facilities if possible. In this way, the operation could either be attached to a menu item, or be accessed from the command line in the same format as any other command in the system.

13.7 Conclusions

The chapter has described a feasible pragmatic approach to the problems of estimating error levels in vector GIS data stored without any. Two approaches are discussed. One is a means of interactively building error models as part of the data capture process, the resulting error model would then be distributed with the database. It would require agreement and standards as to how this might be achieved.

Maybe this is an option more relevant to the twenty-first century. The second approach is simpler and off-line. It seeks to classify line segments into a universal typology based on local line complexity. Three options are then offered:

1. a look-up table that provides relatively crude error information;
2. a look-up table generated from a neural net that takes into account subjective and objective information about the data source, its scale, and quality; and
3. a data or specific neural model that seeks to predict error from available attribute information about the dis-assembled line.

It is believed that both approaches are feasible and provide a comprehensive error simulation strategy. They could also be developed further to handle other types of GIS-relevant error; for instance, near the edges of map categories. There are two remaining related problems one technical and one political. The technical one involves the creation of error database libraries from which to build error models of different levels of sophistication. We have the technology but not the data at present. The second problem involves persuading the GIS world to take error seriously. It is all too often perceived as a weakness of GIS and it may be necessary to wait until either the GIS technology matures or until major catastrophes force a re-think.

References

Ahalt, S.C., Krishnamurthy, A.K., Chen, P. and Melton, D.E., 1990, Competitive learning algorithms for vector quantization. *Neural Networks*, **3**, 277–90.

Brunsdon, C.F., Carver, S.J., Openshaw, S. and Charlton, M., 1990, A review of methods for handling error propagation in GIS, presented at *EGIS '90*, Amsterdam, The Netherlands.

Brunsdon, C.F. and Openshaw, S., 1993, Simulating the effects of error in GIS, in Mather, P. (Ed.), *Geographical Information Handling*. John Wiley and Sons.

Burrough, P. A, Heuvelink, G.B.M. and Stein, A., 1989, Propagation of errors in spatial modelling with GIS, *International Journal of Geographical Information Systems*, **3**, 303–22.

Gray, R.M., 1984, Vector quantisation. *IEEE ASSP Magazine*, April, 4–29.

Hertz, A., Krogh, A. and Palmer, R.G., 1991, *Introduction to the Theory of Neural Computation*, Redwood City, California: Addison-Wesley.

Jasinski, M.J., 1990, *The Comparison of Complexity Measures for Cartographic Lines*, Buffalo, New York: NCGIA Technical Paper 90-1.

Kohonen, T., 1989, *Self-Organisation and Associative Memory* (3rd Edn), Berlin: Springer-Verlag.

McMaster, R.B., 1986, A statistical analysis of mathematical measures for linear simplification. *The American Cartographer*, **13**, 103–17.

Muller, J.C., 1987, The concept of error in cartography. *Cartographica*, **24**, 1–15.

Nasrabadi, N.M. and King, R.A., 1988, Image coding using vector quantization: a review. *IEEE Transactions on Communications*, **36**, 957–71.

Openshaw, S., 1989, Learning to live with error in spatial databases, in Goodchild, M. and Gopal, S. (Eds), *Accuracy of Spatial Databases*, London: Taylor & Francis.
Openshaw, S., Charlton, M. and Carver, S., 1991, Error propagation: a Monte Carlo simulation, in Masser, I. and Blakemore, M. (Eds), *Handling geographic information: methodology and potential applications*. London: Longman, 78–101.
Poulton, E.C., 1974, *Tracking Skill and Manual Control*, New York: Academic Press.
Rumelhart, D.E. and McLelland, J.L., 1986, *Parallel Distributed Processing: Explorations in the Microstructure of Cognition*, (2 volumes), Cambridge: MIT Press.
Tapia, R.A. and Thompson, J.R., 1990, *Nonparametric function estimation, modeling, and simulation*, Philadelphia: Society for Industrial and Applied Mathematics.
Tierney, L., 1991, *LISP-STAT: An Object-Oriented Environment for Statistical Computing and Dynamic Graphics*, Wiley: New York.
Tukey, J.W., 1977, *Exploratory data analysis*, Reading, Mass: Addison-Wesley.

14

An Investigation into the spatial structure of error in digital elevation data

Colin G. Monckton

14.1 Introduction

Error has long been a concern in topographic mapping and more recently in the creation of digital elevation models (DEM). Users are, however, dependent on the published values for error given by the agencies producing the DEM data. Those measures have wholly ignored the spatial structure of error, an omission echoed in much of the research literature. Both the lack of uniqueness and lack of information on the spatial structure in the error measures is now hindering studies which endeavour to propagate DEM error into derived products within GIS such as the viewshed (Fisher, 1991, 1992) or drainage features (Lee *et al.*, 1992). The paramount objective of the research reported here therefore is to examine the spatial structure of error using mapped spot height information to derive a unique error term for an individual DEM. The method used here may be implemented by others without extensive and costly fieldwork.

Error may be defined in a number of different ways. In working with DEM data, the ideal situation is to study and compare the divergence between elevations recorded in the digital version of the terrain, and the elevations found at the same locations on the ground. On the other hand, error has been identified as the difference between elevations stored in an original DEM, and those found in one interpolated from a sample of points in the DEM (MacEachren and Davidson, 1987). In the proposed Standard for Digital Cartographic Data (DCDSTF, 1988), the preferred method of positional error estimation is by comparison with an independent source of higher accuracy, and so is close to the ideal situation identified previously. In the current research error is used in accordance with this definition, and a justification for this assertion is presented below.

The chapter first reviews the existing literature and draws up a list of coherent aims. It then presents the methodology used in the research reported here. Next the results of the analysis are reported, and finally a set of conclusions and recommendations given.

14.2 Literature review

14.2.1 Measurement of error

Error in DEMs is most usually reported as the root mean square error (RMSE) calculated as follows from the n known error values:

$$RMSE=\sqrt{\frac{\sum_{i=1}^{n} d_i^2}{n}} \quad \text{where } d_i^2=Z_{\text{ground}.i}-i_{\text{dtm}.i} \tag{14.1}$$

and where $Z_{\text{ground}.i}$=ground elevation recorded at point i, and $Z_{\text{dtm}.i}$=elevation recorded in the DEM at point i. According to the guide for USGS (1987) 1:24 000 DEM the RMSE is found from deviations at a minimum of 20 test locations on contour lines, bench marks, or spot elevations on the original topographic map. In recent DEM products the actual value is reported, while in the past they were grouped into either RMSE<7 or from 7–15, although a numerical value was still given (Elassel and Caruso, 1983). In the UK, the Ordnance Survey also reports the error as RMSE. The only two paragraphs on accuracy in the user manual focus on contour line accuracy in the topographic maps, and simply state that (OS, 1992, p. 13)

> the DTM creation process, using application software, causes only minimal degradation of the original contour data (RMSE 0·1–0·2 m).

In addition and partly in contradiction, the following statement is included in the header of the file for all DEM products derived from 1:50 000 maps. The only report of error with purchase of a DEM is to the effect that it is not tested uniquely for each DEM which has been supplied but in what testing has been done, the accuracy (OS, 1992)

> varied depending on the nature of the ground. Results so far have ranged from 2 m RMS in a hilly rural area to 3 m in an urban lowland area.

Li (1988) has advocated reporting the standard deviation of the errors in place of the RMSE, thus

$$S=\sqrt{\frac{\sum_{i=1}^{n} (d_i-\bar{d})^2}{n}} \quad \text{where } \bar{d}=\frac{\sum_{i=1}^{n} d_i}{n} \tag{14.2}$$

where the mean error, $\bar{d}$, would necessarily be reported too. The only difference in equations 14.1 and 14.2 is the inclusion of $\bar{d}$ in the latter, and underscores the assumption inherent in using the RMSE of zero mean error, and therefore there is no systematic bias in the DEM (Li, 1988; Imhof, 1982). There is little empirical evidence to support either this assumption or a change to the mean and standard deviation. Torlegård *et al.* (1986), however, has reported small but non-zero mean errors for photogrammetrically collected experimental DEM data. In the production situation the assumption of zero mean error seems inexcusable.

In summary, error reporting for DEM data is underdeveloped, and geared more to the production process, rather than assessment of possible effects of error. Users are not provided with adequate data quality information from which to assess the effects of error within their application.

14.2.2 Predicting error

Many workers have attempted to formulate more complete or acceptable descriptors and predictors of error, but a major handicap is the need for the collection of ground truth data, thus several attempts have been made to develop theoretical models of the error. Reviews of a number of these in (Li, 1988, 1992) identify studies which used Fourier (Makarovic, 1972) and spectral analysis (Frederiksen, 1980), as well as variograms (Frederiksen *et al.*, 1986). More recently this is an area of study that has attracted the attention of those interested in fractals (Polidori *et al.*, 1991). But several of these are based on mathematical models of terrain which it is uncertain are applicable to any real terrain.

Other researchers have turned to the work of Koppe, as reported by Frederiksen (1980), who relates terrain slope, α, to mean vertical error, μ, thus

$$\mu = \pm(A + B \tan \alpha) \tag{14.3}$$

where coefficients A and B should be determined for any particular map.

The use of (14.3) as a predictor of error in DEMs remains poorly tested. In examining interpolated error and sample size MacEachren and Davidson (1987) found a strong relationship between slope and error, generally of the form of (14.3). Indeed it has been asserted that mean slope may be a major determiner of error (Ley, 1986). He notes, however, that further work is required, and since it is based on contour map accuracy, there seems no *a priori* reason that it should be applicable to DEM data - which may be derived in a number of ways.

14.2.3 Modelling error in DEM

Equation (14.3) has been used as a mechanism to predict error when assessing the effect of DEM error in GIS products. Thus Fellerman and Griffin (1990), Mills *et al.* (1992) and Teng and Davis (1992) have all used it in Monte Carlo simulation of DEM error to assess viewshed variability. By contrast Fisher (1991, 1992) and Lee *et al.* (1992) have used the RMSE in propagating the DEM error into viewshed and landscape feature extraction. These last authors have also simulated spatial structure (autocorrelation) in the error model they have used, an error descriptor ignored by the other workers, although its importance is acknowledged. For future work of this kind, it is essential to establish which error model is appropriate to a DEM, and the nature of the spatial structure of the error. This last topic is the subject of the research reported here.

14.2.4 Current objectives

Arising from this review, there is a clear case to be made for every DEM to have ground control information attached which will allow an investigator to establish the level of error according to any particular measure chosen for any subset of the DEM which may be of concern. Indeed, reporting all information used in quality evaluation is required in the proposed standard for cartographic data in the US (DCDSTF, 1988). While it might be optimal for investigators to conduct their own ground truth survey, perhaps based on GPS survey, there is a case to be made that the onus is on the data supplier to deliver quality evaluation data with the DEM. The most convenient information is the spot height data recorded in most topographic mapping programs, and included on maps. The purpose of this research is to investigate the spatial dependence structure of the error in DEM using this data and therefore also the usefulness of such spot height information in characterizing the error in DEM.

14.3 Methods

14.3.1 Data

Two OS DEMs were used in this research, both being derived by the OS from digitized contours from the 1:50 000 Landranger map series. The DEMs are distributed on

20×20 km tiles registered to the national grid (SK02 and SO64), and gridded to a 50 m interval with vertical precision of 1 m (all heights are reported as integers). The DEMs used are not claimed to be representative of either the UK or any other terrain, but do include variable elevations. The DEM SO64 includes the Malvern Hills, a massive ridge of pre-Cambrian rock rising dramatically from a broadly flat plain, and SK02 has an escarpment running east to west past Uttoxeter.

All spot height locations were digitized for this study from 1:25 000 series maps for the two areas. Surveying by OS is done at a number of different base scales, namely 1:50 000 and 1:10 000. Intervening scales are generalized from the larger scale maps, thus the surveys on which the 1:50 000 DEM and 1:25 000 spot heights are based are independent. The surveys are, however, co-registered to the national grid and so the possible complicated process of registering the digitized spot heights to the DEM was trivial. For the Uttoxeter DEM 603 points were recovered and 537 points for the Malvern. The spot height data are not only used in the original maps to report precise elevations, but they are also precisely surveyed in the field, and, in addition, the maps are at a larger scale than the DEM. For all these reasons therefore the spot height data can be regarded as providing a reasonable basis for accuracy assessment (Fisher, 1992, p. 133).

14.3.2 Extraction of error data

At the heart of the research reported here is the interpolation of elevations from the DEM to compare with the spot heights. Interpolating an elevation at a point location, when using a DEM, and when the point may not coincide with the mesh of the DEM, is not simple. In this research two different methods were used.

1. **Simple**: Check the cell in which the spot height falls and compare the spot height value with the value of that DEM cell.
2. **Interpolation**: To reduce the possible effect of the arbitrary selection of elevation cells in the DEM and of flexible positions of the spot heights, the cell in which the spot height fell was taken as the centre of a 5×5 window of elevations. The contents of the window were used to create a polynomial surface from which the elevation at any point could be interpreted. Edge effect problems were avoided by discounting any spot heights within a 2-cell border of the DEM area.

14.3.3 Measures of spatial autocorrelation in the error

The measures of spatial autocorrelation in point data are particularly sparse, but the actual measures are essentially the same as in polygon or line data. For ordinal data such as the error fields used here two main indices exist, Geary's *c* and Moran's *I*. Goodchild (1988) argues that the former of these is equivalent to the variogram, and is sensitive to variations in local mean and variance, while the latter being equivalent to the correlogram uses a global mean and so is insensitive. For this reason, the latter was used here (Goodchild, 1988; Griffith, 1987; Cliff and Ord, 1981) where

$$I=\frac{\sum_{n=1}^{i}\sum_{m=1}^{i} w_{nm}(d_n-\bar{d})(d_m-\bar{d})}{S_2\sum_{n=1}^{i}\sum_{m=1}^{i} w_{nm}} \qquad S^2=\text{sample variance} \tag{14.4}$$

All terms are as before except that w_{nm} indicates the weighting of observation n on observation m. It can be seen that the deviation from the mean in all possible paired values is computed (the covariance), and assessed according to the weighted value.

In a raster display, Goodchild (1988) suggests that the orthogonal immediate neighbours should have weighting 1, and all other possible pairs 0, a scheme commonly used in assessing polygon data. In working with point data the weightings might more commonly be related to the inverse of the distance. Thus

$$W_{nm} = \frac{1}{[x_{n-m}]^p} \tag{14.5}$$

where x_{n-m} indicates the distance from case n to case m, and p can have any real value. This general form of the inverse distance weighting function allows the exploration of the influence of values at different spacings (lags). While a global value is calculated each time, by setting values of $p > 1$ short spaced locations are emphasized, and decreasing the value ($p < 1$) greater spacings are emphasized. The actual spacings having the greatest weightings cannot be determined, but the general form can be observed.

Moran's I value calculation is theoretically standardized for variations in data mean and standard deviation within the dataset and the limits would normally be 1 to -1, where a value of 1 indicates excellent spatial ordering, zero indicates no spatial ordering (i.e. random) and -1 negative spatial ordering (i.e. close neighbouring data points are likely to have dissimilar values). Using (14.5) however, represents a variant of Moran's I and no such fixed I values exist because they are dependant on both the weighting method and the data (Cliff and Ord, 1981, p. 23; Goodchild, 1988, p. 32) however zero remains a value of random spatial order and positive and negative I values remain indicative of the strength of spatial order. Because an I value is calculated using all data points for each p value, the I values presented here represent a smoothing of the overall pattern, but are shown here for their ease of understanding. A wide range of variants were used in calculating I to reflect the spatial autocorrelation at different lag distances. These included varying the selection of points by calculating I based on error values within a specified band of distances therefore deriving an individual I value for each spatial lag and restricting the number of points used with (14.5) to emphasize specific lag distances. These variations are restricted only by the requirement for a statistically sound number of pairs of error values to base the I value on which for these data prevented I values being calculated are distances lower that 250 m. The number of pairs required is a subject of debate, however 30–50 pairs as a minimum is widely accepted as being satisfactory (Rossi *et al.*, 1992).

The use of variogram and covariance were also investigated for the explanation of the spatial structure, but both were found to produce much the same results as the correlogram. Discussion here will therefore focus on the correlogram only.

14.4 Results

14.4.1 Error maps

Figures 14.1 and 14.2 show the location of the spot heights throughout the DEM used and the magnitude of error at those points. The spatial distribution of the error in the Malvern DEM can be seen in Figure 14.1. The grey scale point symbols seem to indicate there may be a broad clustering of larger values towards the south but it is very uncertain. There seems to be no similarity of neighbouring values. In short, no pattern is apparent in the values plotted. Exactly the same observation can be made on the distribution of values in the Uttoxeter DEM (Figure 14.2); neither regionalization nor similarity in neighbours is discernible.

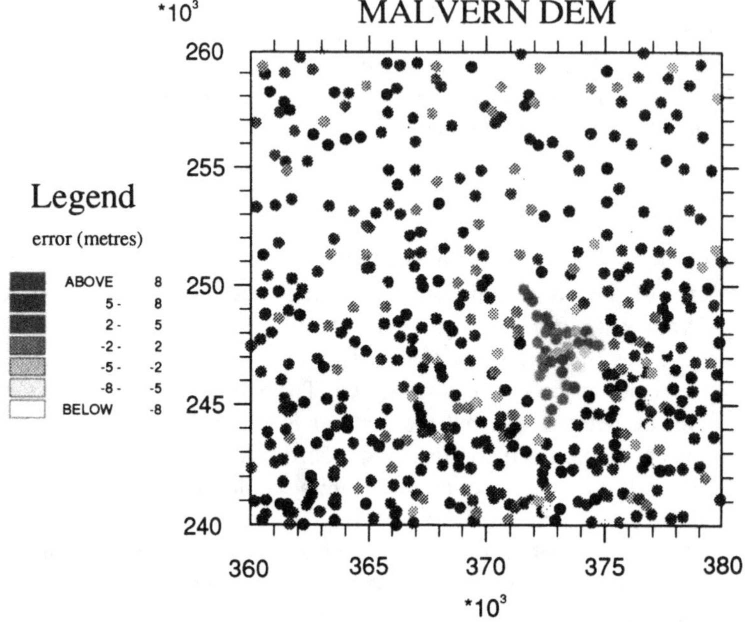

Figure 14.1 Magnitude of error at spot height locations in the Malbern DEM.

The nearest neighbour statistics for the distribution of spot heights show them to be significantly more evenly distributed than random (at the 95 per cent level, Table 14.1). The distributions in both maps therefore show no indication of having been generated by a poisson process, as should be the case if a random sample were collected. This is hardly surprising given the purpose of spot heights, but needs to be born in mind in later discussion.

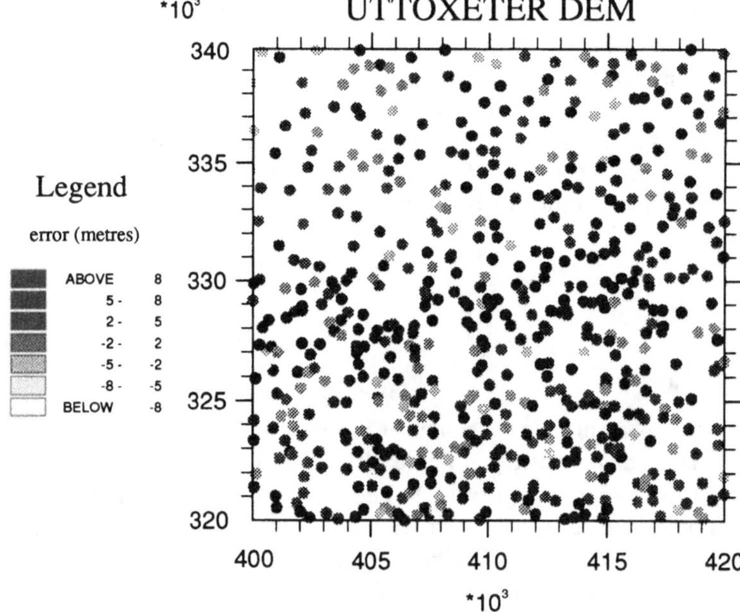

Figure 14.2 Magnitude of error at spot height locations in the Uttoxeter DEM.

Table 14.1 Nearest neighbour statistic for spot height locations

Data set	Points	R	z score
Malvern	537	1.180	7.973
Uttoxeter	603	1.268	12.598

14.4.2 Representativeness

In collecting spot height information, it is apparent that all local peaks are included in the data set, and in several cases the elevation recorded is lower than the elevation of the lowest contour possibly indicating a local pit. Otherwise no characteristics of the distribution of points is apparent. Thus, both sets of points may possibly be biased in favour of peaks, as compared with other types of terrain. This again is unsurprising, but should be considered in later discussion.

14.4.3 Summary statistics

Table 14.2 presents the summary statistics for the two sets of error data. As indicated in the discussion above, various summary statistics are included here. A number of different versions of the data are presented, the reasons for which will become apparent. The mean error is most interestingly always negative, by up to one metre. The consistent overestimate of elevations in the DEM argues against this being the effect of the peaks, the only consistent bias in the sampling identified above. That the distribution is not random, means that the distribution is necessarily biased, but the results seem to be real, and the systematic value of −1 supports Li's (1988) point that the zero mean assumption of the RMSE is presumptive, and may not be valid. Because the means are not zero, the standard deviations and the RMSE are not equal. The RMSE are, however, within the range specified by the OS.

Between the simple and the interpolated data sets the only major difference is in the extreme values, where either maximum or minimum values are different. For example, the minimum in the Uttoxeter DEM are −9 and −13 for the two different methods. For this reason the maximum and minimum one per cent of error values in both data sets were discarded for all subsequent processing. The effect of this is apparent in Table 14.2. Figures 14.1 and 14.2 were prepared with these truncated data sets. Removing these values reduces the RMSE and standard deviation values but not by very much.

The possibility that the spot heights were actually used in the creation of the DEM was of concern in this research. If an exact interpolator were used, as is claimed by the data-provider, and this eventuality were the case, then all the errors should be zero or only different by small amounts due to rounding errors. That this is not the case provides justification for using the spot heights.

Table 14.2 Summary statistics for error at spot height locations (Figures in metres)

Data set	Mean	Std. Dev	RMSE	Maximum	Minimum
Malvern Simple	−0.9685	3.0309	3.1802	9.00	−14.00
Interpolated	−0.8829	2.6641	2.7876	6.6060	−11.6970
Simple with outliers removed	−0.9469	2.7634	2.9212	6.00	−9.00
Interpolated without outliers	−0.8566	2.4358	2.5820	4.600	−8.240
Uttoxeter Simple	−0.9628	2.9112	3.0681	17.0	−9.0
Interpolated	−0.9986	2.8081	2.9803	16.989	−13.233
Simple with outliers removed	−0.9831	2.5256	2.7102	5.00	−7.00
Interpolated without outliers	−1.023	2.6441	2.7876	5.630	−7.240

14.4.4 Autocorrelation

Equation 14.4 was used to generate autocorrelation statistics. Both simple and interpolated data sets were used, however, the maximum and minimum one per cent of distances to nearest neighbour were also removed based on the fact that there should be a certain number of pairs of points at each distance of separation in order for the autocorrelation calculation to be significant. Thus the data sets used and graphed in Figures 14.3 and 14.4 are the Uttoxeter and Malvern DEM with both error and distance outliers removed. The values of p used were in the range 0.2 to 3.8 at intervals of 0.2. A weighting of 3, for example, emphasizing autocorrelation over short lag distances while those weightings less that 1.0 emphasize increasingly long distances. There is little difference between the interpolated and simple error data, therefore the simple data are shown only for both data sets.

These graphs indicate the spatial autocorrelation at lag distances of 250 m and above, values below this distances were unobtainable from this data set due to the arrangement of the spot height data. The x-axis values show the weighting exponent where the shortest distances for which autocorrelation could be calculated are represented at the origin and longer lag distances at the extreme of the x axis. For both DEMs there is clearly a random spatial arrangement of error values at all lags measured in this study. This indicates that the error is likely to be truly randomly generated as opposed to an artefact of the production process and is independent of the nature of the terrain itself. Consequently this indicates that when simulating error fields for these DEMs this random arrangement should be produced at these lags. Below 250 m the spatial autocorrelation structure is unknown. It is intuitive to suggest that some undetermined positive spatial order may exist given

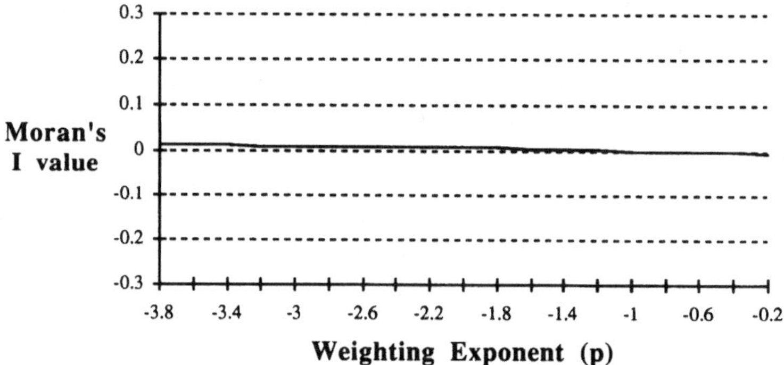

Figure 14.3 Correlogram for the Uttoxeter DEM – simple data set

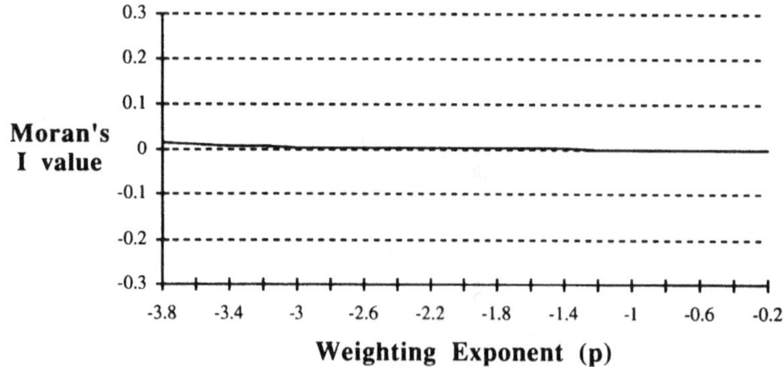

Figure 14.4 Correlogram for the Malvern – simple data set

the nature of the terrain surface, i.e. that a random error arrangement at close distances, if simulated, could produce an alternative realization of the DEM with local peaks and pits uncharacteristic of a terrain surface. It is also the case that positive spatial order that may be created by the production process over short distances (such as from interpolation) is clearly restricted to distances under 250 m. There is an indication in both data sets of a slightly rising I value but clearly this is marginal and significance cannot be attached to this.

The results clearly show that at long and medium lag distances (250m and above) there is a conclusive lack of any spatial structure in the error data, the error is truly random, as indeed should be expected in a well produced DEM. At lag distances below this level there is little of a conclusive nature that can be said using the spot height data as ground truth source data set. It remains that the structure at these close lag distances is a topic for a further investigation. The usefulness of the spot height data has therefore been considerable in providing detailed information about the mid-to long-lag error structure of a DEM.

14.5 Conclusion

Arising directly from this investigation we can make two observations on the nature of error in the DEMs studied which may have wider importance:

1. The statistical impurity of the RMSE was highlighted in this study. The mean error deviated from zero in all cases suggesting support for the argument of reporting the standard deviation and mean instead.
2. Spatial structure. For the two DEMs studied there is conclusive evidence that the spatial arrangement of error beyond lag distances of 250 m is purely random. As such any error simulation should simulate this structure for these DEMs.

Work reported here does not address some major issues due to the unsuitability of the data. It is suggested that at closer lags there may be some autocorrelation, but this does need to be established empirically, from detailed survey. The significance of the study for error modelling cannot be underestimated. It is crucial to establish whether it is necessary to simulate spatial autocorrelation in the error fields, and to what lag the algorithms should be concerned to determine the error (Fisher, 1991, 1992; Lee *et al.*, 1992).

The method adopted in this study of using spot height information has been very successful in the examination of the structure of error in a DEM. It is possible that some amount of bias may be introduced by the location of the spot heights, and this maybe detrimental to their statistical purity for the study of error, but this shortcoming is far outweighed by the advantages:

1. Spot heights allow the user of the DEM product to produce their own analysis of the error structure, and makes them no longer reliant on the use of poor theoretical models for reporting error (such as the RMSE).
2. It allows the determination of a unique error term for a DEM with very little effort, as well as some number of fixed error locations.
3. The data are already available on the topographic sheets, and complete digitizing for a 20×20 km tile can be relatively rapid, if they do not already exist in digital form.

In the light of these three observations, and the conclusions from the empirical study reported here, it is suggested that there is a responsibility upon the data supplier to deliver the spot heights, so that users may have access to a means of a rapid and unique validity check for their DEM.

Acknowledgments

The work reported here was completed as a dissertation for the MSc in GIS at Leicester University, UK. I particularly wish to acknowledge the help and guidance of Pete Fisher and Jo Wood during the research. The ESRC are also acknowledged for Advanced Course Studentship funding. All spatial data retains Crown Copyright, and is reproduced here in accordance with the licence.

References

Cliff, A.D. and Ord, J.K., 1981, *Spatial Processes: Model and Applications*, London: Pion.

DCDSTF (Digital Data Standards Task Force), 1988, The proposed standard for digital cartographic data. *The American Cartographer*, **15**(1), 7–140.

Elassel, A.D. and Caruso, V.M., 1983, *Digital Elevation Models*. USGS Digital Cartographic Standards, Geological Survey Circular 895-B, United States Geological Survey, Reston, VA.

Fellerman, J. and Griffin, C., 1990, *The Role of Error in GIS-based Viewshed Determination: A Problem Analysis*. IEPP Report No. EIPP-90-2, Syracuse, New York.

Fisher, P.F., 1991, First experiments in viewshed uncertainty: the accuracy of the viewshed area. *Photogrammetric Engineering and Remote Sensing*, **57**(10), 1321–7.

Fisher, P.F., 1992, First experiments in viewshed uncertainty: simulating fuzzy viewsheds. *Photogrammetric Engineering and Remote Sensing*, **58**(3), 345–52.

Frederiksen, P., 1980, Terrain analysis and accuracy prediction by means of the Fourier transformation. *International Archives of Photogrammetry and Remote Sensing*, **23**(4), 284–93.

Frederiksen, P., Jacobi, O. and Kubik, K., 1986, Optimum sampling spacing in digital terrain models. *International Archives of Photogrammetry and Remote Sensing*, **26**(3/1), 252–9.

Goodchild, M.F., 1988, *Spatial Autocorrelation*, CATMOG 47, Norwich: Geobooks.

Griffith, D.A., 1987, *Spatial Autocorrelaton: A Primer*, Washington, D.C.: Association of American Geographers.

Imhof, E., 1982, Cartographic relief presentation. Stewart, H.J. (Ed.), translated from the German, Berlin: Walter de Gruyter.

Lee, J., Snyder, P.K. and Fisher, P.F., 1992, Modelling the effect of data errors on feature extraction from digital elevation models. *Photogrammetric Engineeering and Remote Sensing*, **58**(10), 1461–7.

Ley, R.G., 1986, Accuracy assessment of digital terrain models, in Blakemore, M. (Ed.), *Proceedings, Auto Carto London*, **1**, 455–64.

Li, Z., 1988, On the measure of digital terrain model accuracy. *Photogrammetric Record*, **72**(12), 873–7.

Li, Z., 1992, *Theoretical Models of the Accuracy of Digital Terrain Models: An Evaluation and Some Observations*. North East Regional Research Laboratory.

MacEachren, A.M. and Davidson, J.V., 1987, Sampling and isometric mapping of continuous geographic data. *The American Cartographer*, **14**(4), 299–320.

Makarovic, B., 1972, Information transfer in construction of data from sampled points. *Photogrammetria*, **28**(4), 111–30.

Mills, K., Fox, G. and Heimbach, R., 1992, Implementing an intervisibility analysis model on a parallel computing system. *Computers and Geosciences*, **18**(8).

OS, 1992, *1:50 000 Scale Height Data User Manual*. Ordnance Survey, Southampton.

Polidori, L., Chorowicz, J. and Guillande, R., 1991, Description of terrain as a fractal surface, and application to digital elevation model quality assessment. *Photogrammetric Engineering and Remote Sensing*, **57**(10), 1329–32,

Rossi, R.E., Mulla, D.J., Journel, A.G. and Franz, E.H., 1992, Geostatistical tools for modelling and interpreting ecological spatial dependance. *Ecological Monographs*, **62**(2), 277–314.

Teng, Y.A. and Davis, L.S., 1992, Visibility analysis on digital terrain models and its parallel implementation. Technical report CAR-TR-625, Centre for Automation Research, University of Maryland.

Torlegård, K., Ostram, A. and Lindgren, R.A., 1986, A comparative test of photogrammetrically sampled digital elevation models. *Photogrammetria*, **41**(1), 1–16.

USGS, 1987, *Digital Elevation Models Data Users Guide*. Technical Instructions, Data Users Guide 5, National Mapping Program, United States Geological Survey, Reston, VA.

SECTION III

Applications of GIS

Section Three: Applications of GIS

The third and final section of the book is devoted to the treatment of applications. This section is shorter than the other two, as may be expected in a book devoted to research questions. However, it is the applications which are the life-blood of the field and give rationale to its more basic research. Each chapter in this section is not just a description of a particular application area, but a thoughtful appraisal of GIS research in the context of the author's chosen field of interest. The applications are diverse, from governmental and administrative fields to the environment.

The first chapter in this section is written by **Ian Masser** and **Heather Campbell**, from the Department of Town and Regional Planning at the University of Sheffield in the north of England. One of the major benefits to an organization from the acquisition of any information system arises from the ability to share information. This is equally true for GIS but in certain areas has proved problematic. The reasons for this are partly technological, including differing data formats and data transfer between systems, but also institutional and organizational. It is the latter which are discussed in this chapter. The authors consider in detail the implications, advantages and disadvantages, for British local government of the accelerating trend to the introduction of GIS technology. The chapter sets out the premise at the foundation of the case for information sharing in local government as 'the need to regard information as a capital asset'. A logical consequence is the advantage of sharing in terms of the creation of 'economies of scope in the form of multiple applications'. Masser and Campbell have undertaken a survey of British local government, including a telephone survey and detailed case studies, to assess the impact of GIS in this application area. A description of the study and the conclusions drawn make up the latter parts of their chapter.

The next chapter is written by **David Livingstone** and **Jonathan Raper** of Birkbeck College, London University. In arguing for a high-level model of an environmental system, they return us to the beginning of the book, where Smith proposed a high-level framework for scientists to express their scientific models. The authors begin with the assumption that a prerequisite for a well-founded environmental modelling system is a spatio-temporal information system. GIS is thus one element of an integrated system for handling environmental data. As is well known in the database community, problems with integration of heterogeneous components of an information system arise at several levels, including semantic heterogeneity (disparate semantic data models) and physical heterogeneity (hardware disparities and data transfer problems). Usually, the semantic heterogeneity is harder to handle. The authors move on to a general discussion of the computational requirements of environmental system models and categories of environmental data. In this context, they discuss some of the current movements in GIS research, such as spatio-temporality and object oriented systems. They conclude that GIS is becoming well-placed to support an integrated system for environmental information and modelling. They support this proposition by observing that, amongst other advances, GIS are becoming capable of handling different spatial models, are beginning to support spatio-temporal information handling and have a rich collection of analytic tools.

In the final chapter, **Charlotte Norman, Clive Potter**, and **Hadrian Cook**, from Wye College in England, consider the implications of applying GIS technology to optimize the benefits of governmental agri-environmental policies. Recent reforms of the European Common Agricultural Policy have led to a range of measures intended to foster environmentally-friendly practices amongst the agricultural community. Amongst these measures is a long-term set-aside policy, which offers incentives to farmers to take certain land out of production for a period of 20 years. The chapter focuses on a specific application of GIS in East Sussex and Suffolk, which seeks out suitable land for such a policy. Suitability here is measured in terms of soil, hydro-geology, climate, topography, conservation potential and other factors. There is a discussion on data collection in each of these dimensions and their incorporation into a general model of the application. The authors observe that the distribution of suitable land is far from uniform across the areas covered and quite different for the two sample regions. The authors conclude that although there are problems with data support, this approach can provide a useful tool for the implementation of such land policies.

David Miller of the Macaulay Land Use Research Institute in Scotland, proposes the application of an amalgamation of knowledge-based systems and GIS to environmental modelling. In Miller's architecture, the GIS is coupled to the knowledge-base (and thus to the environmental model) by means of an expert system shell. Several constructs derived from artificial intelligence, such as blackboards and frames, are used to manage and represent the knowledge structures. The concept of metadata, ('data about data', for example lineage and error) is important to handle spatial data properly. Output is provided as text or graphics. The author's system is applied to model the distribution of bracken and to predict the spread of bracken in a region of central Scotland. In the discussion, an important point about explanation is made. Expert systems have developed mechanisms by which they can be prompted to explain their conclusions to the human user. This is necessary to give the user confidence in the reasoning of the machine. Such mechanisms are highly pertinent to GIS, where the models and processing is complex. Approaches such as that described in this chapter are at the leading edge of the application of GIS and should provide ideas for other developments.

15

Information Sharing and the implementation of GIS: some key issues

Ian Masser and Heather Campbell

The arguments for information sharing are discussed in the context of British local government. The case for more corporate approaches which maximize the benefits to be obtained from information sharing is evaluated and an alternative case which highlights the operational advantages of departmental approaches is considered. The advantages and disadvantages of these two approaches are examined both in overall terms and in relation to the preliminary findings of a comprehensive survey of GIS implementation. The findings of this survey show that there is a 50:50 split between more corporate and departmental approaches in current British practice.

15.1 Introduction

Data integration and information sharing are becoming recognized as increasingly important issues in the development and implementation of GIS. In February 1992 the National Centre for Geographic Information and Analysis (NCGIA) organized a special workshop on this topic (Onsrud and Rushton 1992) and the subject also featured prominently in the discussion at the Nato Advanced Research Workshop on Modelling the use and diffusion of GIS Technololgy (Masser and Onsrud 1993). What emerged from these meetings is that there are pronounced differences between those who regard information sharing as an essential and inevitable stage in the universal implementation of the new technology and those who argue that there are important impediments as well as incentives which are likely to lead to marked differences in practice. With these considerations in mind this chapter considers some of the key issues that need to be taken account of in the discussion.

It is often argued that the greatest benefits from GIS implementation come as a result of the pressures that are placed on organizations to share information. The findings of the joint Nordic Project, for example, suggest that the returns on investment in GIS range from 1:1 where the automation of basic mapping functions are involved, to 2:1 where GIS is applied to planning activities and are as high as 4:1 where they involve the collective use of databases common to all users (Tveitdal and Hesjedal, 1989). It can also be argued that these benefits will be accompanied by considerable cost savings particularly where a lot of time is spent on inter-agency map exchanges. For example, a study of six agencies in Santa Clara county estimates that overall savings of between 42 and 54 per cent of current costs could be obtained by the development of decentralized and centralized shared AM/GIS facilities respectively (Finkle and Lockfield, 1990).

Nevertheless the theoretical assumptions behind arguments such as these can be questioned from both an economic and a political standpoint. In economic terms, they assume that geographic information can be treated as a public good: i.e. something accessible to everybody which can be shared without loss to any individual. Yet in many respects geographic information must be regarded as a private good, as a commodity which is traded at a price with access being determined by ability to pay (Bates, 1988, Openshaw and Goddard, 1987). It is important also to note that the commodification of geographic information is not restricted to the private sector given the increasing pressures on public sector agencies to recover their costs (Bryan 1992; Rhind 1992).

Similarly these arguments do not take account of the political economy of geographic information: i.e. the extent to which geographic information is used to maintain the existing balance of power within late capitalist society (Mosco and Wasko, 1988). In this perspective geographic information must be regarded as a tool to be used in turf battles between different interests. Moreover, differences in access to information are likely to result in a growing divide between the information-rich and the information-poor within cities and regions while creating circumstances which lead to the emergence of surveillant societies (Pickles 1991).

Even where the public good dimension of geographic information is paramount and there are no conflicts between the agencies involved, the success of information sharing is likely to depend on a wide range of organizational and institutional factors. There have been a number of attempts in recent years to set down maxims or conditions that promote effective implementation (Wellar, 1988; Levinsohn, 1990; Croswell, 1991; Masser and Campbell, 1991). However these are largely directed towards single organizations and must be modified in the context of information sharing.

Some of the key organizational issues involved are highlighted in the reviews by Crosswell (1991) and Masser and Campbell (1991; 1992). Drawing on the work of the Irvine group (Danziger *et al.*, 1982; King and Kraemer, 1985) on the use of computers in local government, Masser and Campbell (1992) contend that three necessary and generally sufficient conditions must be met for the effective utilization of GIS in organizations:

1. the existence of an overall information management strategy based on the needs of users in the agency and the resources at its disposal;
2. the personal commitment of individuals at all levels in the organization with respect to overall leadership, general awareness and technical capabilities; and
3. organizational stability with respect to personnel, administrative structures and environmental considerations.

The effects of moving from a single organization to a multi-organization environment is particularly marked with respect to Masser and Campbell's three conditions. Effective utilization will depend, first on the existence of an overall information management strategy in each of the organizations concerned which takes account of the needs of each set of users and the resources at the disposal of the respective organizations. This assumes a high degree of compatibility between the priorities that are given to these operations where sharing is required and the resources that are set aside by each organization for these purposes. The second condition presupposes a high degree of personal commitment at all levels in all the organizations. For example, given the importance attached in the literature to the role of particular individuals or champions in ensuring effective implementation (Beath, 1991) it will be necessary to think in terms of supra organizational champions in the context of information sharing. However, the biggest obstacle to effective implementation is likely to be with respect to the third condition regarding the need for organizational and environmental stability where information sharing is concerned given that the failure of even one organization in this respect may threaten the success of the whole operation.

It may be argued that the above discussion is unduly negative in tone and that information sharing in most cases involves largely technical operations which are affected to only a very limited extent by economic, political or organizational considerations. It may also be argued that in many cases the benefits from information sharing are sufficiently great to justify special efforts to satisfy the conditions set out above. With these considerations in mind this chapter examines the effect of GIS on British local government with particular reference to information sharing. It examines a case where the circumstances are especially favourable for information sharing: i.e. the local authority environment where information sharing takes place between different departments within the same organization rather than between different organizations. Under these circumstances, given the common corporate objectives of the organizations to which the departments belong and the number of common factors associated with organizational stability it may be argued that the adoption of GIS should be associated with corporate information management strategies which maximize the benefits to be obtained from information sharing.

This chapter first considers the thesis that is implicit in such a standpoint. Then it examines the antithesis which draws attention to the strengths of a departmental standpoint. The third section presents some of the findings from a comprehensive survey of the impact of GIS on local government in Great Britain which are discussed in relation to the assumptions of both the thesis and the antithesis. Finally a number of general conclusions are drawn from the evidence with respect to information sharing as a whole and a number of suggestions are made regarding the further research that is required.

15.2 Thesis and antithesis

15.2.1 Thesis

The basic thesis underlying the case for information sharing in local government revolves around the need to regard information as a capital asset. To exploit this asset an integrated approach is required which maximizes the potential for resource sharing amongst users. This in turn creates economies of scope in the form of multiple applications.

The case for a corporate approach to information resource management in British local government can be developed in a variety of ways. For example, Hepworth (1990) calls for greater efforts by local authorities to mobilize their information capital to improve the territorial management public services. Similarly, Gault and Peutherer (1989, p. 2) argued that 'to manage change in the future, local authorities need to develop and make better use of the information sources at their disposal'. They also argue that GIS represents a useful tool for strategic information management which adds value by integrating diverse data sets. In their view 'a corporate view is essential if the synergy is to be realized which will assist both the management of change and increase operational efficiency'. (p. 13)

The Audit Inspectorate (1990) also note the importance of data integration in their advice to local authorities on preparing an IT strategy. They view with concern the disorganized and uncoordinated approach to acquisition that exists in many local authorities as a result of a preoccupation with short-term considerations. 'Users have invariably been able to demonstrate the short-term benefits of providing their own solutions and have only later come to appreciate such disbenefits as the inability to share corporate data and the consequence of need for expensive duplication'. (p. 4)

The advantages of a corporate approach are highlighted in many empirical studies. For example, Coulson and Bromley's (1990) detailed assessment of user needs in Swansea district demonstrates 'the impracticability of introducing an isolated departmental GIS' (p. 215). In their view four types of benefits would accrue to the local authority as a result of adopting a corporate approach.

1. Maps: 'access to up-to-date urban mapping is a bonus to any organization for reasons of completeness and common usage with other members of the organization';
2. Data: 'the main benefits of GIS are due to better access to data... the corporate approach encourages a steady reduction of the artificial barriers that have grown up between departments over the years';
3. Operations: 'the combination of better mapping and data improvements brought about by a GIS should result in better operations within the local authority, in particular better resource targeting and better forward planning';
4. General benefits 'the most significant is the idea of great coordination, but only if a corporate GIS is adopted. Whilst the management of corporate GIS is clearly more difficult than a departmental GIS the benefits far outweigh the disadvantages'. (p. 216)

15.2.2 The antithesis

It is also necessary to consider the antithesis to these claims. The case embodied in the antithesis revolves largely around matters relating to control and continuity. It can be argued that those most directly concerned with the provision of specific local authority services, that is the individual line departments, need to retain control wherever possible over the technical facilities and the information that they need to carry out their tasks. This presents local government with a fundamental dilemma: 'maintaining sufficient centralization of control in city government that enables oversight, efficiency and accountability, while providing sufficient decentralization to deal adequately with the problems of size and complexity and the demands of neighbourhoods and the special interest groups' (Kraemer and King, 1988, p. 28).

It can also be argued that the importance of continuity in terms of administrative routines and practices is often under-estimated in practice. British local government is highly departmentalized and many departments have their own professional cultures which are reflected in the distinctive character of their administrative practices. Under such circumstances movements between departments within local authorities are relatively rare and it can be argued that the most successful technological innovations are likely to be those which reinforce existing departmental practices.

15.2.3 Evaluation

The arguments underlying the thesis and antithesis can be expressed in terms of pressures towards more corporate approaches and more departmental approaches respectively. The main advantages and disadvantages of these two approaches are summarized in Table 15.1 which is based on Campbell (1992). This shows that the respective advantages of corporate approaches in terms of information sharing, more informed decision making and increased efficiency which were incorporated in the thesis are often offset by corresponding disadvantages in practice. For example, against the benefits to be derived from data integration, there are the potential costs associated with variations in priorities between departments. Similarly, the advantages of improved access to information may be offset by differences between departments in terms of their abilities to exploit the potential offered by GIS.

Table 15.1 The main advantages and disadvantages of corporate and departmental approaches to implementing GIS

Corporate approach		Departmental approach	
Advantages	Disadvantages	Advantages	Disadvantages
Integration of data sets	Variations in priorities between departments	Independence	Departmental isolation
Increased data sharing	Differences between departments in their ability to exploit GIS facilities	Control over: – priorities – the form and accessibility of information – technical specialists – equipment	Lack of support in terms of: – finance – technical specialists – training
Improved access to information	Differences in the level of awareness and spatial data handling skills between departments		
More informed decision making	Inter-departmental disagreements over: – access to information – leadership – data standards – equipment – training	Clear lines of responsibility	Absence of authority-wide including lack of system and data compatibility
Increased efficiency due to reduced duplication leading to time, staff and cost savings		Continuity	Inertia

Furthermore the benefits to be derived from corporate approaches in terms of more informed decision making may be outweighed by differences between departments in terms of the level of awareness and the spatial and data handling skills at their disposal. In addition, the potential savings to be derived from increased efficiency in terms of the reduction of duplicated efforts may be offset by inter-departmental disagreements with respect to access to information, professional leadership, data standards, equipment and training needs.

On the other hand the advantages of a departmental approach in terms of independence, control over resources, clear lines of responsibility and continuity which underlie the antithesis may be offset in practice by the corresponding disadvantages. For example, the benefits to be derived from independence may be outweighed by the costs incurred as a result of the department's isolation within the authority. Similarly the advantages of retaining control over priorities, information content and information processing facilities may be offset by the lack of support in terms of finance and technical expertise. In addition, the benefits to be derived from clear lines of responsibility may not outweigh the lack of system and data compatibility which would enable authority-wide benefits to be derived from departmental activities. Under such circumstances the advantages to be gained from continuity may be offset by inertia which seriously impedes the realization of the potential offered by the new technology.

It is also important to bear in mind that the advantages of corporate and departmental approaches are likely to change over time within any organization. To explore the dynamics of such processes two simple models of information-sharing are outlined in Figure 15.1. These describe information-sharing in terms of exchanges between different departments within an organization or different organizations and in terms of a functional model where one department or agency provides services for the other departments or organizations respectively. Figure 15.1 also shows that the relative benefits and costs to be obtained from collaboration are likely to vary from participant to participant.

The dynamic impacts of these differential costs and benefits are illustrated in Figure 15.2 for both models. This shows that as soon as the costs outweigh the benefits for one of the partners involved the arrangements as a whole are likely to be threatened even though there may be net positive advantages for the other partners.

The most important conclusion to be drawn from Table 15.1 is that both approaches have their advantages and disadvantages. The extent to which the advantages overcome the disadvantages is likely to vary considerably in practice according to the organizational context within which GIS is being introduced. The extent of these variations is explored in the next section of the paper with respect to preliminary findings of the study of GIS on British local government that is being carried out by the authors (Campbell and Masser, 1993).

15.3 The impact of GIS on local government in Great Britain

15.3.1 Research strategy

The findings discussed below form part of a comprehensive survey of British local government which is being undertaken by the authors. A combination of research methods has been adopted as the basis for this project, including a telephone survey of all 514 local authorities in Britain and detailed case study investigations. This chapter concentrates on the results of the telephone survey which was undertaken between February and June 1991. Given the large number of postal questionnaires which have been sent to local authorities and their often partial coverage it was decided to undertake a telephone survey. The 100

1. THE EXCHANGE MODEL

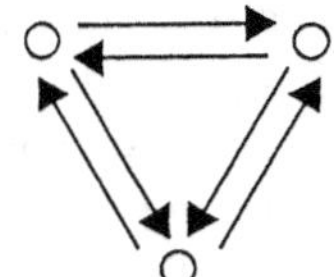

DIFFERENTIAL BENEFITS FROM SHARING

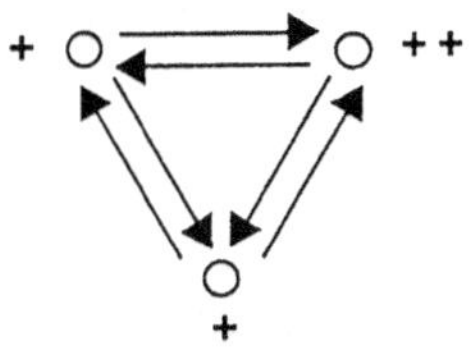

DIFFERENTIAL COSTS OF SHARING

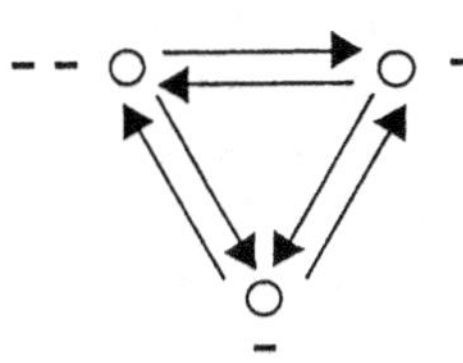

(a)

2. THE FUNCTIONAL MODEL

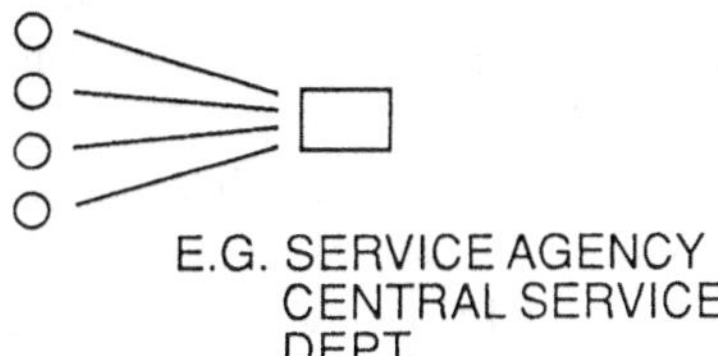

DIFFERENTIAL BENEFITS OF SHARING

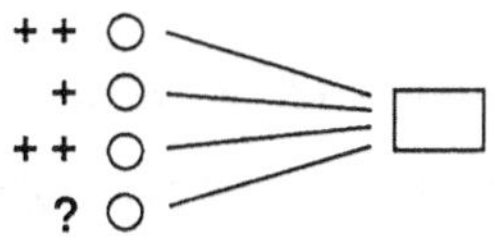

DIFFERENTIAL COSTS OF SHARING

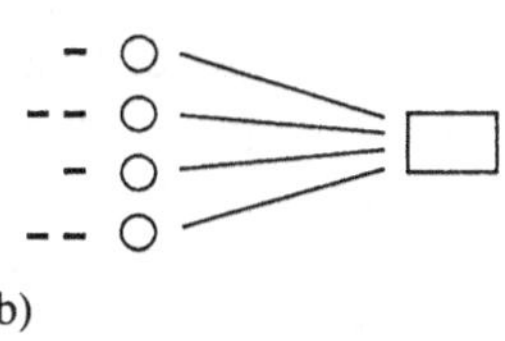

(b)

Figure 15.1 Two simple models of sharing dynamics: (a) the exchange model; and (b) the functional model.

STAGE 1

+/- - ○ ○ -/++

○ +/-

STAGE 2

+ ○ ○ +/-

○ -

STAGE 3

+ ○ ○ -

○ +

STAGE 4

○ ○

○

(a)

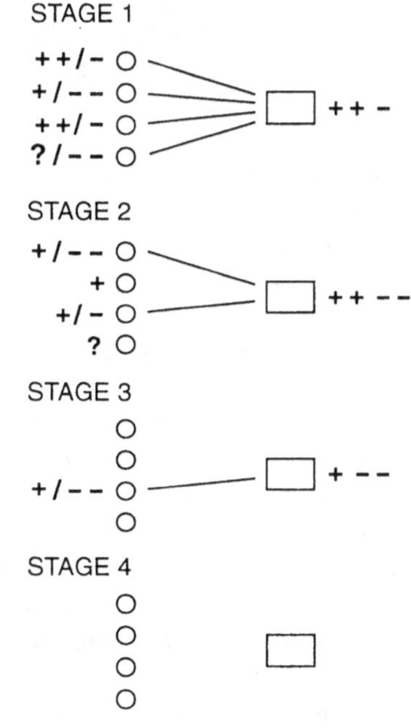

(b)

Figure 15.2 The dynamic processes: (a) the exchange model: and (b) the functional model.

per cent response rate to the survey supports the adoption of this method, and also provided valuable subsidiary information during the interviews. This method also removes the ambiguity which exists in some surveys regarding the definition of GIS, as the capabilities of the software and the precise nature of its use in the authority can be related to operational definitions. The researchers adopted a broad interpretation of GIS including automated mapping/facilities management and land information systems but excluding thematic mapping and computer aided design (CAD) packages. A more detailed discussion of the survey findings with respect to the adoption and implementation of GIS in British local authorities can be found in Campbell and Masser (1993).

15.3.2 Some findings from the survey

The units of study in this analysis are systems rather than authorities. Systems are regarded as a distinct piece or combination of software which one or more departments within a local authority are implementing. The findings of the survey show that 98 systems have been purchased by 85 local authorities. Shire Counties are the most likely to have more than one system within an authority as there are 44 systems in 32 Counties. The only other authority to have adopted more than one system is a large Scottish District.

The presence of more than one system in many of the Shire Counties indicates the extent to which departmental-based approaches are still prevalent in some areas of British local government. In fact there is a virtually even split between departmental and corporate approaches. Given this broad trend there are distinct variations between the different categories of authority. It is nearly twice as likely that a system in a County or Region will have been implemented departmentally rather than corporately. In contrast over three quarters of the systems in Metropolitan Districts and nearly 60 per cent of those in Shire and Scottish Districts were to some extent corporate in nature.

One important factor influencing the type of GIS adopted in many authorities is the division of responsibilities for funding. Although funding is more or less evenly divided between central and departmental sources, there are important differences in emphasis between the different types of authority. In the case of the counties and regions nearly 47 per cent of funding came from departmental sources. In contrast nearly 45 per cent of funding in Shire and Scottish Districts came from central sources. Nevertheless, it appears that the introduction of GIS into local authorities frequently requires a significant commitment to resources by departments even in cases where a corporate approach is adopted.

There are considerable differences between the number of departments involved in the various types of authority where some form of corporate approach has been adopted. Generally, system implementation involves no more than 3 or 4 departments and over 70 per cent of corporate systems are being developed by less than five departments. Only 12 per cent are being introduced in line with the classic corporate approach which would entail all departments in an authority participating in the implementation of GIS. Smaller authorities such as Shire Districts are most likely to favour this framework.

Planning is most likely to be the lead department followed by information technology/computer services where some form of corporate approach has been adopted. The findings suggest that, even in instances where a top down structure is favoured the lead is taken by the central computing department rather than by the Chief Executive's department. However, only in the case of one in four systems is the key coordinating role within the authority taken by one of the central departments. If it is assumed that the lead department is the one with the strongest commitment to GIS, then this is particularly the case with the technical and property related departments. Consequently the introduction of GIS appears to reflect the bottom up demand for such facilities rather than a perceived corporate need by those at the centre. The contribution of planning is most pronounced in the case of the Shire/Scottish and Metropolitan Districts where 47 and 26.5 per cent of systems are led by planning departments respectively. However, planning departments are far less significant in the case of systems being developed by County and Regional authorities with highways and information technology/computer services most likely to take the lead. There is also a high proportion of systems in these authorities for which responsibilities are shared by all participating departments.

These findings suggest that active interest in GIS has not permeated all local authority activities. Large community service departments such as housing, education and social services tend not to have become involved. This may be a reflection of the considerable administrative pressures faced by these departments and the limited resources that are available for new activities associated with data handling.

Departmental systems are associated with very similar types of activity to the GIS that are being implemented corporately. Again technical services such as property management and planning are prominent with a particular tendency for departmental systems to be developed by the engineers and highways authorities.

Overall the findings of the survey indicate that there are very few examples of centrally led fully corporate arrangements for the implementation of GIS in Great Britain at the present time. In practice around half the systems are being developed by a single department while collaborative projects most often involve no more than 3 or 4 departments with a technical service department taking the lead. There is also evidence to suggest that within the counties in particular there may be several systems being developed within one authority.

15.4 Synthesis

The findings from the survey of GIS adoption in British local government must be interpreted from two different perspectives reflecting the thesis and antithesis that was set out in the previous section respectively. In terms of the thesis the fact that half the systems currently involve more than one department points to the extent to which the benefits of information sharing have been recognized in British local government. In this respect it may be argued that the adoption of GIS is having a marked impact on existing administrative practices particularly in the Shire Districts/Scottish Districts and Metropolitan Districts.

At the same time the findings of the survey also indicate the diversity of current arrangements within local authorities where more than one department is involved in GIS implementation. The classically corporate approach led by the central Chief Executive or Computer Services Department and involving most of the departments in the local authority is only being implemented in a relatively small number of cases, primarily, small Shire Districts. More typical are the more pragmatic arrangements that had been made between three or four departments in the larger Shire Districts and Metropolitan Districts. These have usually been instigated by planning or property related departments (Campbell, 1993).

The arguments behind the antithesis are evident in the number of departmental applications of GIS that are being implemented in British local government. This is particularly apparent in the Shire Counties and Scottish Regions. It is especially pronounced in respect of departments such as Highways which already possess in-house technical expertise. Under such circumstances the advantages to be derived from information sharing as a result of inter-departmental collaboration appear to be offset by the benefits of retaining control over GIS operations.

With these considerations in mind it can be argued that these two contrasting standpoints reflect the technological and organizational context within which GIS is being implemented. On the one hand the facilities and technical experience of the County Highway Departments makes it relatively easy for them to adopt and implement GIS without having to make major changes in administrative practices. Under such circumstances there are strong pressures towards a departmental rather than an inter-departmental or corporate approach. On the other hand, the position in many Shire Districts and Metropolitan Districts is entirely different. The more limited experience of the line departments together with their more limited resources tend to favour some form of collaborative venture. In cases of this kind the impetus to innovation often comes from the top initially, even though it may eventually be implemented by Computer Services or the Planning Department rather than the Chief Executive.

A critical factor in both cases is the nature of the geographical information that is being used. Apart from basic map information, it can be argued, for example, that a large proportion of the information required by Highways Engineers is operational information which is specific to the Department itself and the use of strategic information related, for example, to population and social trends is relatively limited in this case. Under such circumstances the case for information sharing is relatively weak by comparison with the advantages of retaining control. On the other hand, it can be argued that, especially in small authorities, property based maps are a key component in the activities of several departments and that there are considerable advantages to be gained by combining efforts. This is often reflected in existing practices prior to the adoption of GIS whereby one department, for example, Planning, already provides a mapping service for other departments. In cases of this kind it is not surprising that collaborative and/or corporate approaches to GIS will be implemented.

To explore these questions in greater depth a number of detailed case studies are currently being carried out by the authors. In addition to providing valuable information about the technological and organizational context within which GIS are being implemented in different authorities, these case studies will also throw light on the dynamics of the GIS implementation process. Until the findings from the case studies are available only interim conclusions can be drawn with respect to the effect of GIS on British local government. However, there is a great deal of evidence from the survey to suggest that the adoption and implementation of GIS has led to the emergence of a variety of corporate approaches which seek to exploit the advantages of information sharing especially in the Shire Districts/Scottish Districts and Metropolitan Districts. On the other hand, there is also evidence of the continuation of department-based approaches, especially in the Shire Counties, despite these apparent advantages. As a result it must be tentatively concluded that the findings from British local government point to the extent to which organizational and political factors apparently offset in many instances, the theoretical benefits to be obtained from structures which seek to promote information sharing.

Acknowledgment

The research described in this paper is funded under the Economic and Social Research Council/Natural Environment Research Council Geographic Information Handling Project. An earlier version of the paper was presented at the NCGIA Workshop on information sharing in San Diego in February 1992.

References

Audit Commission for local authorities in England and Wales, 1990, Preparing an information technology strategy: making it happen. London: HMSO.

Bates, B.J., 1988, Information as an economic good: sources of individual and social value in Mosko, V. and Wasko, J. (Eds). *The Political Economy of Information*, Madison, US: University of Wisconsin Press.

Beath, C.M., 1991. Supporting the information technology champion. *MIS Quart*, 355–72.

Bryan, N.W., 1992; A review of pricing and distribution strategies: local government case studies, *Proc. URISA 92*, **4**, 13–25.

Campbell, H., 1992, Impact of geographic information systems on local government. *Computers Environment and Urban Systems*, **16**, 531–41.

Campbell, H., 1993, GIS implementation in British local government, in Masser, I. and Onsrud, H. (Eds). *The Diffusion and Use of Geographic Information Technologies*, Dordrecht: Kluwer.

Campbell, H. and Masser, I., 1993, The impact of GIS on British local government: some findings from Great Britain. *Int. Jour. GIS*, **7**.

Coulson, M. and Bromley, R., 1990, The assessment of user needs for corporate GIS – the example of Swansea City Council. *Proc. EGIS 90*, Utrecht. The Netherlands.

Croswell, P L., 1991, Obstacles to GIS implementation and guidelines to increase the opportunities for success. *URISA Journal*, **3**, 43–57.

Danziger, J.N., Dutton, W.H., Kling, R. and Kraemer, K., 1982, *Computers and Politics: High Technology in American Local Governments*, New York: Columbia University Press.

Finkle, R.W. and Lockfield, F.M., 1990. *Major Cost Savings in Multiparticipants AM/GIS: the Santa Clara County case study*, JMM Tech. Pub. Service, J.M. Montgomery Inc.

Gault, I. and Peutherer, D., 1989, Developing GIS for local government in the UK, paper presented at the European Regional Science Conference, Cambridge.

Hepworth, M., 1990, Mobilising information capital in local government, *Local Government Policy Making*, **16**, 42–8.

Kraemer, K.L. and King, J.L., 1988, The role of information technology in managing cities. *Local Govt Studies*, **14**, 23–47.

King, J.L. and Kraemer, K.L.., 1985, *The Dynamics of Computing*. New York: Columbia UP.

Levinsohn, A., 1990, Institutional issues in GIS implementation. *Proc. ISPRS Commission IV*. Tsukuba, Japan.

Masser, I. and Campbell, H., 1991, Conditions for the effective utilization of computers in urban planning in developing countries. *Computers Environment and Urban Systems*, **15**, 55–67.

Masser, I. and Campbell, H., 1992, Geographic information systems in organizations: some conditions for their effective utilization, in Lepper, M. de, Scholten, H.J. and Stern, R.M. (Eds). *The Added Value of Geographic Information Systems in Public and Environmental Health*, Dordrecht, Kluwer.

Masser, I. and Onsrud, H. (Eds). 1993, *The Diffusion and Use of Geographic Information Technologies*, Dordrecht, Kluwer.

Mosco, V. and Wasko, J. (Eds), 1988, *The Political Economy of Information*, Madison: University of Wisconsin Press.

Onsrud, H. and Rushton, G. (Eds), 1992, Institutions sharing geographic information Tech Report 92-5, National Centre for Geographic Information and Analysis, Santa Barbara University of California.

Openshaw, S. and Goddard, J.B., 1987, Some implications of the commodification of information and the emerging information economy for applied geographic analysis in the United Kingdom. *Envt and Plg*, **A19**, 1428–39.

Pickles, J., 1991, Geography, GIS and the surveillant Society, *Proc. Applied Geog. Conf.*, Toledo, OH.

Rhind, D., 1992, Data access, charging and copyright and their implications for GIS. *Int. Jour. GIS*, **6**, 13–30.

Tveitdal, S. and Hesjedal, O., 1989, GIS in the Nordic countries – market and technology, strategy for implementation – a Nordic approach. *Proc. GIS 89*, Vancouver, Canada.

Wellar, B., 1988, Institutional maxims and conditions for needs sensitive information systems and services in local governments. *Proc. URISA 88*, **4**, 371–8.

16

Modelling environmental systems with GIS: theoretical barriers to progress

David Livingstone and Jonathan Raper

The integration of dynamic environmental models within GIS is something that is seen by many as desirable but, in comparison to the number of implemented models, is very rarely done. This chapter recognizes the difficulty in reconciling information held in different spatial and temporal data models. It argues that in order to help solve this problem, the semantic data model which characterizes the environmental system under consideration must be defined at a higher level of abstraction than the spatial and temporal data models used to represent the data. This higher level involves the definition of environmental metaclasses which are able to assimilate data held in different data models. Differing process model schema require different spatial and temporal data models as these often reflect the numerical methods used to acquire solutions. These different spatial and temporal data models should be representable in a GIS so as to be able to integrate, represent and compare diverse process modelling approaches within the same system whatever the underlying GIS data model. To be able to do this requires an interface to the GIS application and environmental models/applications at a higher level of data definition so that morphisms between different models can be achieved.

16.1 Introduction

16.1.1 Integration of environmental models with GIS

Aims

Simply stated, spatially and temporally referenced information is fundamental to an information based approach to modelling environmental systems, and in turn spatio-temporal information systems are fundamental to the successful management of this information. Spatial information systems can provide an important part in the construction of integrated systems for storage, analysis, visualization and modelling of environmental data (Raper 1991). Many authors have described methodologies and prototypes for such integrated systems, both in a largely GIS context (Fedra, 1993, Nyerges, 1993; Maidment, 1993) and largely outside the field of GIS (Farmer and Rycroft, 1991). This chapter looks at the way in which GIS in its current form can be incorporated into more generic systems, and at the way that the nature of both environmental information and information systems affect their design.

Fedra (1993) notes that whilst the linkage of environmental models to GIS is becoming increasingly frequent, 'in the majority of cases GIS and environmental models are not really integrated – they are just used together'. The question of when integration is *really* integration is one of degree, however, the scope of each system is a key issue as is the coupling between the components that provide the system's functionality.

Coupling

As indicated above, coupling varies in character. The basics of coupling are determined by the architectures, protocols and structures of information systems and, in those terms, 'openness' is a key design principle. Close coupling implies integration and a high degree of control and functionality. With respect to GIS which is the natural 'focal point' for a closely coupled system (Nyerges, 1993), the degree of coupling can be characterized by how it works (Table 16.1) or what it achieves (Table 16.2).

A well integrated system should give all components of the system access to all of the data in the system and the data should have enough information associated with it to allow it to be used, without manual intervention, by these components. A well integrated system is desirable because it should allow many different problems to be addressed. The data can then be structured in many forms and processes and associated with many different representations or visualizations (Fedra, 1993).

Limitations

The limitations of a particular GIS when used for the data management tasks in an integrated environmental modelling system, are largely caused by the underlying data models that are used by the GIS and the representations derived from these data models. Maidment's (1993) work in integrating GIS and hydrologic modelling routines has caused him to observe that 'the most critical limits are the differences in the data models and in the way relationships between variables are handled in GIS and hydrologic models'. To address this problem, 'it is essential that there be a one-to-one mapping between the portions of their data structures that share common data'. Likewise Nyerges (1993) says 'developing effective process models within a GIS context depends on the nature of the architecture that fosters such development'. For a spatial information system to be able to hold data derived from a variety of different sensors and sampling techniques, and to represent and transfer data between a variety of

Table 16.1 After Nyerges (1993)

Technology	Degree
Data transfer	Low
Reference/cross index	Low-middle
Federated	Middle-high
Integrated/embedded	High

Table 16.2 From Maidment (1993)

Functionality	Applications
Hydrologic Assessment	Mapping, risk assessment
Hydrologic Parameter Determination	Elevations, slopes, watersheds
Hydrologic Modelling Within GIS	Network flows (costs)
Linking GIS and Hydrologic Models	Using grids as input/output to finite difference and element solutions (coded routines)
Object-oriented Linkage	Feature abstractions, feature and model hierarchies

different models it needs to be able to cope with the mapping of data from one model to another.

In a well integrated system the data would be located in the same database and be accessible by both the models and GIS functions (Fedra, 1993). Fedra further envisages a situation whereby the system is perceived by the user to be a kind of modelling workshop where environmental systems can be analysed and the distinction between GIS functions and modelling routines are blurred. This illustrates the point that one of the key objectives of a GIS is to create a spatial, and possibly temporal, model of the data in order to analyse and perform operations upon it.

However at present many GIS do not support a rich enough variety of data models to be able to provide the kind of close coupling required for the kind of integration envisaged by Fedra. Thus the kind of system that is really required by environmental GIS users is not actually currently available. Designing a more integrated, dynamically coupled system poses questions both about the nature of models and of the real systems that they represent. Hence, the actual data models that are required are not just a function of the information system, they are also affected by the nature of the environmental systems that are being represented and the techniques that can be used to analyse them.

16.2 Environmental systems

16.2.1 Environmental systems - problem domain

Environmental systems are amongst the most complex and poorly understood in science (Chorley and Kennedy, 1971). This is due to the difficulty of identifying the relevant functional objects in space or time, their temporally and spatially nested nature, the existence of system thresholds and feedbacks creating complex response to external forcing, the unknown characteristics of the external energy inputs and the precise impact of human intervention and management. These characteristics of environmental systems render the problem domain extremely complex: attempting to hold any one of these factors steady while varying the others is only possible in the laboratory where scaling problems are then involved, or with luck in the field when a system state change occurs during monitoring. The potential richness of representation in an environmental GIS has great potential in this area.

In this chapter examples of the problems associated with the modelling of a coastal geomorphological system are used to illustrate the problem domain for modelling environmental systems using GIS. Work on Scolt Head barrier island in North Norfolk by Bristow, Raper and Allison (1992) and Raper *et al.* (1993) has described the environment of an evolving coastal barrier backed by sand dunes and salt marshes and terminated in spits and ebb tidal deltas. In this environment, GIS are needed for both mapping change (which is rapid), morphology (which is complex), *and* for predicting future development based on wave energy inputs and sediment supply.

16.2.2 Environmental systems - modelling approaches

Spatial data organization

To model environmental systems we need to be able to associate data from field measurements such as borehole samples, ground survey, monitoring records, aerial photography and video with key environmental entities. For example, in a coastal system the entities will be dunes, spits, beaches, marshes and the like. However the behaviour of these entities is tightly coupled with their spatial expression: after any storm their positions will have changed, making it difficult to represent them with fixed geometric data structures. Therefore the system must be dynamic, capable of supporting process models, and representing historical and predictive scenarios. For the data management tasks of the system to be carried out by a GIS, then the GIS must be capable of storing or representing the data pertaining to the environmental entities in the most appropriate way. When environmental modellers design data storage/representation schemes they are often closely linked to the way in which the numerical methods chosen need to access the data, the GIS needs to be able to provide access to the data in a variety of forms. Hence it is instructive to look at the kind of data models used by environmental modellers.

Computer Modelling in the Environmental Sciences (Farmer and Rycroft, 1991) is derived from the conference of the same name held at the British Geological Survey, Keyworth, April 1990 and contains 29 papers, most of which describe the implementation of environmental computer models. Table 16.3 gives a breakdown of a sample of these models in terms of the environmental systems they represent, the numerical methods they use, and the spatial models that are adopted. There are 18 models represented in Table 16.3, 13 of which use some kind of gridded or cellular spatial data model.

The spatial data model used reflects the numerical methods that solve the process model, which in turn can be influenced by computer data structures and storage schemes such as beta trees and arrays. For example since finite difference solutions to problems in the spatial domain are efficiently solved by mapping values onto a regular grid, this approach is well suited to computer methods, and specialized processor architectures have been developed that can make these procedures even more efficient. Geographical data can be collected and have specific spatial sampling constraints imposed upon on it: gridding is a good example of this, e.g. quadrats. Some data is collected in gridded form for reasons

Table 16.3 Comparison of environmental models described in Computer modelling in the environmental sciences (Farmer and Rycroft, 1991)

Environmental model	Numerical method	Spatial model
Tidal-surge	Finite difference (time steps)	Regular grid
Wave model	Finite difference (time steps)	Regular grid
Ocean circulation	Finite difference (time steps)	Regular grid
Hydrodynamic	Finite difference (time steps)	Regular grid 3D
Pollution spill	Gaussian spread (time steps)	Regular grid
Pollution spill	Random walk (time steps)	Regular grid
Pollution dispersion	Langragian/random walk	Line+point vector
Catchment hydrology	Monte Carlo simulation	Gridded planes
Rainfall-runoff	Not specified	Linear reservoirs
Rainfall runoff	Functional flows	Lumped area regular grid
Atmospheric circulation	Finite difference (time steps)	Regular grid
Seismic waves	Propagator matrices/Bessel functions	Anisotropic symmetry/ rotating tensors
Porous media/permeability	Random walk	Conic sections/unit cell
Magnetic and gravity fields	Not Specified	Triangular Facets on grid
Thermo-rheology of crust and mantle	Fourier analysis/finite difference	2D plates/envelopes
Atmospheric circulation	Nonlinear dynamics	Empirical orthogonal functions
Atmospheric turbulence over terrain	FFTs	Gridded terrain/stratified layers
Habitat distribution	Bayes theorem	Gridded cells

of minimizing bias, in other cases it may be because of constraints posed by the instrumentation such as with remote sensed devices. It may even be because of the spatial model for which the data is destined, i.e. in some cases the model exclusively drives the sampling procedure. The spatial structure of the data is therefore a limiting factor in the information that can be extracted from the data and this is the case for both storage and collection.

There is no reason to suggest that one particular spatial data model is suitable for all systems except by appealing to a 'grand unified theory' and so it is important for a variety of models to be supported. A view of geographical space that places primary importance on the type of entities that can be represented and can interact dictates that the entities should define the space that they occupy and not be defined by it.

Categories of environmental data

Looking from an environmental standpoint the key problem when building an environmental database is to classify the world in a way that allows relevant data to be associated with models of environmental entities. This requires a consideration of the types of environmental data available from existing surveys (collected for a variety of purposes), and the types of environmental data collectable.

Raper (1989) described types of objects according to the methods by which they are defined from environmental data. First, 'sampling-limited' objects can be defined with reference to a key parameter, e.g. a discrete boundary on the ground. In this case the detail in the spatial representation is limited by the sampling density. Second, 'definition-limited' objects can be defined with reference to the value of continuously varying parameters or model. Here too the detail is limited by sampling density, but is also affected by the magnitude of the chosen bounding value.

The process of creating objects using the definition-limited and sampling-limited methods is a modelling process. This process requires an input of environmental data, of the kind that has been referred to as 'observable', and produces an environmental model as its output; Raper (1989) calls them 'geo-objects'. The environmental data could be from a variety of sources such as remote instrumentation, destructive sampling, local surveys or even derived from other models. It can be regarded as a pool of information that can be used for model building and as a yardstick to use for verification of models. One of the qualities of such environmental data is that either it is based on direct observation or it is capable of being verified by direct observation and any objects in the database that are used to store such data can be stored distinctly with those objects that represent environmental entities.

For example, it is possible to take borehole samples of sediments and measurements of particle size from any environmental entity. The existence and mapping of such information is not necessarily equivalent to the definition of environmental entities as it may be, for example, produced by routine data gathering methods. However, observable data may become an attribute of an environmental entity, for example by using sampling-limited and definition-limited methods. For instance it may be possible to classify regions of an interpolated 3D grid of values by using criteria based on the particle size statistics, thus creating sedimentary structures as a class of environmental entities.

Observable environmental data, whether model-derived or collected by field instrumentation is a source *for* attribute information that can be assigned to environmental entities. Whilst there is a connection between, for example, the form of an entity and the value of a set of elevation readings taken from the surface of an entity, the elevation readings themselves are primarily attributes of a topographic survey. It is only in the process of defining an instance of an environmental entity such as a spit that the elevations become associated with the form of the entity. This might be through a pattern recognition procedure, a database search specification, an interactive assignment on the ground or a sampling-limited decision made in the field. Sources for observable data are not always of the primary kind: in fact the source for geographical information is often a map of some kind which should be regarded as observable derived data where the modelling assumptions used to derive it from the original measured data may not be known.

Observable environmental data should therefore be a pool of spatially and temporally referenced data associated with information about the source and quality of the data, that can act as a source or sink for assignment to environmental data models. In the type of model being proposed here, observable data is associated with instances of an Environmental Data Class, whereas model data is associated with instances of an Environmental Model Class. Models are built by reference to information, and information can be derived from models, but they are distinct concepts and the data about them should be treated differently.

16.3 Information systems

16.3.1 Space and time – a geographical context

When modelling an environmental feature its form is not the only way in which the feature can be categorized. The feature can also be defined by such factors as its composition (both in terms of materials and any sub-structures/features), its process-response characteristics and the special environment that it is evolving in. Relict features are thus defined by their relict environments but may warrant new definitions in terms of their currently active process-response characteristics. The geometric representation of the feature is a generalized abstraction of its form which enables us to perform quantitative and qualitative spatial analysis. Whilst this is a very important mathematical modelling approach it should not be the driving force behind our feature classification model. Nunes (1992) describes geometry not as a tool for representing space but rather for representing objects that occupy space. The idea that objects, and the relationships between them, divide space is not a new one: 'space, time and colour (being coloured) are forms of objects' (Wittgenstein, 1921), but it is particularly relevant with the advent of object-oriented data modelling and is the basis for our approach to representing environmental features.

Many of the important features that are identified in the barrier island system can be modelled in both the spatial and temporal frameworks associated with the extent of recognizable environmental entities such as a tidal creek, for example, the sediment load of the creek as a function of space and time. In other situations entity recognition may not be so clear cut but instead depend on the location of statistical or threshold boundaries. In the wider context many authors have recognized the importance of tackling the fundamental concepts of geographical space and reasoning when attempting to model geographical systems (Dikau, 1990; Feuchtwanger, 1989; Herring, 1991; Nunes, 1991).

16.3.2 Space and time – a GIS context

While GIS have become widely used over the last few years for the storage and analysis of spatially referenced data. The systems in use are most commonly used for automated mapping (display and interrogation of a screen map) and facilities management (maintenance of a spatially referenced asset register) in the utilities and local government sectors (CAC 1991). This kind of GIS use reflects the software architecture of the most common commercially available systems i.e. the shape/position of features in geographic space are approximated by vector or raster 'geometry', and descriptive 'attributes' are recorded using standard alphanumeric data types (Raper and Maguire, 1992).

Vector systems of this kind have been called position-based systems (Herring, 1991) since this representational scheme is organized by the number and type of positions stored in the geometry store. Position based systems are usually structured in some way: typical structurings are based upon the creation of nodes at line intersections through the 'planar enforcement' process (Goodchild, 1992). A consequence of applying planar enforcement is that the polygons formed by this process are non-overlapping and every locatable point is assigned to one and only one polygon. Raster systems are also explicitly position-based since the cells are mutually exclusive and the cell's value determines whether it is grouped with other cells. Storage of overlapping areas or non-intersecting lines requires the use of separate layers. Commercially available position-based vector GIS contain tools to build structures describing topological relationships, such as connectivity and adjacency, to permit the overlay of different layers graphically and logically, and to update the links between the geometry and the attributes. Raster GIS are also 'layered' since each cell may only contain one value making it necessary to store different geometries in different raster maps. In both raster and vector cases referred to there is a one-to-one relationship between a piece of geometry such as a polygon and a record containing alphanumeric attributes.

Position-based GIS offer powerful functionality for applications where the layers can be readily defined, and where they contain non overlapping features which are static. These conditions are met in many application areas such as fixed asset management and modelling statistical data over administrative areas. However, many organizations handling environmental data find that position-based GIS are inadequate for the storage and modelling of features such as river systems or coastlines which undergo rapid temporal change. Such systems cannot store complex features which are aggregations of geometry (without using grouping operations on the attributes) and must store different states of a feature which is changing its position by placing each state in a separate layer. It is also difficult or impossible to derive change factors for the differences between feature states as, by definition, different geometry is treated as a different feature. Such restrictions apply equally to vector and raster geometric data alike as it is the one-to-one nature of the linkage between the geometric and the non-geometric attributes which is the limiting factor. These limitations have acted as significant restrictions on the use of GIS in hydrological, geomorphological and marine applications (Raper, 1991).

The arrival of 'object' approaches to GIS offers some powerful new solutions to these limitations of position-based GIS (Worboys *et al.*, 1990; Worboys, 1992). Object approaches in GIS are focused on the types of features or objects being represented. This means that the database is constructed from sets of related object classes, where an object class is defined as a phenomenon having defining characteristics which may be grouped to higher levels and broken down into lower ones. These characteristics define each object class in the system data model, and instances ('objects') of the object class are stored in the system database. In an object GIS the descriptive characteristics are not differentiated between geometric and non-geometric types at the interface level. Different object GIS employ different mechanisms to unify storage of the geometric and non-geometric attributes: some systems employ object databases with unified storage, others use an object-oriented language to manipulate the data storage in a relational database.

Object GIS have flexible representational tools, hence objects in any object class can have associated geometries which overlap each other to any extent necessary. Interaction of objects in and between object classes are defined by interaction rules which may be governed by topological criteria. Operations which act only on the object class can also be stored with the instances, making it possible to manipulate particular object classes in a variety of ways.

16.3.3 Object-oriented data modelling

The object-oriented paradigm incorporates concepts which can be applied to both the description of real world systems and to the definition of computer systems through a programming language. This appears to offer a good opportunity to represent concepts in both of these types of systems; object-oriented systems incorporate concepts that translate well into the language of real world systems, for example sub-classing, inheritence, encapsulation of data and behaviour, and aggregations; these concepts are analagous to similar terms used to describe the nature and classification of physical systems. In environmental systems we perceive such things as hierarchical relationships (Oertel,1985), when we classify the world into typologic and taxonomic schema, and aggregation relationships, features made up of agglomerations of smaller features. Therefore a data model that also incorporates these concepts is a promising vehicle for the storage and processing of environmental data.

The object-oriented data model offers potentially more powerful semantic concepts than does the relational model which is fundamentally mathematical in nature. However, unlike the relational model, there is no set of 'commandments', akin to Codd's (1970) principles, that define the object-oriented model. This problem is being approached by standards bodies such as the Object Management Group (Soley,1990) but in the meantime the effect has been to limit the uptake and development of object-oriented databases. The lack of a standard terminology is a limitation on the efficacy of object-oriented data modelling, because one of the major advantages of an object-oriented approach is the tendency to dissolve the boundaries between programs and databases, allowing structures and parameters for the numerical models to be persistent objects in a managed database. This is exactly the kind of situation that we require for a GIS-based environmental modelling database. One of the ways that GIS vendors have sought to get around the problem of storing persistent objects in a non object-oriented fashion, is to map the object-oriented model onto a relational one and then to store the objects in a relational database. Examples of such databases used in a GIS environment are the POSTGRES database and Smallworld's Version Managed Datastore both of which access the records in the database tables as if they were objects (in the object-oriented sense). This allows us to design systems using an object-oriented model and then to implement them in a more traditional way.

In order to design an object-oriented system, in the absence of a definitive object-oriented model, a consistent set of semantics needs to be adopted; this is often based on models used by particular programming environments such as Smalltalk, C++ and Smallworld Magik. However, from the standpoint of system analysis and design, it is not desirable to design a system on the basis of the mode of implementation, and so a more generic set of semantics is required. Booch (1991) describes a set of semantics and methodologies for the design of object-oriented systems which draws upon the histories of object-oriented programming languages some of which have been available for over twenty years. The usages and descriptions of object-oriented concepts in this chapter are based upon the Booch terminology.

16.4 Modelling environmental systems in a GIS context

16.4.1 Environmental data models – a GIS perspective

Herring (1991) describes morphisms, the mapping of one model to another, as the means which enable a GIS to implement a variety of spatial models and suggests that object-oriented and relational models have the necessary tools to do this as well as to create a natural language interface based upon their semantic concepts; such an interface is akin to the the kind of situation envisaged by Fedra. This kind of user environment where the language of the spatial data types and functions are meaningful to the user in terms of their application is a feature of the function mapping facility as defined in the UGIX project (Raper and Bundock, 1992) where the entities and functions are defined at a higher level than the GIS.

In order to be able to model geographical space as a property of entities (Nunes, 1991) the definitions must be made at this higher level. This is the process of defining an environmental semantic data model. Since the aim is to devise a model whereby entities and their spatial attributes define space, the situation will arise where different representations and therefore different ways of dividing space are most suitable for different purposes. It is desirable to avoid the situation whereby different instances of the same entity have to exist for different types of representation. Since it is still the same entity that is being represented, the definition of the entity needs to be at a higher level of abstraction than that of its representations. Any morphisms that map between the GIS data models that represent the entity should therefore be available at this higher level.

A question that also needs to be considered is whether more than one object can occupy the same part of geographical space at the same time. Since the definition of the entity is at a higher level than its representation such a constraint would not preclude different representations overlapping and so a principle of object exclusion would go further than the technique of planar enforcement which ensures that polygonal geometries cannot overlap. The principle of exclusion is of prime importance in modern physics and Nunes (1991) refers to such principles to argue for a geographical space defined by the relationships of discrete physical objects. Whilst such arguments are less clear for more abstract geographical notions, e.g. demographic statistics and associated spatial patterns, they can be used for the construction of environmental objects in aggregated hierachies with relationships between and within hierarchies.

One of the problems with a project such as this is the continual temptation to try to define a metaphysical model of the universe precisely because of the complexity and scale of environmental processes. To some extent these issues must be faced and opinions formed about what might be termed environmental metaclasses. Ultimately decisions about metaclasses will be the most important as regards the flexibility and durability of the system since these classes will determine what objects it will be possible to represent. In object-oriented parlance metaclasses are those classes whose instances are classes in themselves, and they can be used to define attributes and methods that are common to all the instances of the class that they create. These instances are the models of environmental entities using a particular definition and representation required by the GIS data model and deriving their data from a pool of spatially and temporally referenced environmental information.

Instances of the environmental model class can thus derive their attributes from instances of the environmental data class whether derived, from other models, or observed, directly by experiment or sampling. Both of these different class categories (environmental data and environmental model) have associated metaclasses that are independent of the mode of representation but are concerned with the behaviour and relationships of the class categories and the morphisms required to get the data into a usable form.

16.4.2 Proposed design of an environmental modelling system

As shown in Figure 16.1 there are four major components to the system; the spatial

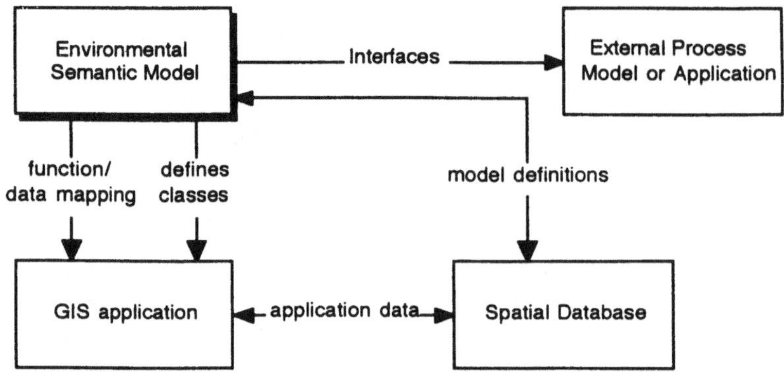

Figure 16.1 Proposed system structure.

database, the GIS application, the environmental semantic model and the external process models/applications. The important features of the system are:

- all the data is held in the same spatial database;
- different levels of data are held in the database;
- environmental metaclasses:
- are the structure of the semantic model
- integrate external models with the GIS
- reconcile different representations within a GIS using morphisms defined as methods
- allow entities in the database to be defined in an implementation independent way
- allow operations on these entities using generic functions
- provide different class interfaces for models/data from each phase of the modelling;
- the encapsulation of data and methods.

This encapsulation allows the simulation of dynamic response using self modifying methods in response to external or internal factors.

The spatial database acts as a repository for data at different levels of data definition. The concepts of data definition levels are set out in the BS ISO/IEC 10027: 1990 framework for Information Resource Dictionary Systems (IRDS). Instances of entities that occur in the GIS occur at the application level whereas the definitions of these entities occurs one level above, the Information Resource Dictionary Level. The object-oriented concept of class and instance spans two levels, in IRDS parlance this is a level pair, the level pair that occurs between the IRD level and the application level is the application level pair. The environmental semantic data model is the component of the system which integrates external models with the GIS application and reconciles different representations. It is defined at a higher data definition level and forms an applications interface for application level pair data in accordance with the UGIX model (Raper and Bundock, 1992) and is shown in Figure 16.2. Incorporating the concepts of data definition-levels into the system design also allows the tackling of the requirements stated by Feuchtwanger (1989) for application-geometry independence and in fact goes one stage further by specifiying 'model-application-geometry' independence as an aim.

Classes defined in the Environmental Semantic Model operate as metaclasses to the environmental entities in the GIS, a metaclass being a class whose instances are classes themselves. The requirement that entities should be capable of being represented by a different spatial and temporal data model is tackled by associating such models with the entity classes and providing methods on the metaclass which will perform morphisms

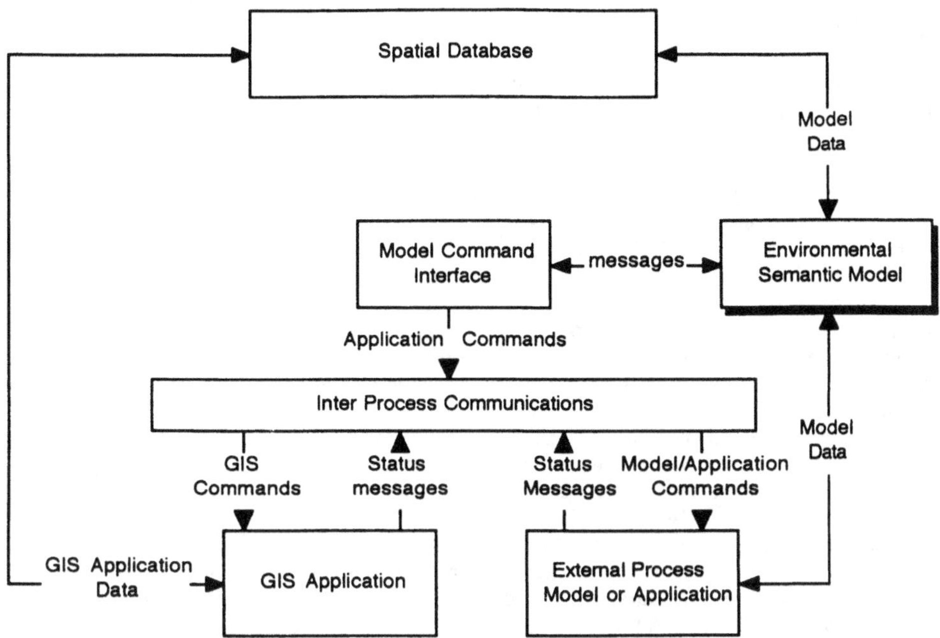

Figure 16.2 Integration of GIS and environmental models via an applications interface.

between the models. These models and methods are associated with the metaclass and through this to the lower level class definition.

16.5 Conclusions

This chapter has demonstrated that GIS has great potential as the core of an integrated environmental information and modelling system due to its evolution as a spatial database management system capable of handling different spatial data structures, its rich suite of associated analytic tools and its migration on to open computer platforms. Whilst acknowledging that there is an increasing tendency by vendors to provide customizable interfaces and object-oriented type programming tools, the authors believe that for GIS to fulfil its potential in the field of environmental modelling it needs to be in an application-independent environment so as to maximize the number of models that can be supported and to minimize 'blind alley' implementations. Obviously this requires a commitment by the vendors to an open systems approach and the development of standards for translations between spatial and temporal data models.

The authors also see object-oriented data modelling methods as the best way of tackling the issues raised and believe that the adoption of true object-oriented databases with a standard set of semantics is vital to this approach; this would greatly aid the goal of interchangeability between spatial data models. Even without the frameworks described above being in place, systems available today can still tackle the implementation of an integrated environmental information and modelling system. This chapter has put forward a design approach which can concentrate on producing a system that uses semantics and methods derived from the language and procedures of environmental data and environmental modelling.

References

Booch, G., 1991, *Object-oriented Design with Applications*, Redwood City: Benjamin-Cummings.

Bristow, C.S., Raper, J.F. and Allison, H.M., 1992, Sedimentary architecture of recurved spits on a macrotidal barrier island, Scolt Head Island, Norfolk, England, *Courier Forshungsinstitut Seuckenburg*, **154**, 14.

CAC, 1991, *GIS: Requirements and marketing opportunities in W. Europe*. London: Computer Applications Consultants.

Chorley, R and Kennedy, B., 1971, *Physical Geography: A Systems Approach*, London: Prentice Hall.

Codd, E.F., 1970, A relational model of data for large shared databanks. *Communications of the ACM*, **13**(6), 377–8.

Dikau, R., 1990, Geomorphic landform modelling based on hierarchy theory. *Proceedings, 4th International Symposium on Spatial Data Handling*, Zürich, 23–27/7/90, 230–9.

Farmer, D.G. and Rycroft, M.J., 1991, *Computer Modelling in the Environmental Sciences*, Oxford: Clarendon Press.

Fedra, K., 1993, GIS and environmental modelling, in Goodchild, M.F., Parks, B.O. and Styaert, L.T. (Eds) *Environmental Modelling with GIS*. Oxford: Oxford University Press, pp. 35–50.

Feuchtwanger, M., 1989, Geographic logical database model requirements. *Proceedings, Auto-Carto 9*, Baltimore, MD, USA, 2–7/4/89, 599–609.

Goodchild, M.F., 1992, Geographical data modelling. *Computers and Geosciences*, **18**, 401–8.

Herring, J., 1991, The Mathematical modelling of spatial and non-spatial information in Geographic Information Systems, in Mark, D.M. and Frank, A.U. (eds) *Cognitive and Linguistic Aspects of Geographic Space*. NATO ASID 63, Las Navas del Marqués, Spain, July 8–20, 1990, Dordrecht: Kluwer, 313–50.

Maidment, D.R., 1993, GIS and Hydrologic Modelling in Goodchild, M.F., Parks, B.O. and Styaert, L.T. (Eds) *Environmental Modelling with GIS*. Oxford: Oxford University Press, pp. 147–167.

Nunes, J., 1991, Geographic Space as a Set of Concrete Geographic Entities, in Mark, D.M. and Frank, A.U. (Eds) *Cognitive and Linguistic Aspects of Geographic Space*. NATO ASID 63, Las Navas del Marquás, Spain, July 8–20, 1990, Dordrecht: Kluwer, 9–34,

Nyerges, T.L., 1993, Understanding the scope of GIS: its relationship to environmental modelling, in Goodchild, M.F., Parks, B.O. and Styaert, L.T. (Eds) *Environmental Modelling with GIS*. Oxford: Oxford University Press, pp. 75–93.

Oertel, G.F., 1985, The barrier island system. *Marine Geology*, **63**, 1–18.

Raper, J.F., 1989, The 3-Dimensional Geoscientific Mapping and Modelling System: A Conceptual Design, in Raper, J. (Ed.) *Three Dimensional Applications for GIS*. London: Taylor & Francis.

Raper, J.F.; 1991, Geographical information systems. *Progress in Physical Geography*, **15**(4), 438–44.

Raper, J.F. and Bundock, M.S., 1991, UGIX: A layer based model for a GIS user interface, in Mark, D.M. and Frank, A.U. (Eds) *Cognitive and Linguistic Aspects of Geographic Space*. NATO ASID 63, Las Navas del Marqués, Spain, July 8–20, 1990, Dordrecht: Kluwer, 449–75.

Raper, J.F. and Maguire, D.J., 1992, Design models and functionality in GIS. *Computers and Geosciences*, **18**(4), 387–94

Raper, J.F., Bristow, C.S., Livingstone, D. and Köppen, A., 1993, Reconstructing 3D sand body geometry with IVM: A case study from a Barrier Island Spit Bar. *Dynamic Graphics Applications Journal*, **1**, 43–51.

Smith, M.B. and Brilly, M., 1992, Automated grid element ordering for GIS-based overland flow Modelling. *Photogrammetric Engineering and Remote Sensing*, **58**(5), May 1992, 579–85.

Soley, R.M. (Ed.) 1990, Object management architecture guide. Object Management Group document 90.9.1, 4/9/90.

Wittgenstein, L., 1921, *Tractatus Logico-Philosophicus*, Routledge & Kegan Paul 1961. Translated from *Logisch-Philosophicus Abhandlung*, Annalen Der Naturphilosophie.

Worboys, M.F., 1992, A generic model for planar geographic objects. *International Journal of GIS*, **6**, 353–72.

Worboys, M. F., Hearnshaw, H. and Maguire, D.J., 1990, Object-oriented data-modelling for spatial databases. *International Journal of Geographical Information Systems*, **4**, 369–84.

17

Coupling of process-based vegetation models to GIS and knowledge-based systems with reference to vegetation change

David Miller

This chapter describes the coupling of a process-based vegetation succession model with GIS using a knowledge-based system. Knowledge of the components of the vegetation model and of spatial data-handling are encoded to facilitate the graphical and statistical expression of possible changes in distribution of a candidate vegetation type (Bracken – *Pteridium aquilinum*). A frame structure is used for the representation of knowledge and attributes of spatial data. Analysis of the changes in vegetation distribution with respect to the surrounding physical environment and the vegetation model is used to describe and explain the predicted change in distribution.

17.1 Introduction

The use of knowledge-based systems (KBS) in association with GIS has been suggested by several authors (Ripple and Ulshoefer, 1987; Robinson *et al.*, 1987) but with the emphasis on data management. Rapid development of GIS has enabled the incorporation of modelling of different forms: geographic (Fisher, 1991); inductive (Aspinall, 1992); rule-based (Wright and Morrice, 1988) and knowledge-based (Peuquet, 1984). In the mapping of vegetation, GIS maybe employed to present spatial data and aspatially represented knowledge (Lee *et al.*, 1987; Lowell and Astroth, 1989).

Attempts to integrate models and GIS within single packages currently leave the user with little or no control over the operation of the model. However, the user who has an interest in the understanding of scales of ecological phenomena in time and space might benefit from a tool which provides some guidance on spatial analysis. GIS and digital image processing systems provide the functions for spatial data analysis but the techniques for their use may not be explicit in the package's user interface or documentation. As an example of what might be done, this paper describes a prototype knowledge-based system for predicting where changes in semi-natural vegetation will occur. Botanical and ecological knowledge of interactions between vegetation types, specifically with respect to bracken, provide a process-based vegetation model to couple to a GIS.

17.2 Knowledge-based system

17.2.1 SBS expert system shell

The expert system shell, SBS (Baldock *et al.*, 1987), provided the framework within which the GIS was coupled to the knowledge-base and thus to the environmental model. The development of this system made use of a model-based approach which permits an interaction between high and low level models to direct the needs of the lower level processing.

Separation of types of knowledge into different groups of sources (Tailor, 1988) helps modularity in system development. Using a blackboard system (Tailor, 1988), each source may have its own type of knowledge and processing. A blackboard system comprises a global database with knowledge sources, data structures and control structures (Tailor, 1988). It is well suited to application in problems that require a variety of input data and a need to integrate diverse information, with independent or semi-independent pieces of knowledge (Tailor, 1988). The knowledge sources are represented as procedures, sets of rules or logic questions. They modify only the blackboard or control data structures. The knowledge sources react opportunistically to changes on the blackboard, which lead incrementally to a set of acceptable solutions.

EXPERTs are structures within which certain rules are encoded and links made to algorithms encoded outwith the system. Their communication is with the blackboard, not with each other and their role is to address the goal at the top of the goal list. A numerical priority is assigned to each EXPERT which determines the order in which its capabilities will be tested against the goal list contents. The selection between different EXPERTs which can undertake the same generic function is determined by one of two ways: different prerequisites before a goal can be tackled; or different priorities set for each EXPERT. The body of the EXPERT is the code which is executed if the goal matches. Two alternatives are possible.

1. RULEs which are used in a production rule system (analogous to the operation of the EXPERT). The RULEs are executed in order of the numerical priority associated with each and those priorities are alterable within a system run.
2. Procedures which are directly executed without prior reference to any other code, as with the procedure to calculate sun elevation and azimuth before any illumination dataset.

A scheduler is used to select the knowledge source appropriate to the current task, which utilizes the control structure information. This shell uses the POP-11 language (Barratt *et al.*, 1986) for the encoding of rules, procedures and the scheduling of EXPERT routines, which retain system-wide information. Information about individual EXPERTs is held in a frame structure (described below, Lynn-Usery, 1988) for access by the scheduler. Access to the raster format, spatial datasets is by means of FORTRAN routines initiated by EXPERT routines.

17.2.2 Frame representation of knowledge

Frame-based representations of objects allow 'class-subclass,' 'is-a' and 'has-a' relations (Lynn-Usery, 1988). Attributes are assigned to these objects which, if changed, have a 'trigger-system' for recalculating other associated attribute values possibly within other objects. For example, changes in the area of one object will necessitate a change in the area of neighbouring objects. The frame is used to represent a group of entities with attendant facts which are the frame slots (Lynn-Usery, 1988). For example, the frame called soil_mapunit (illustrated below) contains slots for soil map units characteristic of the original soil map (scale, source, classification), these slots are, in turn, frames containing details of individual soil map units. Associated vegetation types are included as lists in order of dominance and include woodland categories which may be found in addition to grassland communities. Landform has the subjective estimate of slope and rockiness in its list.

The example frame shown below is a representation of the soil map key, including observed bracken presence.

```
frame soil250 isa soil_mapunit;
  slots soil_mapunitno = [250]                  ;;; soil map unit
    soil_type = [hip bfs hg pp]                 ;;; textual description with dominant
                                                    soil type first
        soil_vegetat = [[acid grass] [rush pasture][hazel oak birch]
                                                ;;; vegetation types
    soil_wetness = freely                       ;;; drainage level
    bracken_presence  =  [[observed] [fair 0·6]] ;;; recorded presence and assessed
                                                    as 0·6 likelihood present
endframe;
```

17.2.3 Meta-data

In order to effectively manage, share, and use spatial data, it is important to have information about the data. This is the meta-data (Townshend and Justice, 1987), which contains details of processing steps and parameters plus quantitative and qualitative judgements about the data. Any close coupling of the environmental model and the GIS requires a sufficiently open GIS architecture to provide all the linkages and interfacing necessary. For example, in the frame listed above, the nature of the vegetation types associated with the soil map unit are listed without any information on the proportions of the soil map unit covered in each vegetation type. Information can be computed directly from the basic spatial data (for example, biomass estimates from satellite imagery for polygons of land cover) and may or may not be stored for the duration of the system run.

Meta-data includes the history of the processing on the data and the history of requests the system has had to tackle. Assess is required to the meta-data for query and update, where appropriate, for use in data processing procedures. The meta-data for each data set is preserved and accompanies a data transfer automatically. This includes transferral of the characteristics of a dataset prior to processing and an explanation of why the processing step was performed.

```
frame data_history;
  slots data_title = land_cover
      scale_capt = 50 000                ;;; denominator
      scale_maxuse = 20 000              ;;; denominator
      data_type = raster                 ;;; raster, vector, text, table
      data_source = classific            ;;; classif. or digitiz
      data_dimension = [500 601]         ;;; eastings, northings
      data_reliab_type = probability     ;;; reliability datasets associated with dataset:
                                            Perkal epsilon error
      data_acquired = [26 july 1976]     ;;; day month year
      data_currency = [50 1]             ;;; years
      locn_os = [240.0 740.0]            ;;; nw east, nw north (km)
```

```
    locn__cent = [265.0 710.0]          ;;; approx. geometric center of dataset
    locn__acc__av = 0.8                 ;;; absolute
    routine__history = [[satellite mss][geocorr nn 1.4][illum 135 32][maxlike]];
                          ;;; routine identifier added when called
endframe;
```

The example frame shown above lists some of the contents of the header information associated with each data file. If the data is raster scanned, the scale of capture is taken from the empirical relationship between scale and maximum resolution (Fegeas *et al.*, 1992). For the purposes of this application the limit of use is taken as 2.5 times the source scale. The possible values of data__source are 'classifed' (from satellite imagery), 'satellite' (raw data) or 'digitized' (vector digitized and rasterized). Data__currency relates to the estimated length of time for which the data are relevant. Although the figure of 50 years was selected, in practice the land cover classification will be relevant from the time of its collection for analysis at all future dates. The stability of the data will be shorter term: annual changes are assumed so a one year estimate is given.

17.2.4 Error estimates

Each EXPERT routine adds the goal assess error to the goal list (see Figure 17.1, in the Blackboard) before it exits. One EXPERT has assess error as the goal to satisfy. The returned value restricted in its range of options to one which is quantifiable or an expression of reliability based upon knowledge derived from the literature which gives the user some indication of the veracity of the output.

The EXPERT routine consults the contents of the data history frame associated with each dataset. The data reliability slot indicates which types of error are to be considered. All datasets comprising discrete classes have a boundary error band associated with them and the description of each class within its frame structure contains information on the heterogeneity of each map unit.

17.2.5 Output

The nature of the output from this system is restricted to two forms: textual (including numerical summaries) and graphical (a map or graphical representations of spatial data). Textual output includes explanations from the system of what has been done.

Each processing step undertaken on the data is recorded in the data history file. Along with this information is an explanatory statement of what the routine was doing and why it was being done. A text string is written out for each rule and procedure which is called. The prerequisite and condition results are reported whenever one is triggered such that each failed test is also recorded and can be reported.

Graphical output is of the form of raster datasets in formats suitable for display on digital image processing systems (Erdas, 1992). These output datasets - which can hold intermediate results rather than final output - are for inspection or printing, and analysed in their own right at a later date. Graphical output of histogram or scatter-diagram plots have been left to those facilities provided by the image processing packages.

17.3 Schematic enquiry

An example enquiry on the spread of bracken (*Pteridium aquilinum*), in the Trossachs area of central Scotland, is used to describe the nature of information used and some of

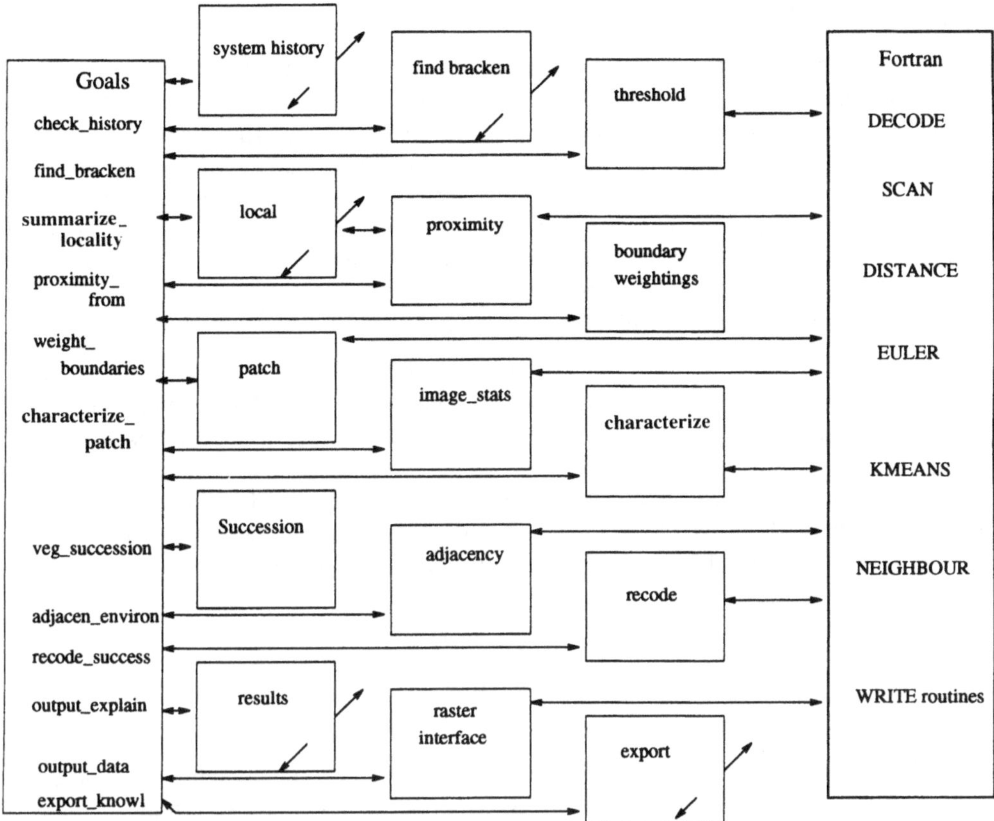

Figure 17.1 Schematic diagram of the EXPERTS involved in predicting bracken change.

the GIS requirements. The enquiry initiates the sequence of responses which attempt to satisfy conditions regarding the current or historical position (or how that might be established), what information is required to assess the potential for spread, and finally to predict where the spread may occur (in a 20-year time period).

In response to a system-prompted enquiry which requests the estimates of acceptable accuracy, percentage errors for bracken identification and area of bracken spread are entered. A look-up table approach is used to match error in area with resolution and thus to sensor type (for example, 20 m resolution for SPOT and 50 m for Landsat MSS) and scales of available datasets.

Figure 17.1 shows one set of linkages between the goals to be attained by the EXPERT routines and the FORTRAN routines that may be required to process the spatial data. In between the two groups are the EXPERTs which may be called. This forms part of the coupling of the environmental model and the GIS. The 'triggering' of an EXPERT routine is not restricted by the order presented in Figure 17.1. Any change in the order or detail of the goals will cause a different sequence of EXPERT routines.

Using the frame representation of each type of raster data file permits multiple-overlay simultaneously, without each file having to be consulted by each relevant rule. The instances are formed from geometric, coincidence and adjacency models of the environment around the patches of bracken. Operations on the slot values are then undertaken using these models to assess the likelihood of bracken spread from its current patch.

17.3.1 Predict bracken spread

The model of changes is based upon vegetative reproduction of bracken and as such requires first, a bracken presence and second, changes along the boundaries of that existing bracken. Contextual information of the vegetation cover may be lost with interpretation of polygonal information without boundary information. The mature hinterland to bracken may locally show signs of degeneration which may be as low growth or discoloration similar to deficiency disease. Regression in bracken may be due to patch aging, climate or pollution. Locally, soil conditions and micro-climate may promote dieback. These changes become obvious when bracken cover reveals gaps.

The stages of assessment of likelihood of spread involve the checking of:

1. bracken heterogeneity as measured by the number of bracken polygons in an area minus the number of holes in those polygons (the Euler number) – this is calculated from different probability levels to give a measure of the susceptibility to bracken-spread according to the rules in the POP-11 database;
2. adjacent environment to that occupied by bracken – the altitude, slope, aspect, soil types, exposure levels and land cover types adjacent to existing bracken are summarized and each bracken pixel is assigned a score of suitability of its neighbouring environment to bracken spread;
3. which types of temporal change are being observed to identify active and static edges to the bracken stands;
4. the vegetation succession model (Miles, 1988) which allocates likelihoods of spread according to the neighbouring vegetation type.

If a satellite image classification is appropriate it uses a maximum likelihood algorighm which outputs the first three likelihood levels for every pixel rather than just the top level. The likelihood of different land cover classes present at each pixel provides information on how well discriminated the first likelihood level is from the second. The structure of the variation within a polygon-representation of land cover using satellite imagery is used in conjunction with ecological observations on vegetation succession (Miles, 1988) which relates to the measurements of dieback within the bracken patch. Once separated into the most likely class at every location, the probability surfaces are usually treated as contiguous classes. The adjacency of land cover types and the rate of transition between vegetation types at the boundary are two of the inputs required to run a simulation of the succession model.

If there is evidence of a hole in a well established bracken stand, the active edge may be a bracken front retreating rather than spreading. However, further testing is required of the ecological models to assess the validity of this hypothesis.

Information available pertaining to conditions associated with bracken presence which were encoded as rules (5 out of about 200 rules) are as follows:

1. Not present on waterlogged soils because oxygen deficiency sensitive. The implication from this was to mask the peatland out of the analysis. By relaxing the interpretation of bracken presence from 'none' to 'low', bracken was predicted to be present. Investigation of the location showed bracken present on 'dystrophic peat'. (Dystrophic peat has low biological activity and nutrient status where there has been surface drying.)
2. Soil acidity of between pH 3·0 and 7·6. Information about the soil maps is contained within the map key and accompanying handbook. No evidence is available from the soil maps of peatland deposits comprising dystrophic peat, therefore, using the soil maps alone for a source of soils information will not address these misclassifications.

   The error bounds used to represent the boundary accuracy of the mapped soils do not permit any distinction between boundaries which have gradual changes between neighbouring soil types across several tens of metres and changes which occur over distance of two or three metres. For example, a boundary between soil map units identified as Peat, Peaty Gley and Peaty Podzol and Peaty Podzol, Peaty Gley and Peat (Macaulay Institute for Soil Research, 1984) is not definable within the accuracy the cartographic representation of a single line suggests. The difference between the two classes is in the proportions of the same component soil types in each of the map units. A description of the differences between the two map units has been obtained from the soil maps handbook. That was used in the explanations of different susceptibility of those soils to bracken spread. The locational data model soil parameter has five components: soil code; dominant soil type; number of component soil types identified; soil wetness and soil pH. These components were dealt with independently of their soil map unit number and thus reduce the significance of the boundary location on the mapping of bracken. No information is available to indicate the accuracy of boundaries between soil map units. To test the impact of the interpretation of soil boundary on predicted changes in bracken distribution, the data model of the local environment was checked against the soil type slots to check for similarity between component major soil sub-groups. The boundary definition was relaxed by up to a further 200 m when considering adjacent soil types. Two options were tried:

- different orders of occurrence of the same com_ponent soil types, e.g. [Peat Peaty_Gley Peat_Peaty Podzol] compared to [Peaty_Podzol Peaty_Gley Peat] boundary relaxed 100 m either side;
- changes of one component soil types (apart from the first), e.g. [Brown_Forest_Soil] [Brown_Forest_Soil Brown_Forest_Soil_Gleying] boundary relaxed 50 m either side.

3. Bracken has a competitive-plant strategy. There is some debate among plant ecologists over the theoretical basis of this rule but it has been taken from Ripple and Ulshoefer (1987). The importance of assessing the context of bracken with its surrounding environment is twofold. First, the errors in representation of the surrounding environment impact directly on the assessment of the distribution of bracken. Second, the extent to which bracken is already present within the niche of available environmentally favourable conditions determines the limitations of its future encroachment on neighbouring land.
4. An invading front may advance up to 0.5 m per annum and in ideal conditions up to 1.5 m per annum. A feedback loop calculating transition matrices based on an incremental inclusion of examples of change provided a means of learning what changes to expect and their spatial extent.
5. Main spread is by rhyzomes because the establishment by spores is rare. The vegetation succession model (Miles, 1988) being used in this study indicates those vegetation types which are expected to be succeeded by bracken. The knowledge of bracken spread by vegetative means (provided by the model) indicates that the distribution of vegetation requires assessment in the context of its neighbouring vegetation types. Context is not restricted to first order measures but also rates of change of categorical and non-categorical data, for example, soils and land ownership.

The land cover parameter of the locational data model has four values for each location: first and second most likely land cover types and the probability levels associated with each land cover class likelihood. There are also two measures of adjacency and proximity built in: proportions of each adjacent land cover type and distance from the nearest different land cover type where none is immediately adjacent. The changes and rate of changes at each pixel contribute to the assessment of the dynamic or static nature of a vegetation boundary but these measures have directions associated with the changes. A data model for each pixel will require some means of assessing the influence of direction. The simplest form of direction in this work is that which is away from the bracken patch, therefore directional influence is inherent in the dataset of proximity from patch boundary.

17.3.2 Explanations of subject of enquiry

The history files are updated after every processing step which includes the EXPERTs, rules employed in the system run and any new data output going to the data file. The pro-forma of the explanations follows a sequence of: what was mapped; where was the bracken; what were the likely errors, e.g.:

> The object mapped was bracken. The information was for July, bracken is fully grown.
>
> The distribution of bracken was on predominantly south and east facing slopes, in an altitudinal range of 5 to 440 m.
>
> The distribution of bracken was on predominantly humus iron podzols and brown forest soils.
>
> The environmental niche available for bracken was 27 631 ha (36·7%).
> The area occupied by bracken was 2364 ha (3·1%).
>
> The scale of remotely sensed data was not ideal, satellite not aircraft.
> A masking satellite classification solution was used.
>
> The scale of soils digital terrain model woodland were suitable.
> The bracken has 6% of area susceptible to soil map boundary error.
> The bracken has 4% of area susceptible to woodland map boundary error.
> The specified accuracy was met.

The same approach was employed in explaining bracken spread as for that of distribution. The (partial) explanation reported below includes reasons why bracken would not spread in certain areas as well as why it could spread.

> The object spread was bracken. The information was for July, bracken is fully grown.
> Time period was 20 years.
>
> Patches of bracken large (av. 2.35 ha). Spread was by rhizomes into heather (15%) and grass (8%). 6 ha previously bracken now grass or heather.
> Conditions of soil, topography, land use good for further spread.
> 41% of Forestry (6865.35 ha) present on bracken suitable land prevents some bracken spread.
>
> No accuracy was specified.

17.4 Discussion

In the context of the GIS, knowledge of the scales being used to collect, analyse and display the data is vital to valid use of the spatial data. If the preferred environment of an image processing or GIS facility is available, the user has the facilities such packages offer independent of the knowledge-based vegetation mapping system. Such independence is one advantage of the loose coupling of GIS or image processing packages with environmental models. The value of tighter coupling is that the procedural knowledge is stored in the KBS and such knowledge is used in alerting the system if use of the data is likely to be inappropriate.

Using a frame-based system for representation of storage of data (attributes of spatial data or aspatial data) and knowledge permits advantage to be taken of the characteristics of inheritance which it offers. Hoever, this approach would benefit from the use of object-oriented software and an object-oriented database.

The range of functions which a GIS is required to undertake cannot normally be supported with only one data structure. The translations between field to object and object to field as part of the processing provides for this requirement only when the data has been interpreted for the data model. Viewing the data in the context that the data model was designed to provide, adds value which can be taken advantage of during the processing. There is a computational overhead associated with the processing for the data model. However, more powerful computing facilities, and efficiencies in coding, should offset some of this. Further developments may include incorporating linear networks and triangulated irregular networks (Aspinall, Chapter 10, this volume).

The allocation of a priority label to each EXPERT provides a mechanism for controlling ordering and selection of knowledge, and execution of GIS functions according to the governing principle of appropriateness, a fundamental aspect of which is measures of error. Using this mechanism, an estimate of accumulated error is instrumental in selection of methodology, selection of function and explanations of processing used and results obtained.

The building of the history file to track the procedures undertaken is an extensive part of the total length of code. However, it provides the basis for explanation of the routines considered and those activated. For an exploratory session, this facility is more important than any preliminary results. It provides the basis for a user to undertake the suggested tasks outwith the hardware configuration used for the knowledge-based system. It also fulfils a role of pointing the user towards the possible deficiencies in the task being undertaken. An extension of the simplification of processing by reference to a history file would be to learn from the steps undertaken in previous enquiries of the system.

The facilities for describing and explaining what processing has been done and why, plus an interpretation of the results, improves the use of GIS for environmental applications. Guidance is available for the non-expert user of digital spatial data towards the processing applicable for the task. Further, the interpretation of the results in the context of the data available and the goals set, demonstrates the potential for a feedback loop governing the coupling of models and GIS, and changing the emphasis of that coupling. The emphasis is on evolution of the application during processing rather than the tackling of technical requirements prior to processing. Considerably more work will be required in building extensive knowledge-bases and providing the range of generic GIS functions before the potential for this approach is beneficial to a wider applications audience.

As the vegetation changes are taking place, so the applicability of the vegetation succession model begins to change. If changes are significant enough over time for the mix of vegetation types to change, where each vegetation type reacts individually to changes in the environment, current observation-based models of vegetation succession will be inappropriate. There is insufficient evidence to estimate the validity of predictions beyond another 40 years. Identification of what threshold of land cover mixes beyond which the model ceases to be reliable may be done in two ways. Either land cover could be monitored at sites and the succession model tuned through time, or sites studied where the mix of vegetation types varies enough to indicate different stages of succession.

Two opposite development directions could be taken. The first is adoption of higher-level expert system shells which build rule modules and require less writing of code in languages such as Prolog or Pop-11. The second option is to write more routines in languages such as C++ where structures such as frames are replaced within an object-oriented software environment implying that the programmer is responsible for more of the system encoding.

References

Aspinall, R.J., 1992, An inductive modelling procedure based on Bayes' Theorem for analysis of pattern in spatial data. *International Journal of Geographical Information Systems*, **6**(2), 105–21.

Baldock, R.A., Ireland, J. and Towers, S.J., 1987, SBS User Guide. Medical Research Council and Population Cytogenetics Unit, Edinburgh, internal report.

Barrett, R., Ramsay, A. and Sloman, A., 1986, *POP-11: A Practical Language for Artificial Intelligence*, Chichester: Ellis Horwood Limited.

Erdas Inc., 1992, *Erdas Imagine, Version 8.0.1 Documentation*, Erdas Inc., 2801 Buford Highway, NE, Suite 300, Atlanta, Georgia 30329-2137, USA.

Fegeas, R.G., Cascio, J.L. and Lazar, R.A., 1992, An overview of FIPS 173, The Spatial Data Transfer Standard. *Cartography and Geographic Information Systems*, **19**(5).

Fisher, P.F., 1991, Modelling soil map-unit inclusions by Monte Carlo simulation. *International Journal of Geographical Information Systems*, **5**(2), 193–208.

Lee, T., Richards, J.A. and Swain, P.H., 1987, Probabilistic and evidential approaches for multisource data analysis. *IEEE Transactions on Geoscience and Remote Sensing*, **25**(3), 283–93.

Lowell, K.E. and Astroth, J.H., 1989, Vegetative succession and controlled fire in a glades ecosystem – a geographical information system approach. *International Journal of Geographical Information Systems*, **3**(1), 69–81.

Lynn-Usery, E., Altheide, P., Deister, R.R.P. and Barr, D.J., 1988, Knowledge-based GIS techniques applied to geological engineering. *Photogrammetric Engineering and Remote Sensing*, **54**(11), 1623–8.

Macaulay Institute for Soil Research, 1984, *Soil and Land Capability for Agriculture 1:250 000 Survey: Organization and Methods*. Macaulay Institute for Soil Research, Aberdeen: Aberdeen University Press, 81.

Miles, J., 1988, Vegetation and soil change in the uplands, in Usher, M.B. and Thompson, D.B.A. (Ed.) *Ecological Change in the Uplands*. Special publications series of the British Ecological Society, No. 7, 57–74.

Pequet, D.J., 1984, Data structures for a knowledge-based geographic information system, in *Proceedings of the International Symposium on Spatial Data Handling*, Zurich, Switzerland, 372–91.

Ripple, W.J. and Ulshoefer, V.S., 1987, Expert systems and spatial data models for efficient geographic data handling. *Photogrammetric Engineering and Remote Sensing*, **53**(10), 1431–1.

Robinson, V.B., Frank, A.U. and Karimi, H., 1987, Expert systems for geographic information systems in resource management. *Artificial Intelligence Applications in Natural Resource Management*, **1**(1), 47–57.

Tailor, A., 1988, MXA – a blackboard expert system shell, in Engelmore, R. and Morgan, T. (Eds), *Blackboard Systems*, Addison Wesley, 315–34.

Townshend, J.R.G. and Justice, C.O., 1987, Selecting the spatial resolution of satellite sensors required for global monitoring of land transformations. *International Journal of Remote Sensing*, **9**(2), 187–236.

Wright, G.G. and Morrice, J.G., 1988, Potato crop distribution and subdivision on soil type and potential water deficit: an integration of satellite imagery and environmental spatial database. *International Journal of Remote Sensing*, **9**(4), 683–99.

18

Using GIS to target agri-environmental policy

Charlotte Norman, Clive Potter and Hadrian Cook

There is growing concern in the UK about environmental problems such as soil erosion, aquifer and river pollution, wildlife and habitat conservation and flooding. At the same time the emergence of new agri-environmental policies under the Common Agricultural Policy will make available public subsidies and incentives which may be used to encourage changes in farm practices and land-use on affected land. A critical consideration will be how such measures can be targeted to ensure maximum environmental value for money. In this chapter we look specifically at a GIS methodology which aims to target land to be diverted from agricultural production under a long-term set-aside scheme. Spatial and statistical results are presented for the counties of East Sussex and Suffolk in the UK. Thoughts are also discussed about the policy implications of such an exercise.

18.1 Introduction

Agricultural change over the last 25 years has been dominated by the twin processes of intensification and specialization. This creates a situation where land-use is often incompatible with land capability (Green,1991). As a consequence of these and other processes, increasing by significant areas of farmland are now subject to: soil erosion, which contributes to water pollution; are affected by periodic flooding; and/or have a conservation potential. Targeting these problems for environmental land diversion policy provides an ideal application for GIS.

18.2 Soil erosion

Although the full extent and scale of soil erosion is known only in outline, it is clear that erosion of agricultural land in certain areas is occurring regularly and some of the rates are sufficiently high to give cause for concern (Boardman, 1990). It has been estimated that some 73 800 km or 9 per cent of the total agricultural area of England and Wales has been affected by erosion over the last decade (Evans and Cook, 1986). Evans (1990) on the basis of a classification of soil associations, estimates that 5.9 per cent of surveyed land in England and Wales is at high or very high risk of erosion. Water erosion especially by rills, is fairly common in England and Wales while sandy and peaty soils in areas like the Vale of York, the Fens and the Suffolk sandlings are vulnerable to erosion by wind (Boardman, 1988).

18.3 Agro-chemical pollution

The concentration of nitrates in UK waters is largely due to intensive use and untimely application of fertilizers to agricultural land (Croll and Haynes, 1988). Nitrate in drinking water has long been perceived as a health hazard. The World Health Organization recommended levels are 50 ppm for babies and 100 ppm for adults, set, in the former case, to reduce the risk of methaemoglobinaemia, a very rare condition, and in the latter, to reduce the risk of stomach cancer. The EEC Drinking Water Directive (1985) sets a maximum admissible concentration of 50 mg of nitrate per litre in drinking water supplies (DOE, 1986). Supplies of drinking water from both surface and groundwater frequently exceed this limit in some areas. Britain in 1992 was in breach of the EC Standards in 28 supply zones in England alone (Hornsby and McCarthy, 1992). Uncertainties surround the precise prediction of future nitrate concentrations in groundwater but many sources are likely to exceed 100 mg NO3/1. The long-term trends in river nitrate concentrations in Central and Southern England show rises at varying rates until 1976. Since then values have in general remained constant (DOE, 1986).

Pesticide levels in drinking water have on several occasions breached EC limits introduced in 1985 of 0.1 mg/L with a maximum of all pesticides plus metabolites of 0.5 ppb. A large majority of the pesticides detected are accounted for by weed control; only a small proportion is associated with agriculture (Eagle, 1992). The behaviour of pesticides in soil is greatly affected by its chemical nature and is poorly understood (OECD, 1986).

18.4 Conservation potential

In England and Wales in the period of mechanized agriculture since 1945, 70 per cent of the original peat lands, 58 per cent of the ancient forests and 40 per cent heathlands have disappeared (Agra Europe, 1991). Hedgerows have suffered similar levels of loss as a consequence of modern farming techniques. The Institute of Terrestrial Ecology (ITE, 1991), concluded from a comparison of 1984 and 1990 data that the rate of hedgerow removal is greater than that between 1978 and 1984. In addition, there has been an overall decline in the intensity of hedgerow management, leading to an increase in the boundary type defined as relict hedgerows (defined as a line of shrubs or trees showing where a hedge had once been).

Concern about habitat reduction and fragmentation has fostered the belief that this loss can to some extent be compensated for and the environment be improved, by re-creating new semi-natural areas which resemble the old (Buckley, 1989). English Nature's objective, in the light of changes in agricultural policy, is to conserve all remaining semi-natural habitat and land with good wildlife value. Changes in agricultural policy also create a role for habitat reconstruction or recreation. The aim here is to maintain, defend but also enhance the small areas of high quality habitat scattered amongst the predominately cultivated landscape. Land adjacent to these high quality habitats could be diversified through natural colonization and appropriate management. The potential for re-establishing wildlife habitat in the farmed countryside is very large whether the areas withdrawn are whole farms, single fields or strips alongside field margins.

18.5 Flooding

Periodic flooding is a further environmental hazard. Many parts of the UK, usually floodplains and grazing marshes, have been given protection against flood damage by costly tidal barrages. (Bowers and Cheshire 1983), reviewing these investments, questions whether the supposed agricultural benefits which follow from agricultural intensification, which often includes areas of high conservation value, really justify many such schemes. Their thesis is that too much social capital has been diverted to subsidize these large areas which are still subject to high return periods from flooding and therefore should be targeted for land diversion. Re-establishing a low intensity pastoral agriculture in such areas under a land diversion programme would create an agricultural system base which is both environmentally benign and better adjusted to the level of flood risk. Such restoration of arable to grass or other types of vegetation might do much to restore wildlife lost through agricultural intensification.

Around 720 000 ha of coastal land may be affected by flood or salination as a result of sea level rise (Boorman *et al.*, 1989). These one-time coastal wetlands, a proportion of which were drained for agriculture, could be re-created by allowing the area to flood. Water authority engineers look at this as one alternative for a cheaper flood alleviation strategy bearing in mind farming extensification and diversion schemes and the prospect of sea level rise (Newbold, 1989).

18.6 Agricultural policy under the CAP

Reform of the European Common Agricultural Policy (CAP) provides a new opportunity to address a number of these problems. The European Commission has in its recent reforms agreed to move towards a more streamlined and environmentally sensitive CAP, essentially decoupling farm support from the quantities farmers produce. Compensation for price reductions will take the form of direct payments to individual farmers. To qualify for payments cereal farmers have to set aside a prescribed portion of their arable land base.

As part of the CAP reform the European Community also agreed an Environmental Regulation (2078/92). These measures are to include a package of direct payment schemes which governments are to offer farmers in return for a range of 'environmentally-friendly' management practices. The forms of land management covered include taking land out of production for at least 20 years for environmental purposes. This long-term set-aside scheme has a target of 10 000 h but all types of farmland, not only arable, are to be included.

18.7 UK conservation reserve

One of the more promising components of the package is the long-term habitat improvement scheme (HIS). The objective of this scheme is to withdraw agricultural land from production for the long-term in carefully selected areas for the creation or improvement of a range of wildlife habitats - a Conservation Reserve in embryo. Commentators have been arguing for some time for an American-style Conservation Reserve in the UK. In the US the Conservation Reserve Policy (CRP) is used to target areas at risk from soil erosion. A UK CRP would bring about a carefully targeted and long term diversion of cropland for the purposes of habitat recreation, soil erosion control, water protection, and flood alleviation (Potter *et al.*, 1991). In the US, one of the principle lessons, after 50 years experience with Federal Soil Conservation Programmes, is that targeting the land to be taken out of production in this way maximizes the net social benefits of government programmes by channelling resources and money into areas or onto farms where they will generate the greatest possible environmental benefit per dollar spent (Griffen and Stoll, 1983). Targeting allows policy makers faced with limited budgets to set priorities and direct resources in ways which maximize the social and environmental return on every pound spent. It is likely to be most effective where problems are geographically specific and where levels of programme expenditure are so large that even small gains in policy efficiency saves money (Potter *et al.*, 1993). However, the technical ability to target such a programme is still poorly developed in the UK, with methodologies and procedures for identifying vulnerable land still relatively crude by US standards. Even the simplest and crudest targeting of land would seem to be frustrated by the lack of data, though previous research suggests a number of ways forward.

18.8 Previous research

Potter *et al.* (1991) developed a potential surface analysis model using map overlays to target land diversion policy at 10 km squares in England and Wales. The model used map overlays based on the degree of mismatch between existing agricultural use and land capability, vulnerability to water and wind erosion, aquifer contamination, conservation potential and conservation designations. The model adopted even weighting between the superimposed map layers. The resultant map highlights areas of greatest potential for land diversion. The map consistently highlights chalk and sandstone outcrops, locations where agriculture is unsustainable in the long term.

Cook (1991) refined this methodology to a micro-targeting level for nitrate pollution, identifying possible target areas for a risk assessed, incremental programme of land diversion in south east England. In this study, pedologic, climatic and hydrogeological information was assimilated in a GIS to map areas where nitrate pollution to the unconfined aquifer is likely to occur. The combined model identifies areas on the basis of their vulnerability to the unconfined aquifer, ranking land areas and thereby permitting a sequential approach to land diversion by removing the most vulnerable intensive arable or horticultural areas from production first of all.

18.9 New methodology

The aim of the present project was to build on this work by developing a GIS methodology to assess the potential and scope for targeting long term set-aside more precisely. The overriding aim was to combine available geographic data on soils, hydrogeology, climate, topography and conservation sites through spatial modelling to target areas of environmental vulnerability and conservation potential for agricultural land diversion. The hypothesis guiding this exercise was that the incidence of locationally-specific environmental problems and conservation opportunities overlap, creating scope for targeting these areas for land diversion. This type of application lends itself to GIS which offers a flexible means of synthesizing a large amount of spatial data (Figure 18.1) from a variety of different sources, scales and data formats, allowing the integration of environmental policy objectives (Norman *et al.*, 1991).

This methodology adopts a study area approach based on 10 km squares. Ten study areas were selected from clusters of the highest ranking squares in the Potter *et al.* (1991) study. This provides a regional scale methodology. The approach was then to assess the scope for targeting environmentally vulnerable land with conservation potential within these areas. Eligibility criteria selected for each of these areas are as shown in Table 18.1. Any GIS methodology which attempts to integrate a wide range of datasets will face problems of spatial data availability. The spatial data required to develop this targeting exercise are highly diverse and available from a variety of agencies at varying scales and formats. Data needs are clearly dictated by the selected eligibility criteria and include information relating to soils, hydrogeology, climate, topography, habitat designations and agricultural cropping patterns. A review of available datasets highlights the incompleteness of data coverage (Table 18.1). Soil survey data is a key source when attempting to target land on environmental criteria, though complete coverage is only available in England and Wales at a scale of 1:250 000. Soil information at this scale is presented in terms of 'soil associations' (geographically associated soils) rather than true soil series (describing

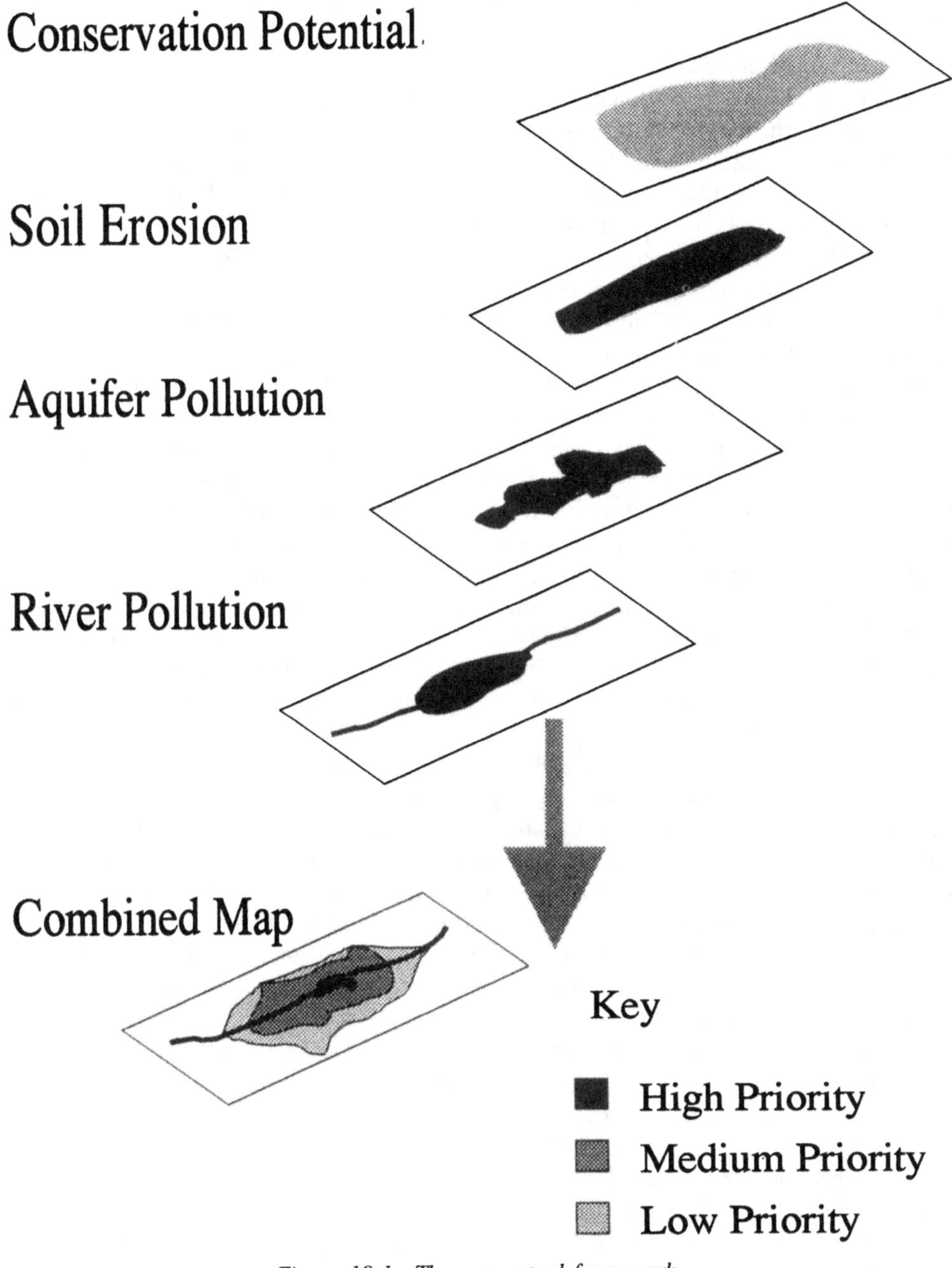

Figure 18.1 The conceptual framework.

Table 18.1 The conceptual framework.

Environmental vulnerability
Erosion vulnerability: Areas with soils liable to erosion, > 10 degree slope, >200 metre slope length, high rainfall intensity.
Pollution vulnerability: Areas with soils liable to leaching, presence or absence of unconfined aquifer, high effective rainfall, river buffer zones.
Conservation buffers
Areas in proximity to SSSIs, areas in proximity to woodland.
Flood vulnerability: Areas liable to river flooding, areas below 5 metre contour.

soils alike in soil characteristics and behaviour). Land use capability class maps are also potentially useful for targeting, particularly in identifying areas where the current farming uses are mismatched with physical land capability, though here to coverage is presently only available at a 1:250 000 scale and then only for a 4 per cent sample area of England and Wales. The agricultural land classification series produced by MAFF may be an acceptable substitute here, with universal coverage at 1:63 360 scale.

Hydrogeological information covers most aquifers at a scale of 1:100 000. Topographic is available from Ordnance Survey maps at an increasing scale from 1:50 000 forming a complete dataset for England and Wales. Flood data from Section 24 maps can be obtained from the National Rivers Authority. Information on cropping patterns is available from the Ministry of Agriculture Fisheries and Foods (MAFF) Annual Agricultural Parish Census Summaries which are complete for England and Wales for 1988.

Perhaps the most problematic data to obtain in terms of compatibility relate to conservation potential. Information about the type and distribution of habitats is not available at a national scale, although sources like the national red data book provide data on the distribution of rare or endangered species on a 10 km square basis. Data on the distribution of habitats is increasingly well documented through the Phase 1 Habitat Survey presently being conducted at a county level under the Wildlife and Countryside Act. Maps are now available for completed counties but between-county comparisons are difficult as some of the data is more detailed than others. English Nature have digitized maps of conservation sites and reserves which for this exercise have proved useful in identifying important habitats for protection and extension.

The actual modelling procedure involves the identification and reclassification of the above datasets for importation into GIS to produce maps for each of the environmental criteria (the most suitable available to satisfy the criteria). To satisfy the criteria of soils liable to leaching, the soil association categories were reclassified according to the degree of erosion hazard based on a classification developed for the NERC Erosion Study (1990). Each study area was classified according to its average Rainfall Intensity using the Morgan Rainfall Index. To define those areas susceptible to soil erosion by virtue of their slope characteristics distance between contours and contour intervals were measured from a 1:50 000 Ordnance Survey map.

To define soils liable to nitrate leaching to ground and surface waters soil association information has been classified using the Burnham (1991) classification. Each soil association is grouped into one of four weighted categories from low risk – very poorly-drained soils, waterlogged to within 40 cm of the surface for more than 180 days in most years through to high risk – dry stony or shallow soils, liable to summer drought. Aquifers were classified, from hydrogeological maps, according to the degree of protection afforded by drift geology and bedrock cover. Geology and drift geology maps provided the presence and thickness of drift material above the aquifer. Effective rainfall was obtained, from

Annual Average Rainfall maps, by adjusting the rainfall isohyets by the average figure for annual potential evapotranspiration for each study area obtained from the Agricultural Climates of England and Wales MAFF report (Smith, 1984).

To spatially define areas of conservation potential and to protect existing semi-natural areas, both Sites of Special Scientific Interest (SSSIs) and woodland boundaries were buffered to a distance of 0.5 km. Also from the 1:50 000 Ordnance Survey maps, rivers and major drainage dykes were digitized and buffered to a distance of 0.2 km and 0.5 km to act as a retentive zone for both potential pollutants and eroded particles. These datasets have been captured in GIS (SPANS) to produce map layers. Each map layer (or overlay) distinguishes areas of high priority from areas of low priority or presence or absence in accordance with the eligibility criteria. Through a process of matrix overlays maps contributing to each category (Table 18.1) are then superimposed under an equal weighting rule to produce both an environmental vulnerability map and a conservation potential map of areas at highest risk. These two models are then aggregated to produce a combined model which shows a grading of agricultural land according to its degree of potential environmental vulnerability and conservation potential. The combined map has the potential to produce 23 classes, these are categorized into high, medium and low risk. The identified environmental hazards are rarely a problem unless arable land is present. Therefore an overlay of the percentage of arable land within parish boundaries is cross-tabulated with the combined model to enable a statistical analysis of land, which is both environmentally suitable and in arable production (target areas). Target areas are arable areas in which the highest number of attributes coexist at the same location. Target areas do not contain a homogenous number of attributes as it is the combination of six classes (12–18) which individually combine an identical number of attributes which are not homogenous in type.

18.10 Results

The combined maps for each study area resulting from this targeting GIS methodology clearly show that environmentally vulnerable land is far from evenly distributed across the land base of England and Wales and is therefore highly susceptible to a targeting approach. The percentages of target land within and between each study area are very unevenly distributed, varying between 0 per cent of the total arable land area in the Brecon study area and 71 per cent of the total arable area in the East Sussex study area. To help explain the variation in results two study areas, East Sussex and Suffolk, will be described in more detail.

The East Sussex study area (Figure 18.2 OS sheet reference TQ30/40)) is downland and is nearly entirely designated as the South Downs Environmentally Sensitive Area (ESA). This contrasts well with the Suffolk study area which is a lowland fen. The parent material in East Sussex is chalk and associated drift. The soil type is shallow chalky soils which are well drained. The relief of the area is dominated by the slopes of the South Downs; the land use is mainly arable. Average annual rainfall is greatly affected by the topography of the area ranging from 762–952.5 mm. The River Ouse is the only main river running through the east side of the area. As Table 18.2 shows, target land makes up 71 per cent of the arable land base in this area. By targeting highly vulnerable arable land 4674 h is vulnerable to erosion. This is mainly due to the presence of high-scoring erodible land in an area with steep slopes, subject to heavy rainfall. The same land area would also be highly vulnerable to agro-chemical leaching to the unconfined aquifer. Leaching to surface waters is not an important problem in this area, accounting for less than 3 h. There are also some 243 h (5 per cent) of land with a conservation potential of which 83 per

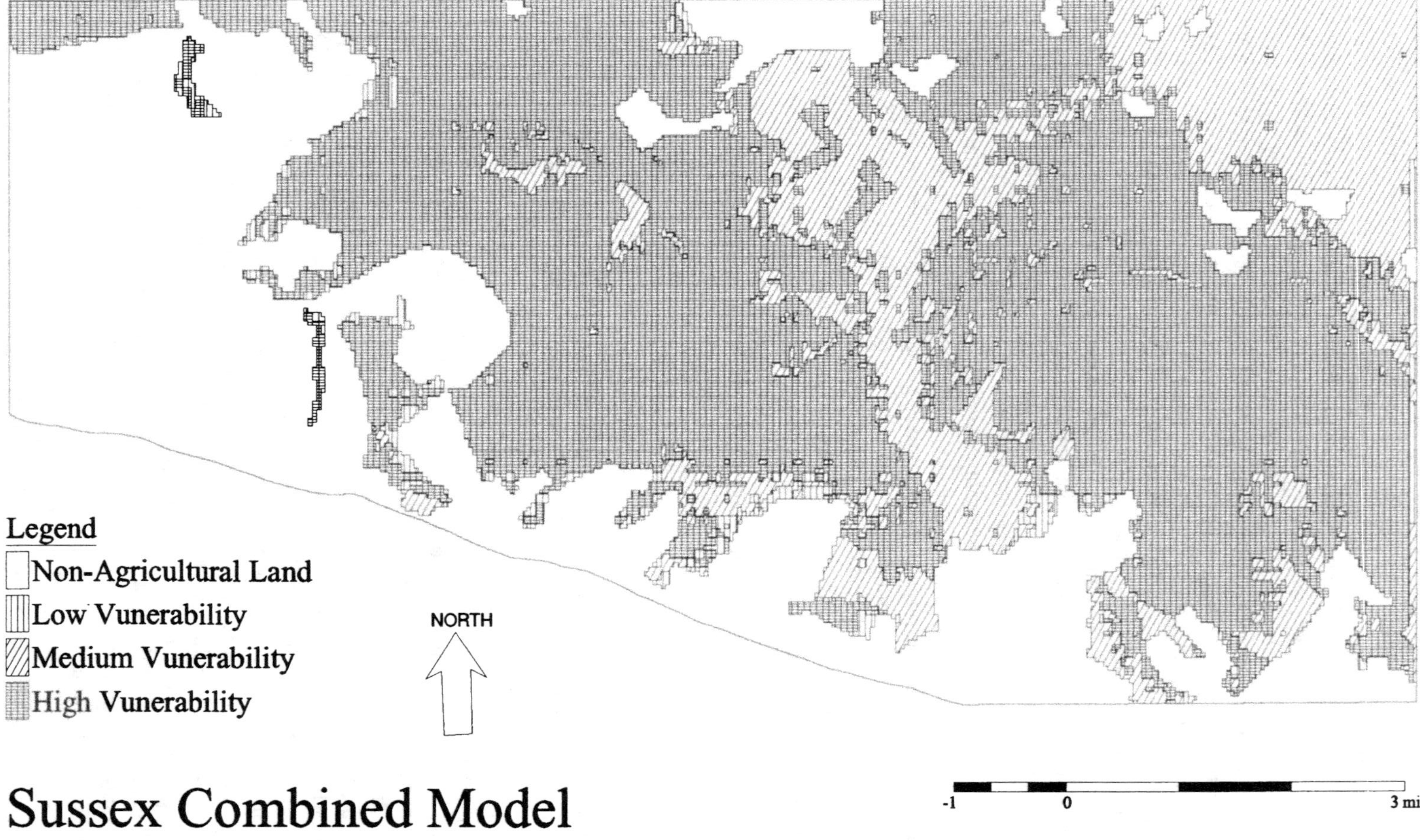

Figure 18.2 Sussex combined model.

cent is flood-prone, 5 per cent in woodland buffer zones and 12 per cent buffering Sites of Scientific Interest.

In contrast the Suffolk area (Figure 18.2 OS sheet reference - TM34/5) is a fenland with an intricate drainage system at low altitude. The soil parent material is dominated by clay. The associated soils are mainly gleys with some calcareous brown earths. This peninsula area is a flat or gently undulating plain, known as the Sandlings. The area is predominately in agricultural use and is for the most part intensively farmed. Rainfall is lower than in East Sussex with an average annual fall of 525 mm. Besides an intricate drainage system the area includes some of the course of two main rivers, the River Deneden and the River Adle. This target area comprises 55 per cent of the total arable area (Table 18.2). Soil erosion is also an important factor in Suffolk, although to a less extent than in East Sussex, despite the absence of highly vulnerable slopes. Soil erosion is a problem on all this vulnerable land due to vulnerable soil types combined with a high rainfall intensity. The same area is highly vulnerable to agro-chemical leaching but only to surface waters as an aquifer does not underlie this area. About 2130 h of this area has conservation potential, 56 per cent being woodland buffer zones, 7 per cent buffering Sites of Special Scientific Interest and 46 per cent having potential for flooding. Diverting land only running alongside rivers would reduce the target area to just 10 per cent of the total arable area.

18.11 Policy implications

This regional scale targeting exercise reveals a high concentration of highly suitable arable land in locationally specific areas, adding considerably to the idea that environmentally vulnerable land can be identified and is thus susceptible to targeting. These results have a number of implications for policy design. This targeting methodology, impossible without GIS mapping techniques, could be extremely effective in assisting policy makers to increase the environmental and conservation benefits from the new CAP reforms, especially the long term set-aside scheme. Previously untargeted set-aside schemes were unlikely to withdraw those areas which offer the maximum environmental benefit and conservation potential.

In an ideal world the government would start from scratch by drawing up a Conservation Reserve Programme (CRP) as proposed by Potter *et. al.* (1991), designed to divert land and resources in some locations to tackle the nitrate problem and the still underestimated problem of soil erosion on farms. In practice a CRP in the UK is unlikely given the incremental nature of policy change and the fact that various (short-term) set-aside schemes have already been agreed. Having said that, the long-term set-aside introduced through the new agri-environmental package under the CAP reform is a signficant step towards diverting land from productive agriculture for environmental purposes and should be a first candidate for a targeted approach based on GIS. A targeting approach should now be brought to bear to increase policy effectiveness and clarify policy objectives.

18.12 Conclusion

In this chapter we have introduced the idea of using GIS as a tool to test the potential for targeting long-term set-aside in order to tackle soil erosion and nitrate pollution and identify areas in which to reconstruct particular habitats and landscapes. We can conclude that although data is severely limited in availability and patchy in places, there is considerable scope for deploying existing datasets to identify broad regional scale target

Figure 18.3 Suffolk combined model.

Table 18.2 Land areas for vulnerability classes.

	Study areas	
	East Sussex	Suffolk
Total arable area	6598	6970
Low vulnerability (ha)	164 (2%)	568 (8%)
Medium vulnerability	1760 (27%)	2607 (37%)
High vulnerability (ha) (target land)	4674 (71%)	3795 (55%)
	Percentage of target land	
Soils liable to erode	100%	100%
Soils liable to leaching	100%	100%
Slope	19%	–
Aquifier	100%	–
Woodland buffer zone	4%	56%
SSSI buffer zone	12%	4%
River buffer zone	0·05%	0·03%

areas (see Table 18.3). As these data needs are met, further work would be beneficial, thus increasing the detail of the eligibility criteria and the precision of the target areas.

Acknowledgments

The funding for this research was based was provided by the Economic and Social Research Council, and finance for essential equipment by the University of London, Central Research Fund.

References

Agra Europe, 1991, 'Agriculture and the Environment – How Will the EC Resolve the Conflict?' Special Report No 60.

Boardman, J., 1988, Public policy and soil erosion in Britain, in Hooke, J.M, (Ed.) *Geomorphology in Environmental Planning*, Chichester: Wiley, 33–5.

Boardman, J., 1990, 'Soil erosion in the South Downs: a review.' in Boordman, J., Foster, J.D.L. and Dearing, J.A. (Eds) *Soil Erosion on Agricultural Land, Chichester, Wiley.*

Boorman, C.A., Goss-Custard, J.A., McGroty, 1989, *Climate Change, Rising Sea levels and the British Coastline*. Institute of Terrestrial Ecology. HMSO.

Bowers, J.K and Cheshire, J., 1983, *Agriculture, the Countryside and Land Use: an Economic Critique*. London: Methuen.

Buckley, P., 1989, Biological Habitat Reconstruction, London: Belhaven Press.

Burnham, 1992, personal communication.

Cook, H F., 1991, Nitrate protection zones: targeting and land use over an aquifer. *Land Use Policy*, January, 16–28.

Croll, B.T and Haynes, C.R., 1988, Nitate and water supplies in the UK, *Environmental Pollution*, **50**, 163–87.

DOE, 1986, *Nitrate in Water*, Pollution Paper 36, London: HMSO.

Eagle, 1992, Leaching of pesticides to surface and groundwater, *SEESOIL*, **7**.

Evans, R., 1990, Soils at risk of accelerated erosion in England and Wales. *Soil Use and Management*, **6**(3), 125–31.

Evans, R. and Cook, S., 1986, Soil erosion in Britain, *SEESOIL*, **3**.

Green, B., 1991, cited in Marsh, J., *The Changing of the Common Agricultural Policy*, London, Belhaven Press.

Griffen, R. and Stoll, J., 1983, 'Evolutionary processes in soil policy.' *Land Economics*, **60**, 30–9.

Hornsby, M. and McCarthy, M., 1992, EC convicts Britain of poor quality tap water, *The Times*, 26 November.

ITE, 1991, Changes in hedgerows in Britain between 1984 and 1990, Cumbria: Merlewood. Institute of Terrestrial Ecology.

MAFF, 1993, Consultation Papers for England and Wales – Agriculture and the Environment.

Newbold, C., 1989, Semi-natural habitats or habitat recreation: conflict or partnership? in Buckley, G.P. (Ed.) *Biological Habitat Reconstruction*, London: Belhaven Press.

Norman, C., Potter, C.A. and Cook, H.F., 1991, Targeting environmental policy - developing a geographical information system methodology, in Rideout, T.W. (Ed.) Conference Proceedings on Geographical Information Systems and Urban and Rural Planning, The Planning and Environment Study Group of the Institute of British Geographers.

OECD, 1986, Agricultural Policy Reforms and Public Goods.

Potter, C.A., Cook, H.F. and Norman, C., 1993, (in press) The targeting of rural environmental policies. *Journal of Environmental Planning and Management*.

Potter, C., Burham, P., Edwards, A., Gasson, R. and Green, B., 1991, *The Diversion of Land, Conservation in a period of Farming Contraction*, London: Routledge.

Smith, L.P., 1984, Agricultural climates of England and Wales, area of averages 1941–1770, Technical Bulletin 35, Ministry of Agriculture, Fisheries and Foods, HMSO.

Index

abstract relation 18
access to data 45, 75, 81, 86, 92–3, 242–3
  information sharing 218, 220–2
  integration of systems 232, 236
  spatio-temporal 12, 15–16, 19–23, 105, 108, 114
accuracy 26, 32
  DEMs 120, 204
  remotely sensed data 70, 72
  vegetation modelling 245, 247, 248
aerial photography 66–7, 177, 231
aggregation 110–11, 126, 131, 132, 236
agriculture 66, 71, 215, 251–61
agri-environmental policy 215, 251–61
aircraft 66, 67
aliasing 67
algorithms 9–11, 49, 101, 134, 195–8
  classifications 150
  concepts 131–2
  database support 21–3
  data conversion 28–30, 32–3
  error simulation 190–5, 196
  generalization 54, 56–7, 60–1
geometric complexity 178–9
  information handling 75, 82, 83–7
  viewshed function 120
  Amazon watershed 14, 23
ArcEdit 47
ARC/INFO 47, 76, 93, 96, 161, 197
archiving 25, 39, 93, 114
area-value data 149–58
areal features 8, 53–4
artificial intelligence 119, 124–5, 216
atomicity 22

back-propogation 195–6, 197
Bayes Theorem 28, 119, 139–40, 141–5, 170
Bidirectional Associative Memories (BAMs) 135
binary skeleton 29
binary viewsheds 169, 170, 172
bioclimatic mapping 119, 139–46
black and white pattern search 133–4
blurring 67
Boolean viewshed 162–4, 174
bracken (*Pteridium aquilinum*) 216, 241–9
Brunsdon's conjecture 189
buildings 54–62, 68, 72

C (language) 53, 56, 121
  error simulation 192, 197, 198
C++ (language) 236, 249
cancer 130
CCD scanners 27, 28
chaining 29, 31–3, 95–7, 101, 110
change propagation 21
chaos theory 136
Chief Executive of local authority 224–6
choropleth maps 120, 149–53, 156
class relationships 78
classification 149–53, 192, 195
  neural nets 186–9, 196–7
  spatial autocorrelation 151–8
climate monitoring 2, 119, 140–5, 246
clusters 129–30, 192, 205, 255
  identification 141–2
  length-angle indices 183–5
  neural nets 186–7
coastal regions 253
  environmental modelling 231–2, 234–5
  map conversion 26–7, 37
coherency checking 21
colinearity 26
Common Agricultural Policy (CAP) 215, 251, 253, 259
complexity 177–9, 188–9, 194–5
computational geometry 83–5
Computer Aided Design (CAD) 8, 25, 224
computer assisted tools 43–4, 46, 53, 150
  data capture 8, 39–51
  digitizing 43–4
  evaluation 46–51
computer science 1–3, 124
  information handling 75–87
computer services departments 224–6
computer vision 119, 125, 135
concentric ring theory 127, 128–9, 132, 134
concept approach 2, 69–70, 89, 90, 119, 123–37, 162
conceptual domains (C-domains) 13, 16–17
concrete relation 18
conservation 251–7, 259, 261
Conservation Reserve 253–4, 259
contours 40, 54, 257
data conversion 25, 26–7, 37

DEM error 202–3, 207
corrupted data 22
costs 65, 75, 85
data capture 39–42, 49
information sharing 217–18, 220–3
coupling 230–1, 241–9
coverages 66–7
see also overlays

database administrator (DBA) 72–3
database and modelling system (DBMS) 11–15, 20–3, 79, 99
data capture 1–2, 7–8, 10
automatic 8, 25–37, 40–1
computer assisted 8, 39–51
costs 39–42, 49
errors 41–2, 45, 49, 120
interactive 42–3
remotely sensed 65, 67
spatio-temporal 106
data conversion 7–8, 26–7, 39–40, 66
automatic 8, 25–37, 39
data incorporation 7–10
data integration 217, 221
environmental modelling 229–31, 232, 236, 239
data transfer 215–6
deductive systems 76–7, 81–3, 89
digital elevation models (DEMs) 15, 19, 120
errors 163, 164, 201–10
viewsheds 162–4
digitization 25, 37, 40–4, 46–51
errors 8, 177–9, 181–4
manual 39–42
dis-assembling line segments 185–9, 199
display consistency 70, 71
distance decay 127, 130–1
viewsheds 165, 166–7, 172

Earth Observation System (EOS) 7, 11–12, 14, 22–3
edge effect problems 204
element relations 18
embedded SQL 53, 56
encapsulation 77
end points artefacts 37
environmental information 3, 139–40, 142, 145, 218
agriculture 251–61
modelling 1, 215–16, 229–39, 249, 254–5
Environmentally Sensitive Aras (ESAs) 257
erosion
soil 251, 254, 254, 256–7, 259, 261
vectorization 29–31
errors 8, 22–3
data capture 41–2, 45, 49, 120
data conversion 32–5, 37
DEMs 163, 164, 201–10
estimation 189–90
remotely sensed data 65, 72, 74
simulation 177–99
spatial analysis 3, 119–20, 139, 141–5
vegetation modelling 244–9
events 108–9, 114
exchange models 222–3
EXODUS 98, 114
EXPERT 242, 244–6, 248–9
expert systems 60–2, 77, 81, 215–16, 242
see also EXPERT
exploratory data analysis (EDA) 139–46
extended relational database 98, 113

farm distribution 145
Fastrak digitizer 40
Fifth Generation 9
First Law of Geography 125–6, 151, 152
flooding 251–4, 256, 259
forestry 68–9, 90, 106, 252
FORTRAN 82, 121, 142, 242, 245
error simulation 192, 197
Fourier transform 134, 180–1, 182–3, 202
funcional models 222–3
fuzz 119, 120, 189
data conversion 34, 36, 37
viewshe ıs 161–70, 172–4

GAM-K cluster searcher 130
Geary's c 204
GemStone 114
generalization 7, 8, 78–9, 110, 189
automated 53–63
concepts 119, 127–31
errors 144, 178, 189, 204
GENESIS 114
genetic programming 135
GEO system 113
Geographical Analysis machine 141, 142
geographical hypotheses generation 139–46
geomorphology 231, 235
geo-relational model 90, 92–7
granularity of time 9, 91
graphics 8, 153, 154, 191–2, 244
grey scale images 27, 29, 205

Haar functions 179–80, 182–3
hab itat 140, 145
conservation 251–4, 256, 261
Habitat Improvement Scheme (HIS) 253
highways departments 225–6
homomorphisms relation 18
Hopfield nets 135
Hough transform 134
Human Computer Interface (HCI) 87
hydrogeology 254–7

hydrology 12, 106, 230, 235
hypermedia coordination 105

iconic representations 68–9, 71
indexing 107, 114
inductive learning 119, 140, 241
Information Resource Dictionary Systems (IRDS) 238
information technology departments 219, 224–5
inheritance 17, 77, 78, 98, 110, 236, 249
interpretation of data 65–74
intersection artefacts 35–6, 67

knowledge-based systems 241, 242, 248, 249

lambda notation 191, 193, 194
land terrier system 9, 89–97, 99, 101
land use 139, 145, 215
  agri-environmental 251–6, 259
  remotely sensed data 66, 67, 72
  temporal dimension 91–2, 94
LANDSAT 65, 66, 67, 245
landscape viewsheds 161, 165, 168, 169, 174
languages 2, 14–23, 121, 125
  environmental modelling 235–7, 239
  error simulation 190–8
  expert systems 242, 246, 249
  generalization 53, 56
  information handling 76, 77, 79–82
  query 2, 9, 70, 89, 99, 101, 102
  spatio-temporal modelling 11–16, 18–23, 105, 113–14
Leicestershire electoral wards 152–3
length-angle indices 183–5
library of spatial patterns 133–6
lineage tracking 21, 23
line-following tool 43–4, 47–50
Lisp (language) 191, 193–8
list-based approach 193–4, 196
local government 3, 215, 234
  information sharing 217–26
  land terrier system 89, 90, 93
logic-based approaches 2, 7, 9, 119, 121
  information handling 75, 81–3

machine learning 119, 135
Magik 80, 236
Malvern Hills 120, 163, 204–8
MARIS (map recognition input system) 40
massively parallel systems 121, 123–37, 198
Medial Axis Transform (MAT) 29
metadata 17–18, 87, 216, 243
meta-objects 55–62
modelling
  agri-environmental 254–5
  bioclimatic 140–1
  conceptual 2, 89, 90
  database language (MDBL) 11, 14–23
  database system (DBMS) 11–15, 20–3, 79, 99
  environmental 1, 215–16, 229–39, 249, 254–4
  errors 120, 177, 197–9, 203, 209
  information handling 75–82, 86
  information sharing 217, 222–3
  mathematical 123, 126–7
  metadata 17–18
  probability 120, 139, 161–5, 170–4
  procedure 141–4
  scientific 1–3, 7
  spatio-temporal 11–23, 90, 92, 97, 101–2, 105–14
  vegetation 241–9
Monte Carlo simulation 134, 141–4, 163, 203
Moran's *I* 120, 149, 151, 153–7, 204–5, 208
multiple nuclei pattern 134
multiple sector model pattern 132–3

National Centre for Geographic Information and Analysis (NCGIA) 1, 217
nearest neighbours 189, 206–8
negentropy 144, 145
neural computing 120, 177–99
neural nets
  algorithms 195–6
  dis-assembling 185–9, 199
  error 189–90, 193–7, 199
  feature identification 196–7
neurocomputing 119–21, 130, 135
nitrates 252, 254, 257, 259, 261
noise 28–9, 37, 135
Nordic project 217
null hypothesis 126, 128–30

O2 114
object-oriented systems 2–3, 7, 9–10, 21, 216
  environmental modelling 234–9
  generalization 55
  information handling 75–81, 83
  integration 230
  spatio-temporal 89, 90, 97–102, 106, 108–14
  vegetation modelling 249
  views 69
ObjectStore 113
ONTOS 113, 114
Openshaw's conjecture 189
Oracle database 57
Ordnance Survey maps
  agri-environment 256–7, 259

data capture 39–40
errors 202–4, 207
generalization 53–63
viewsheds 163
ORION 113
OSGEN 8, 56–62
overlays 66–9, 71–2, 74, 235, 245
agri-environment policy 254, 257
data capture 40–50

patterns 140, 141, 205
classification 150–1, 155
recognition 124–37, 183, 234
pesticides 252
physical heterogeneity 216
planning departments 217, 224–6
pollution 91, 246
agri-environment 251, 252, 254, 256–7, 259, 261
polygons
capture tool 44–6, 49, 51
choropleth maps 120, 150
DEM error 205
environmental modelling 235, 237
information handling 79, 83–6
spatio-temporal modelling 13–14, 16, 19
street blocks 56–8
temporal dimension 96–7, 98–100
vegetation modelling 243, 246
polymorphism 78
POP-11 (language) 242, 246, 249
POSTGRES 20, 23, 98–101, 113, 236
probability models 120, 139, 161–5, 170–4
PROBE 113
process-based models 241
PROLOG 53, 56, 57–62, 81, 249
Prospero 20
prototyping 13, 22–3, 56–62, 79, 113–14

quantitative analysis 123–4
quantization 186
query languages 2, 9, 70, 99
information handling 76, 79, 81
SQL 53, 56, 102
temporality 89, 101, 102, 105, 113, 114

randomness 129–30, 172, 174
rasters 8, 66–7, 71, 98, 161, 205, 235
data capture 39–40, 42–5, 49
data conversion 27, 40
vegetation modelling 242, 244, 245
re-drafting 25, 37
regression models 126–7
relational databases (RDBs) 7, 9, 18, 76, 81
extended 98, 113
models 76–7, 236, 237
temporal dimension 89, 90, 107, 114
see also geo-relational model
remotely sensed data 9, 41, 65–74, 233
aerial photography 66–7, 177, 231
see also satellites
representation 7, 76, 91, 140
environmental modelling 230, 232, 237
iconic 68–9, 71
remotely sensed data 70–1
spatio-temporal 106, 114
representational domains (R-domains) 11, 13–18, 23
response time 92, 95
retrieval of data 20–1, 45, 92–3
see also access to data
rivers 54–5, 179
pollution 251, 252
remotely sensed data 67–8, 71
roads 9
generalization 54–62
remotely sensed data 65–8, 71–2
root mean square error (RMSE)
DEMs 164, 201–3, 207–9
rule-based systems 60–2, 130–3, 136, 241

SAFE systems 41
satellites 7, 9, 11–12, 14, 22–3, 65–7, 177
vegetation modelling 243–4, 246, 248
see also remotely sensed data
SBS (expert system) 242
scale 8, 132, 134, 231, 248, 256
DEM error 204
generalization 53–63
remotely sensed data 68, 71–3
spatial autocorrelation 126, 131
vectorization 34, 36
scanning 25–7, 28, 39–41
Scotland
bracken 216, 244–9
wildlife 119, 139, 140–6
semantics 8, 10, 13, 16, 81
environmental modelling 229, 236–9
heterogeneity 216
inheritance 77
temporality 97, 101, 105, 106, 107–12, 114
Sequoia project 1
set-aside 215, 251, 253–4, 259, 261
shared objects 97, 98
Sites of Special Scientific Interest (SSSIs) 256–7, 259, 261
Smalltalk 236
Smallworld 10, 80–1, 236
smoothed length-angle indices 183–5
social trends 127–33, 226
socio-economic data 139, 145
soil 12, 242–3, 246–7, 256, 259
erosion 251, 254, 256–7, 259, 261

SPANS 257
spatial association 126–7, 131–2, 149–58
spatial autocorrelation 133
  classification 151–8
  concepts 125–7, 131
  DEM error 203, 204–5, 208–9
Spatial Data Transfer Standard (SDTS) 17, 21, 114
spatio-temporal systems 9–10, 216, 229
  information handling 76, 78–80
  modelling 11–23, 90, 92, 97, 101–2, 105–14
  POSTGRES 98–101
  see also temporal dimensions
species distribution (wildlife) 119, 140–6
SPOT 65, 67, 245
SQL (query language) 53, 56, 102
Standard for Digital Cartographic Data 201
storage 18, 25, 56, 150, 248
  cartographic data 45
  environmental modelling 229, 232–6
  information handling 75, 76, 82, 85
  spatio-temporal 12, 20–1, 92, 101
Suffolk 257, 259, 260, 261
Sussex 257–9, 261
Synthetic Aperture Radar (SAR) 29

T displacement 35, 36
template matching 133, 134
temporal dimensions 7, 9–10, 70, 72, 89–102, 105–14
  information handling 76, 78–81
  see also spatio-temporal systems
text capture tool 45, 51
thresholding 27, 28
thinning 26
  vectorization 29–35, 37
time slicing 93–4
time stamping 92, 93–5, 110
  POSTGRES 98–100
time values 69
topographic maps generalization 53–63
TOSQL 101
transformation 8–9, 119, 180–1
  spatio-temporal modelling 11, 13–19, 21
triangulation 83–6

UGIX 237, 238
unitary skeletons 30
updating 25
  spatio-temporal 105–7, 110–14
urban areas 66, 127–33, 220
  map generalization 54, 56–62
utility companies 25, 39, 234
Uttoxeter 204–8

VAX/VMS 56
vectors 26–7, 66, 71, 161, 235, 244
  data capture 40, 41, 43–6
  data conversion 28, 32–7
  erosion 29–31
  errors 120, 177–99
  inside bias 37
  temporal 90, 99, 106
  thinning 29–35, 37
vegetation 12, 216, 241–9
  see also forestry
versioning 21, 99, 101, 105, 106–7, 109–14
views 68–73
viewsheds 120, 161–74, 201, 203
visual analysis 153–6
visualization 119–21, 125–6, 149–58, 229–30
VMS DecWindows 45

weights matrix 126
wildlife 91, 119, 139, 140–6, 251–6
Windows 191, 193
Worcester Beacon 164, 165–73

X destruction 35, 36
XLISP-STAT 190, 191–3, 197–8

Zenith object-oriented model 101

For Product Safety Concerns and Information please contact our EU
representative GPSR@taylorandfrancis.com
Taylor & Francis Verlag GmbH, Kaufingerstraße 24, 80331 München, Germany

www.ingramcontent.com/pod-product-compliance
Lightning Source LLC
LaVergne TN
LVHW061220100826
845148LV00004B/810

* 9 7 8 0 3 6 7 5 7 9 8 3 8 *